D1597483

Growing Trees from Seed

Growing Trees from Seed

A practical guide to growing native trees, vines and shrubs

Henry Kock

with Paul Aird, John Ambrose and Gerald Waldron

FIREFLY BOOKS

Published by Firefly Books Ltd. 2008

First printing

Publisher Cataloging-in-Publication Data (U.S.)
Kock, Henry.
Growing trees from seed : a practical guide to growing native trees, vines and shrubs / Henry Kock ; with Paul Aird ; John Ambrose ; and Gerald Waldron.
[280] p. : ill., col. photos., maps ; cm.
Includes bibliographical references and index.
Summary: A guide to growing native trees, vines and shrubs from seed. Includes information on identifying, gathering, storing and germinating the seeds, along with transplanting, protecting and nurturing the seedlings.
ISBN-13: 978-1-55407-363-4
ISBN-10: 1-55407-363-4
1. Tree planting. 2. Trees – Seedlings.
3. Climbing plants. 4. Shrubs. I. Aird, Paul.
II. Ambrose, John. III. Waldron, Gerald. IV. Title.
635.977 dc22 SD391.K635 2008

Library and Archives Canada Cataloguing in Publication
Kock, Henry, 1952-2005.
Growing trees from seed : a practical guide to growing native trees, vines and shrubs / Henry Kock; with Paul Aird, John Ambrose and Gerald Waldron.
Includes bibliographical references and index.
ISBN-13: 978-1-55407-363-4
ISBN-10: 1-55407-363-4
1. Tree planting. 2. Trees — Seedlings. 3. Shrubs.
4. Climbing plants. I. Aird, Paul L. II. Ambrose, John D.
III. Waldron, Gerald IV. Title.
SD391.K63 2008 635.9'77 C2008-900806-5

Published in the United States by
Firefly Books (U.S.) Inc.
P.O. Box 1338, Ellicott Station
Buffalo, New York 14205

Published in Canada by
Firefly Books Ltd.
66 Leek Crescent
Richmond Hill, Ontario L4B 1H1

Produced for Firefly Books Ltd. by
International Book Productions Inc., Toronto
Cover and interior design by: Dietmar Kokemohr
Edited by: Barbara Hopkinson

Printed in China

The publisher gratefully acknowledges the financial support for our publishing program by the Government of Canada through the Book Publishing Industry Development Program.

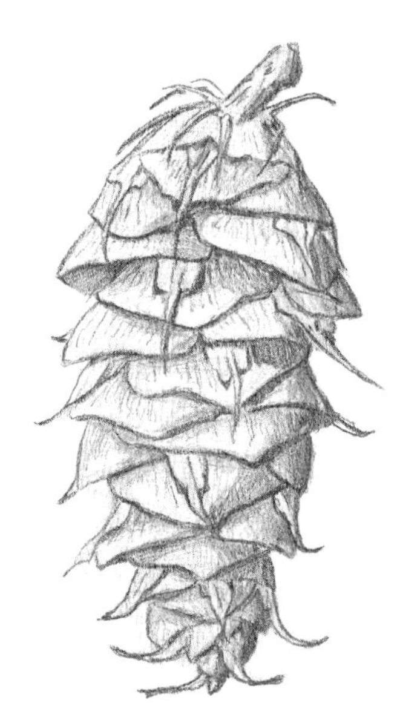

Table of Contents

To my mother and father –
for teaching me how to grow plants
— Henry Kock —

Preface

There is much satisfaction to be gained in gathering seeds from an old oak or elm or witch hazel, and propagating them, knowing that you are playing a role in preserving the genetic diversity of important local trees and shrubs. But how do you know which seeds to collect, and from which tree, how best to treat them so that they will germinate, or how and where to plant them?

In this book, my aim is to pass on my knowledge to you about how to do these things. This means helping you learn what kinds of native woody plants to look for, showing you how to sharpen your senses to identify plants, even how to "think like a seed," how to plant the seeds, germinate them and protect the seedlings, and what species are not native but invasive species.

To help you identify the fruit of trees and other woody plants to the species level, especially if closely related exotic species exist in the wild, extensive descriptions and illustrations are provided in Chapter 6. While this book presents procedures found successful for propagating the woody species of the Great Lakes bioregion, the techniques apply as well to the same species where they grow naturally outside this region, and to other species in other regions.

My other aims in writing this book are to inspire the gathering and growing of seeds; to help restore and sustain the precious diversity of our natural heritage; to nurture a deeper appreciation of the natural diversity of landscapes; to help everyone to understand plants as members of communities of plants and animals rather than as isolated specimens; and to understand how lands and forests become fragmented and what this means.

Covering many thousands of square miles, the Great Lakes bioregion is the ecological community of plants, animals and microorganisms encompassing the largest group of freshwater lakes in the world: Lake Ontario, Lake Erie, Lake Michigan, Lake Huron and Lake Superior. All of Michigan, Wisconsin, Illinois, Indiana, and Ohio, and parts of Ontario, New York, Minnesota, and Pennsylvania belong to this bioregion. The meeting place of three vast life zones — the boreal, mixed and deciduous forests (including prairies) — it contains a rich diversity of woody plants — trees, shrubs and vines — arising from the convergence and interaction of species from these three very different life zones. In addition, the birds migrating from as far away as South America — insectivores, seed-eaters and their avian predators — have also had a profound influence on the distribution and growth of the woody species in the Great Lakes region. Likewise, the Great Lakes bioregion has had a profound influence on avian populations.

Regrettably, it is one of the most heavily impacted bioregions in all of the Americas. It has been significantly fragmented by intensive human settlement; now, in some rural areas, the amount of coverage with native plants is less than 2 percent. The disruption of the natural landscape caused by land clearing, paving and extensive planting of exotic (non-native) species has resulted in a climatic, hydrologic, vegetative and faunal imbalance that dramatically restricts the natural evolution of the region's native species.

Planting native tree, shrub and vine species to address this massive injustice to the land is a nearly sacred act.

Leaves of the Shumard oak (Quercus shumardii)

Chapter 1

Seeing the Trees in the Forest

The secrets are out. The bracken tips have unfurled and baby birds are squawking and flapping among the dense foliage. The trees are fully dressed, brilliant and "spandy" in their new clothing put on with an imperceptible and silent push. There is nothing so strong as growing.

— Emily Carr —

Trees, shrubs and many vines are known as woody plants because some woody tissue survives above ground through the non-growing season. Trees usually have one central trunk, shrubs usually have multiple woody stems, and perennial vines have fiber bundles or woody stems to support growth along the ground or to elevate themselves by climbing on other plants or objects.

Most north temperate woody plants flower, although not always with showy flowers (the alternate reproductive structures are the cones of conifers and berry-like structures of the conifer relatives the yews). The flowers produce fruits, and there are fantastic botanical terms for describing all the different kinds of fruits. The familiar fruits are berries, winged keys, nuts and capsules. A fruit is an expanded part of the original flower and contains the ovary, or ovaries, which become the seeds. The fruit of each species has evolved in fascinating ways to assist in the dispersal of the seeds — wings to catch the wind, hooks to hold on to fur, sweetness to attract animal carriers.

Looking for and finding seeds is easier than ever before because trail systems now wind their way through watersheds and woodlands in every county. Along with better access to natural ecosystems, some of the best trail guides indicate the plant species that can be found along the way.

However, before looking for seeds to grow trees, you will need to know something about the woody plants themselves. At first the mass of large trees, bushes, small trees, shrubs, thickets and vines may look like an undifferentiated green wall. Give yourself time to learn about the woody plants. Bit by bit, you will learn to identify individual species. Joining a local field naturalist club, or going along on outings organized by conservation groups, can help you learn to identify plants. Visiting an arboretum or botanical garden is another way to learn about woody plants. These institutions often offer workshops on both tree and shrub identification.

As you walk through the landscape, familiarize yourself with the common trees and shrubs first and learn their names. Don't try to remember everything and become intimidated in the process; learning how to see and what to look for is more important than memorizing plant names.

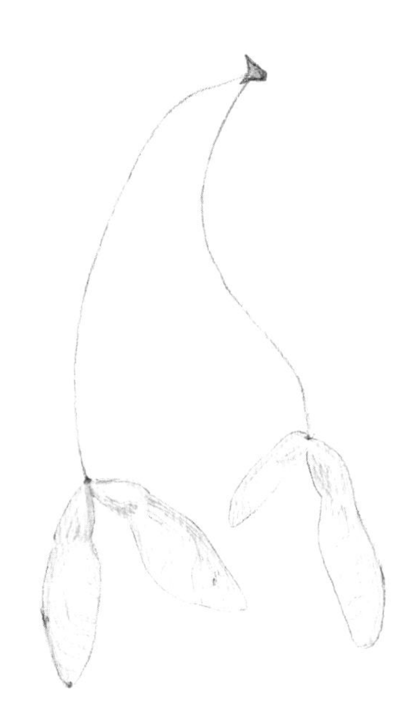

The fruit, called keys, of the red maple, a common native species. Occasionally only one of the pair develops, as shown by the pair to the right.

What Is a Native or Indigenous Plant?

Native, or indigenous, plant species are those species that were growing in North America prior to colonial times. A plant that is native to the Great Lakes region may be native to only part of the region. If planted well outside its natural range it may be called an "exotic" or an "alien."

A native species can cover an extensive area of North America. Red maple, *Acer rubrum*, for example, is native to eastern North America, with an extreme range from Newfoundland to North Dakota and south to Texas and subtropical Florida. A red maple from either Georgia or central Ontario will not grow well in northern Ohio because this tree's genes are finely tuned to day length. But a red maple seedling from a Wisconsin seed source would likely adapt to eastern Ontario because the day length is the same in both locations.

Some indigenous species will grow in a range of habitats. You will find alternate-leaved dogwood growing in moist woodland shade and along open exposed fence lines. Bur oak thrives in wet woodlands at the edge of swamps and on open, dry, gravelly slopes. Red oak is found on deep sand, clay embankments and granitic rock outcrops.

It's a safe bet that seeds from an alternate-leaved dogwood growing in an open fence line will provide the best plants for growing on an exposed site, and that seed from an old bur oak in a swamp may not be the best choice for planting on dry, gravelly soil. By the same token, seed from a red oak growing in sandy soil may not produce a 300-year-old tree, let alone a 40-year-old one, if planted on a clay site. Woody plants are able to tolerate a slight change in habitat, but are less likely to survive if there are two or more differences in climate zone, soil type or soil moisture between the original site and the new planting site.

As you walk through the landscape, note how much variation there is in the location of species. You will soon notice what is called the adaptive range of trees and shrubs. Paper birch occurs frequently on sites where it can thrive (optimum sites), infrequently on sites where it cannot thrive well, and not at all on non compatible sites or south of its climatic range. Your observations will help you find the best places to gather birch seeds to suit your local conditions. Plants grown from seeds collected from a local source growing on an optimum site can themselves be expected to thrive on optimum sites.

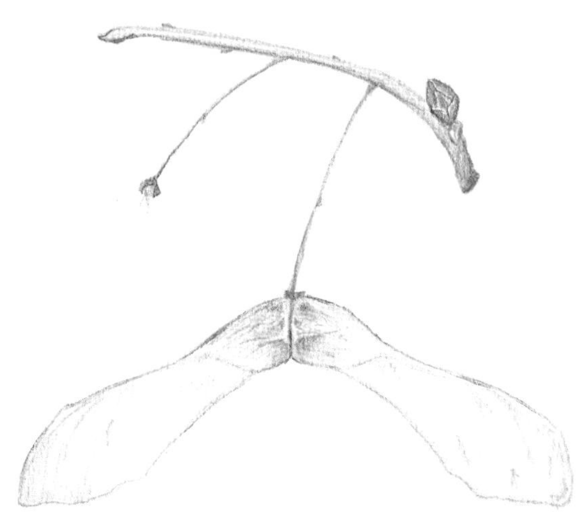

Norway maple keys. This invasive species has become a concern for natural forest health and yet it is still being planted.

Unwanted Visitors: Exotic Trees and Shrubs in Nature

Horticulturists and foresters have planted non native plants, also called exotic plants, in the Great Lakes region for the past 200 years, often giving them preference over native plants. Exotic plants are able to grow in many locales because they came from a similar climate in Europe or Asia. Some of these species have become naturalized, meaning they can reproduce naturally in the wild. Eurasian species have gone through a much longer period of evolutionary adaptation in their natural ranges to sites degraded by humans than our native plants, thus giving them a huge advantage in growing and reproducing on degraded sites in North America.

Several exotic species are so well naturalized that they are beginning to replace the indigenous flora. Exotic plants may become invasive because the insects that feed on them or diseases that affect them are not found in North America, they are able to spread or colonize rapidly, or they produce tremendous seed crops with high rates of seedling survival.

But this does not mean you have to give up your fascination with exotic plants. Most exotic food and garden plants are not invasive. However, there are many species that have become a threat to local biodiversity by either smothering native plants or reducing seedling survival. Tatarian and Amur honeysuckles form dense understory thickets in parts of New York, European buckthorn prevents oak regeneration in parts of Ontario, and Norway maple and tree of heaven are filling ravine lands throughout eastern North America. The list is long and includes herbaceous plants like garlic mustard, *Alliaria petiolata*, and periwinkle, *Vinca minor*. In the interest of promoting the health and diversity of indigenous species, we should end the planting of the known invasive exotic species listed in Appendix I.

Plant identification is absolutely critical to ensure that you do not grow one of the many hundreds of exotic plants now naturalized in the Great Lakes bioregion. When reading field guides, go beyond the picture and read the text, especially the habitat section. Exotic plants such as European birch, European highbush cranberry and Asian bittersweet are so similar

Autumn Olive (*Elaeagnus Umbellata*)

Each year the U.S. spends billions of dollars to control and eliminate non native invasive plants. In Canada, it is essentially a volunteer activity. Overall, exotic species invasion (plants, fish, aquatic organisms, insects, etc.) is rated second to urban sprawl (also a form of exotic invasion) for destroying native plant and animal habitats.

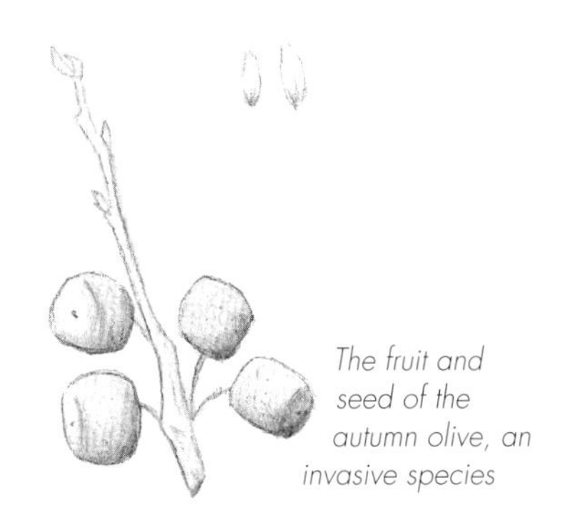

The fruit and seed of the autumn olive, an invasive species

What can you do about invasive species?

- Become informed about the region's most invasive species.
- If a newly acquired exotic plant starts spreading rapidly in the second year, compost it.
- Avoid the disposal of any plants over embankments.
- Do not donate invasive plants to fundraising plant sales or neighbors.
- Use the list in Appendix I to avoid purchasing invasive plants, and let the garden center manager know why you won't buy them.
- Look for exotic species on your property and replace them with non invasive species.
- Volunteer to remove invasive plants in parks and conservation lands.

to our North American species that their seed is often inadvertently collected from the naturalized exotic species. Information on how to tell closely related species apart is provided in Chapter 6.

To avoid what I call "botanical racism," I don't advise removing exotic species that seem relatively benign — European elderberry, daphne, Japanese yew, sweet cherry and horse chestnut come to mind. But if a species is known to be a significant invasive plant and has just become established, efforts to remove it should be made before it becomes well established. Landowners can consult with field botanists or regional foresters to find out what options are available to them.

Where extensive colonization has already occurred, the cost of removal must be weighed against the very likely possibility that the exotic species may recolonize if controlled burns or the removal of nearby seed banks is not undertaken. If the site is left alone, some exotic species may ultimately decline due to succession forces that essentially prevent any species from prevailing to the exclusion of others. At the University of Guelph Arboretum in the 1980s, I was alarmed by the discovery of an established dense understory of glossy buckthorn, but in the early nineties significant die-off was evident in one area and today dead trunks are almost all that remain.

In Toronto, hundreds of the most invasive species (buckthorn, white mulberry, Norway maple and tree of heaven) have been dug from gardens by well-meaning citizens and planted along the banks of the Humber River valley. These species are already managing very well in our landscape and do not need any further assistance, be it through misidentification or misplaced good intention. In urban areas, exotic species make up almost 100 percent of the vegetative cover on abandoned industrial sites and other urban "brownfields."

The introduced dog rose, Rosa canina (above), is found on moist ground, while swamp rose, Rosa palustris (below), is found in wet areas.

Plant Identification

My first foray into native plant identification was observing wildflowers. I initially enjoyed flipping through the pages of a Peterson field guide, but later discovered the incredibly simple key in *Newcomb's Wildflower Guide* — unexcelled for identifying wildflowers for the novice and professional alike. Unfortunately, there is no book with a simple key like Newcomb's that covers all the trees and shrubs in our region.

This book will help you with plant identification to the species level only. Chapter 6 provides identification of the fruit by illustration and description, especially if closely related exotic species exist in the wild. Appendix IV provides a list of identification guides for woody plants along with a few of the best herbaceous guides. And, if you are really stumped by a plant, you can send a dry specimen, donation and self-addressed envelope to the botany or forestry department at the university or college closest to you, or consider hiring a field botanist or forester to identify any plants that you can't figure out.

What's in a Name?

Although most gardeners don't use botanical names for their plants, it's worth knowing a little bit about the conventions of naming plants and why they are useful.

Using the common names for plants is a problem since they frequently change as you move from region to region. For example, in some areas the paper birch (*Betula papyifera*) is known as the white birch or the canoe birch, so common names simply aren't common enough to ensure that we are all talking about the same plant — although alone their regional quirkiness can be of interest! Another, more serious problem is that many different plants often share the same common name. For example, there are two different plants commonly known as ironwood — *Ostrya virginiana* and *Carpinus caroliniana* — though efforts to use unique common names are gaining ground. Nevertheless, the formal naming of plants gives us something better to distinguish them.

The basic structure for classifying and naming plants is founded on Carl Linnaeus's system for classifying plants published in 1753. It is based on reproductive structures and a two-part name (a binomial). Each of the following categories represents a group of plants that share similar characteristics.

Quercus velutina
(black oak)

I. **Family** — plants that share a common general appearance or characteristics; usually grouped according to similarities in reproductive characteristics. Most family names end in "-aceae" and are based on a type genus, such as Asteraceae based on aster (the exceptions being older names such as Compositae, the synonym for Asteraceae).

II. **Genus** — clusters of plants within a family that have strong common characteristics such as leaves, buds, roots, branches or stems. The genus (plural: genera) is what we think of as a particular type of plant — *Rosa* is the genus name for roses, *Pinus* the genus name for pines.

III. **Species** — the grouping within the genus that contains those plants that share essential features that are used to identify a plant before us, such as color, shape, size or habit. *Pinus rigida*, pitch pine, is a species of pine. There often are recognized variants within a species, but generally when they come together they are able to freely interbreed. The idea of a species as a breeding unit becomes blurry when we see that many distinct species can hybridize; their distinctness may be due to a long period of geographic isolation.

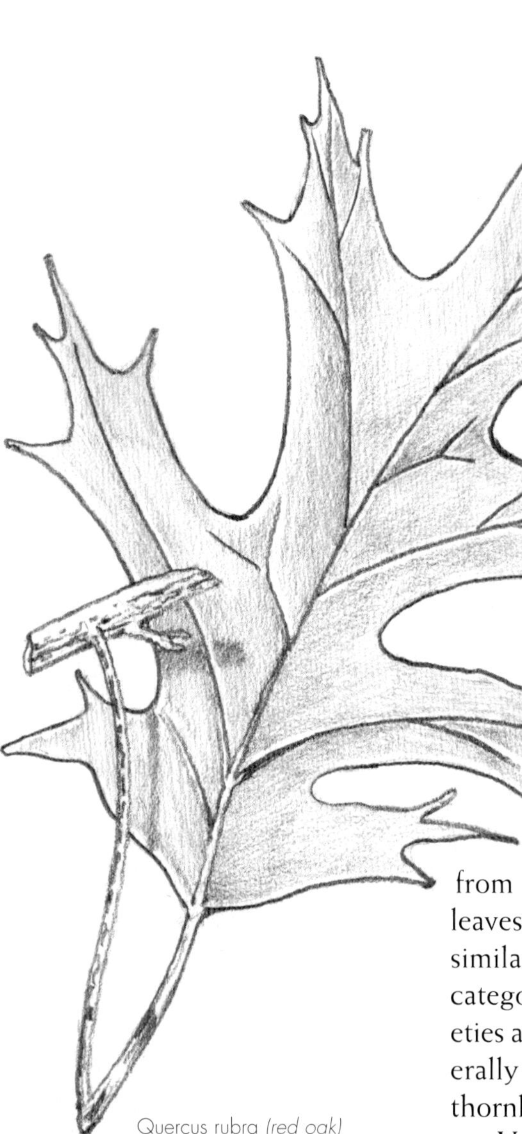

Quercus rubra *(red oak)*

The black oak (see illustration on the previous page), *red oak* (above) *and Shumard oak* (see p. 8) *should have been called savannah oak, red oak and swamp red oak to better describe their habitats instead of being named after a person or the color of their leaves or wood.*

IV. **Subspecies** — groups plants together that have common traits that set them apart from the plants in the species, such as the downy, droopy leaves of black maple and its slightly different habitat to the similar sugar maple subspecies. There are other subspecific categories that will sometimes be encountered, such as varieties and forms. Form is the lowest of the distinctions and generally based on only a single character difference, such as thornless honeylocusts (*Gleditsia triacanthos* f. *inermis*).

V. **Cultivar** — a cultivated variety, i.e., a garden selection of a species. This can originate either from a selection of a superior offspring from hybridizing different species or a selection of a variant found in nature. They are generally reproduced clonally to maintain their genetic distinctness, or, in the case of annuals, from an inbreed line.

Cultivar names are capitalized and in plain type and are noted either by single quotes around the name ('Yukon Gold' potato, or *Solanum tuberosum* 'Yukon Gold') or with the abbreviation cv. Cultivar naming has its code just like botanical names. Examples are that the name should be in the vernacular language (so not in Latin) and not exaggerate its qualities ('Best Ever').

The code of botanical nomenclature is updated at the International Botanical Congress, held every five years. The foundation of the system, going back to the time of Linnaeus, consists of using mostly Latin to create unique names. Sometimes Greek terms are used, as in *Rhododendron* (rose + tree).

Many of the botanical names are also used as the common name, such as *Iris*, *Hosta* and *Gladiolus*.

The basis of the Linnean system is the binomial — two names to define the species. It consists of a unique generic name (e.g., *Cornus* for the dogwoods) and a defining specific name (*Cornus alternifolia*, the alternate-leaved dogwood). As just shown, botanical names are printed in italics (in older texts they may appear underlined), with the first letter of the generic name capitalized.

Developing Your Plant Identification Skills

It is a good idea to carry a shallow plastic container on your walks to carry plant specimens home. A specimen should include leaves attached to the stem, and preferably the flower or fruit. With that small amount of material, it is possible to identify any woody plant.

Some groups of native species are difficult to tell apart if all you have are the leaves. So noting where a plant grows is an important identification tool. In the red oak group, for example, the leaves are similar but the habitat is distinct. *Quercus velutina* (black oak) is a tree of dry oak savannahs, *Quercus rubra* (red oak) is a tree of moist but well-drained soils, and *Quercus shumardii* (Shumard oak) is a swamp-edge species. If there are two or three similar species in a genus, they often segregate out by habitat or geographic range. When you are gathering the fruits of an interesting plant, record its habitat if you are not familiar with the species.

When looking at the plant itself, always examine the underside of the leaf. Its color, veins and hairiness can often tell you more about the plant's identity than the leaf shape. Look for glands on the leaf stalk and how the fruit is attached.

Use all of your senses: smell the bruised leaves of cedar, spicebush and bayberry; listen to the fluttering leaves of aspen, the moan of wind through the conifers, or the rattle of the fruits of bladdernut on a windy winter day. Notice the large hairs on the stems of staghorn sumac and the minute hairs on the leaf petioles of black maple when seen from a distance in bright sunlight. Look for winter catkins on members of the birch family, or overwintering flower buds on red maple, flowering dogwood and pawpaw. These experiences help you to remember the plant and where you found it.

Plant identification, like bird or insect identification, is not so much a task as an adventure. Depending on your experience and available time, it will take several years of observation with the help of guidebooks before you become comfortable with the names and habits of plants and associated wildlife. The best time to get to know a species is after you've first discovered and

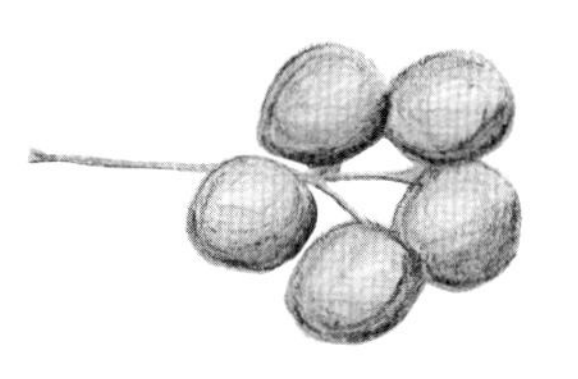

The fruit of the American cranberry, Vaccinium macrocarpon (top), *is very different in taste to that of the highbush cranberry,* Viburnum trilobum (above).

identified it. If you read about a new tree or shrub as soon as you get home, you're more likely to remember it.

Learning about the plants and animals and their interrelationships in a landscape is one of the most rewarding aspects of learning to identify and grow some of your own indigenous plants.

Poison ivy leaves

Poison Ivy Alert

When you are out collecting seeds in the woods, keep an eye out for poison ivy, *Rhus radicans,* which can give you an irritating, itchy rash. Poison ivy is both a low, colony-forming shrub and a tree-climbing vine with branches spreading 6 feet out from the trunk. The much less common poison sumac is a swamp- and marsh-edge shrubby tree with white fruits in open, pendent clusters that persist through the winter.

The nearly mythical "poison oak" is not a poisonous oak tree but a different plant that is similar to poison ivy. It has deeper-lobed or slightly toothed leaves resembling oak leaves. It occurs in British Columbia and the western and southern United States.

Natural Variation and Hybrids

The names of plants lead us to believe that a species is the same wherever it is found. But evolution is never simple. Variation is natural in many species. Sassafras, for example, has smoothish leaves and stems in the north and relatively downy leaves and stems in the south, with "intermediates" where their ranges overlap. A range of variation is frequent in the botanical world, especially in leaf size and shape, hairiness or plant height. Where other characteristics may be indistinguishable, the fruit is generally considered the most useful key for identification. This is particularly true in the oaks, hickories, maples, spruces, pines and other complex genera.

The designation of a group of plants as a species is usually assigned to groups of plants exhibiting significant variation in just a few ways, such as differences in leaf margin, large or small flowers and growing on wet or dry soils. This doesn't mean that they don't still occasionally hybridize. I have seen trees with white oak leaves and bur oak acorns, and red oak-sized acorns with black oak banding on the nuts. *Speciation* is the term used to describe the evolution of a new species, and hybridization is part of this process.

Botanists who trek through the wilds of the world or sit in the international plant registry office at Kew Botanical Gardens in London fall into two main groups. The "splitters" consider variation as the basis for species designation, and the "lumpers" are content with grouping plants with similar characteristics, especially if hybrids are common. I'm a lumper.

There are several genera in our region in which hybrids are unusually common. Each species is relatively similar across its range. Where their ranges overlap, their hybrids can be found. Implicit in the definition of a species is an inability to mate with another species. What does this say about the way botanists determine what is and is not a species? After all, if two species commonly hybridize, they are technically not separate species. The easy way around this conundrum has been to call the two species and their intermediates a hybrid complex or hybrid swarm. The following genera are notorious for hybridizing: *Amelanchier* (serviceberry), *Carya* (hickory), *Cornus* (dogwood), *Crataegus* (hawthorn), *Populus* (poplar), *Quercus* (oak), *Rubus* (blackberry) and *Salix* (willow).

In *River Notes*, Barry Lopez aptly describes the problem: "These are dry willow leaves of some sort. There are so many willows, all of which can interbreed. Trying to hold each one to a name is like trying to give a name to each rill trickling over the bar here and making it stick. Who is going to draw the lines? And yet it is done. Somewhere this leaf has a name, *Salix hookeriana, Salix lasiandra*."

Since hybridization is a frequent occurrence in complex genera, it may not always be possible to identify a plant to the species level. There is nothing wrong with growing natural hybrids that are locally adapted and doing well. It may even be very important, since we haven't a clue which gene pools are going to survive climatic change or pest and disease migration.

I was once asked to define intelligence. After some time and a couple of pints of beer, I came up with this definition: "Intelligence is the ability to resolve a problem with the minimum of waste." I should have simply answered "plants" or, better yet, "evolution."

Seed-Rain

I am constantly amazed at the volumes of fruit produced throughout the year, for example, when I see juneberry branches hanging low with ripe fruit or the ground covered with acorns. I have watched cedar waxwings topple, drunk, from their perch while feasting on black cherry fruit, and chipmunks scamper down their holes with cheek pouches filled with the seeds of red maple, beech and oak.

The natural attrition of seeds and seedlings is very high. In some years, the entire seed crop is consumed by seed-eating mammals and birds or lost to weevils or inhospitable conditions for germination and seedling development. This is as it should be. Fruits and seeds have co-evolved with seedeaters in order to be dispersed widely. In any case, unfavorable germination conditions ensure that a species does not grow where it ultimately can't survive.

Most of our shrubs, trees and vines produce an annual seed crop, but a significant number of tree species have evolved to produce intermittent seed crops, often in unison across large areas. These trees tend to rest from flowering for intervals of two to several years. This fascinating adaptation allows the trees to conserve energy for the next seed-crop cycle while reducing the number of seedeaters associated with that species so that when a seed crop is ultimately produced, seed germination and seedling survival are more likely. How this survival mechanism is triggered remains a mystery; it is only partly explained as a response to drought and other severe climatic events.

For long-term survival, any species must accomplish just one thing. Each individual must replace itself (on average, of course) with one plant reaching sexual maturity. Why, then, does each plant produce many millions of seeds over its lifetime? Clearly this illustrates how extremely vulnerable are seeds and young plants. In woodlands and grasslands and along fence lines, nature is always being replenished by the relatively few survivors of millions of seedlings that germinated from the seeds that rain down each year.

Every now and then an older shrub or tree displays a tolerance to more adverse conditions due to its chance germination in a different place. Plants are always being pushed by nature. Droughts, snow load, wind, insects, diseases and herbivores all play a role in thinning the ranks. The old survivors are worth noting as potential seed sources.

Seen clockwise from the top are a black walnut seed, a red oak seed and American plum seeds that have been chewed on by mice.

The Natural Dispersal of Seeds

The presence of young plants in close proximity to older plants is not surprising, but when a species is found in widely scattered habitats throughout its range, my curiosity is piqued.

The long-distance movement of seed is associated with wind, water and animal movement. The seeds of species such as white oak are locally dispersed by squirrels, chipmunks and jays. Blue jays occasionally carry acorns far from the original tree. It is likely that, years ago, passenger pigeons and native travelers played a role in the movement of acorns right across the Great Lakes barrier itself. Birch seed is dispersed throughout the winter and can easily be blown across a frozen pond or lake, even occasionally as far as across Lake Erie.

The best way to learn about seed movement is by examining animal droppings, called scat, or observing birds feeding on ripe fruit. I have seen coyote scat completely filled with wild plum seeds just after the plums have fallen, and, later in the season, filled with pear and apple seeds. In recent years, the scat has been filled with seed of the very invasive Oriental autumn olive, which was foolishly planted extensively as food for wildlife. Every animal that passes intact seeds — black raspberry, wild plum, bittersweet or juneberry — through its gut is a seed distributer.

Plants have evolved to hitch that all-important ride. In certain plants, a highly nutritious covering has evolved around the tougher seeds to appeal to and be carried by fruit-eaters. In others, the seed itself is small enough to stick to the feathers, feet or fur of a bird or animal traveling from one watershed to another. Humans are also agents for the natural movement of plants. Humans have been moving plants ever since the first hominid ate raspberries and traveled for a day before defecating in the woods.

This brings to mind one of the key questions in the current debate on the movement of seed — how far should we move a species beyond its historic range? All of the botanists and naturalists I have met over the years garden. Non native plants abound in their yards. Humans are like wildlife in this respect. We usually don't stop at the edge of a plant species' range.

One day while I was collecting seeds in southwestern Ontario, a farmer asked, "How did the pawpaw get here?"

"Mammals," I replied. "Bear, opossum, human."

When I said "human," he remembered something about his land. "The Navajo Trail ends at a winter camp at the junction of two creeks on my farm, near the pawpaw grove. The Navajo came here to trade with the Stony Point Indians."

I had heard of this trade, but it was more likely the Potawatomi were the traders, not the Navajo. If they arrived in mid-October, they could have carried pawpaw fruits, which usually ripen in September, with them from what was essentially the same climate zone in Michigan.

The farmer went on to describe another group of people who had lived on his land long ago. The remnants of shelters that had housed slaves escaping on the Underground Railroad also are located in the woods near the pawpaws.

"Could they have brought pawpaw with them?" he asked.

"Possibly," I replied.

Pawpaw is the northernmost member of a huge tropical family, and the slaves would have been familiar with this type of fruit. On their journey north to Canada, they would certainly have taken food — perhaps the nutritious pawpaw — with them. A word of caution if you decide to try a pawpaw: eating unripe pawpaws causes stomach upset.

Our propagation of indigenous plants today partly compensates for the loss of historical seed movers — indigenous peoples, passenger pigeons and, in many locations, the black bear. It also helps to compensate for seeds that fall and germinate naturally but then are destroyed by farm machines and domesticated grazing animals.

The gathering of seed, the storage of seed, the sowing of seed, the germination of seed and the planting of plants requires us to think like a seed. To think like a seed we must first listen to the seed's story as it is written by both the land where the seed is formed and the animals, winds and waters that distribute it. Thinking like a seed means honoring the natural heritage of each landscape by growing and planting its native species from seed derived from its local plants — the plants that have evolved to survive and flourish together in that place.

Chapter 2

Searching for Seeds

Convince me that you have a seed there, and I am prepared to expect wonders.

— Henry D. Thoreau —

One mid-July, I was hiking along a path connected to a section of the Bruce Trail near Inglewood, Ontario. I had been on the path before and knew where several downy juneberries grew (they should be called july-berries in the north). Since robins and cedar waxwings had been feeding for about a week in a small grove of juneberry trees at the University of Guelph Arboretum, I was looking forward to picking a few fruits from young plants along the trail. But before I reached the juneberries, I was elated to find fully ripe fruits on the ground under two huge Allegheny serviceberries. Arriving at the height of the feeding frenzy, when fruits are inadvertently dropped by birds, provided me with a rare opportunity to obtain seeds from old, healthy trees whose branches were way beyond reach.

Once back home, I set the precious fruits out to dry. Glancing outside, I noticed birch seeds raining down from seedlings I had planted nine years earlier. Goldfinches were busy tearing into the mature but still green catkins. Gray squirrels were just as busy gnawing through the hard seed coats of the unripe fruit of alternate-leaved dogwood. In another two weeks the fruit would be ripe and it would be time to gather seeds from a site I knew where some very hardy dogwood grows on a dry exposed embankment.

At certain times of the year, seeds and fruits are so abundant that success in finding them is almost guaranteed.

Seed Crop Cycles and Forecasting

Most shrubs flower at three to ten years of age and begin to produce annual seed crops. Trees flower at from eight to ten years of age but may take up to 30 or 40 years before they produce seed. Many tree species bear fruit on alternating years. Some, like the sugar maple or white pine, are periodic, flowering and producing large crops of fruit every three to

Seed Competition from Squirrels

Squirrel populations are often at abnormally high levels in and near urban areas due to predator decline and ample winter food from bird feeders. This may make it difficult to obtain seeds from hazelnut, oak, beech and other species preferred by squirrels.

four years. With the exception of the green ash, all ash trees fruit heavily only at five- to seven-year intervals.

A number of tree species provide evidence of a potential crop well in advance. The immature fruits of white pine and species in the red oak group, for example, may be seen up to 15 months prior to seed dispersal. The flowering times for each species are indicated in Chapter 6. Use this information to time your seed collection or to confirm species identity, especially in difficult genera such as *Amelanchier* (serviceberry).

Keep notes on the locations of species and times that you have observed them in flower or with developing fruit. Familiarize yourself with the seed dispersal calendar (Appendix II), and make a note in your daybook to plan a hike to collect seeds of the species you are interested in when they are ripe. Then, if that species is not on your regular walking route, the only thing left is to remember where you have seen it before.

The Source of the Seed Matters: Preventing Site Mismatches

Barb Boysen of the Forest Gene Conservation Association in Ontario says, "The most expensive planting is a failed planting." Shrubs will show signs of stress from a site mismatch within a few years. In trees, it can be immediately obvious or it might take 40 to 50 years before the mismatch becomes evident. Years ago, to halt desertification, white, red and jack pine were planted on calcareous (chalky) soils in Ontario. These trees are now in decline. One could say that much time and effort had been lost, or, conversely, that these ill-fated conifers performed a service by stabilizing the site for the eventual return of local hardwoods.

So how do you make sure that your plants will thrive on your chosen site? Professionals working on urban forest and reforestation projects have learned a great deal over the past few decades about matching plants to the local conditions — and not just the local climatic conditions. A plant cannot thrive if it is not growing within its adaptive range of soil, moisture, temperature and light conditions. The most significant factor restricting the range of a species is climate, followed by soil and moisture conditions, and then method of seed dispersal.

At present, experts suggest selecting seed sources from within your climatic zone. Native plants often survive well north of their known natural distribution area, but that does not mean they will reproduce in the new location. An individual plant may not be self-fertile and the growing season may not be long enough to mature the fruits — two very important factors limiting the migration of some species. If it is a species with fruit that matures late in the season, such as redbud or

Kentucky coffee tree, you will sometimes find the seeds shriveled because they have frozen while still immature.

How far the seed moves from the parent tree — the natural dispersion distance — is another important aspect of growing native plants. The vast majority of seed of any species is locally dispersed. It is the rare seed that survives a long journey, but it happens.

Seed movement ensures the maintenance and enhancement of existing genetic diversity. It is nature's life insurance policy. Genetic diversity is, without question, essential for organisms to evolve and survive through climate change, pests and diseases. Clonal propagation of plants by division, cuttings, grafting or tissue culture will help to maintain genetic diversity, but it does not increase genetic diversity. In fact, clonal propagation by itself brings evolution to a grinding halt.

Scientists have studied genetic diversity by collecting seeds of the same species from dozens of different places and planting them in the same location. Observations based on historical plantings indicate that some species have a relatively uniform genetic makeup throughout their range. These plants are called generalists (red pine comes to mind). Plants such as eastern flowering dogwood have very diverse gene pools throughout their range and are called specialists. The specialists have a narrow adaptive range, meaning that plants from seed grown in one place will perform best in that place.

You may be interested in growing a few trees in your backyard, hundreds in a community nursery or thousands in a commercial endeavor. You choose where the seed comes from; there are no plant police watching. Keep in mind that, for the specialist species, how long a plant lives and how well it regenerates naturally in the landscape are directly related to how well the seed source is matched to the planting site.

Since we know the adaptive ranges of only a handful of trees and shrubs (this information, along with my own observations, is provided in Chapter 6), seed source guidelines should be treated simply as that — a guide. In fact, due to our limited knowledge of adaptive ranges and the unpredictability of climate, it may be advantageous to plant seed from two or three regional seed sources of the same species on certain sites, especially where native plants of those species are absent. This is akin to not putting all our eggs in one basket.

There is much satisfaction to be gained in gathering seeds from an old oak or elm or witch hazel, knowing that you are playing a role in preserving the genetic diversity of local trees and shrubs. However, avoid collecting from isolated plants, because they must self-pollinate, which can reduce germination vigor and longevity of their offspring. Also avoid native plants that are mass-produced by cuttings or grafting, which results in very

Seeds and Drought

A moisture shortage with severe droughts over the past few decades in the Great Lakes region has created stress for every species, with little time for adaptation. Given these conditions, it probably makes sense to collect seeds from drier sites, where the plants have a greater tolerance of drought.

Buying Trees and Shrubs from the Garden Center

Most commercial garden centers do not track the seed source of the "native" trees and shrubs they sell. More to the point, the plants are usually mass-produced by cuttings or grafting. This means that every plant of a horticultural cultivar, such as *Rhus aromatica* 'Low Grow' or *Acer saccharum* 'Green Mountain', is genetically identical whether growing in Minnesota, Ontario or Maine. This genetic uniformity is unacceptable in the natural world.

The seed certification program established in Minnesota provided a model on which the Forest Gene Conservation Association of Ontario has established a seed and stock certification program. Designed to provide information on the sources of woody plants, it allows the buyer to make choices based on the climatic characteristics of the seed source. A number of small growers in Canada and the United States now offer source-identified stock. Ask your supplier for the seed source information of the plants you want to buy and decide for yourself if the source is appropriate to your needs.

A mature cone of the white pine, which the tree produces only once every three to four years

uniform plants, since they usually originate from only one parent plant. Old roadside and street trees, on the other hand, are an ideal source of seeds because they were usually obtained from the local woodlands as the town or city expanded. These old trees represent important seed sources for new urban trees.

Seed collectors for large naturalization projects that use native plants to restore or rehabilitate environments need to take a wider view. To sustain genetic diversity, it is important they collect seeds from the population rather than the individual. This thinking goes against the horticulturalist's and, to some extent, the forester's historical tendency to focus on outstanding individuals; however, the most desirable seeds are those collected from several plants at each of several groves.

How to Tell When the Fruit Is Ready to Collect

Make sure that you collect ripe or mature fruit, not the aborted and empty "premature drop." Many species drop some fruit before it has matured (these are noted in Chapter 6). Immature fruit is usually green. Fruit maturity is associated with a shift in color, usually toward reddish or yellowish green. Overly mature or ripe fruit is indicated (at times on only part of the fruit surface) by a full shift in color to red, blue, brown or black, and occasionally to yellow or orange.

For many species, gathering the mature fruit is acceptable or even advantageous; gathering ripe fruit is always acceptable. The distinction between mature and ripe fruit, as well as the fruit-ripening time, is described for each species in Chapter 6.

There will be times when you are in a location that you will not return to and the fruit is still green but approaching maturity. On two occasions, I collected the green fruits of both *Viburnum* and *Amelanchier*. First, I picked a few short branches with the fruit still attached, removed half of the leaves, and placed the branches in a moist plastic bag. Once home, I put the branches into a vase filled with water, and used a plastic bag as a "greenhouse" to enclose the whole works. Then I placed the vase out of direct sunlight and changed the water every few days, producing mature and ripe fruit within three weeks.

When gathering from large-seeded species, check for seed soundness by cutting a few fruits open to see if they are filled with seeds and see how sound the seeds are. Species like oak, ash and hickory may produce a full crop of fruit that has been entirely consumed by weevils. As a general rule, sound seeds are bright white and crisp inside. The exceptions are maple seeds, which are green inside, and pawpaw, which are unusually discolored and streaked with brown.

The cut test is a simple way to check the contents of larger seeds. Hold the seed lightly between your thumb and forefinger and cut straight through the narrowest dimension using a good pair of pruning shears.

Seed-Collecting Techniques

Seed collectors use many creative techniques. Fruits can be picked by hand and tossed into an open umbrella or onto a drop cloth. Sometimes fruits will fall onto a tarpaulin if the shrub or tree is given a shake or if a rope is thrown over a branch and lightly jerked. Occasionally, seeds can be swept from the ground or from a tarp that has been placed downwind of a seed-dispersing tree. Gone are the days when a tree was simply cut down to allow handpicking from the crown.

Fruits that squirrels drop to the ground can be picked up if your timing is right. The high crowns of some interior forest trees — red maple, beech, tulip tree, oak, butternut, hemlock, pine and spruce — make it difficult to obtain seed without their help. Squirrels begin their harvest about a week prior to the color shift that indicates maturity. I have often obtained good seeds from these barely mature fruits by spreading them out and monitoring their dryness as described in Chapter 6.

Can You Eat the Fruit?

I lived for a while with a wonderful dog, a cross between a border collie and a husky. He was never fed table scraps. When I took him on wilderness trips, he packed in and ate his own food. But once, during the height of blueberry season, I was amazed to see him carefully eating individual ripe berries. How did he know that canines can eat blueberries? Was it genetically coded? Did he smell blueberries in the feces of his cousin the wolf?

Even in humans, gathering wild seeds or fruits often triggers a memory of ancestral ways. It's an instinctual survival response. There are very few of us who can pass raspberries or blueberries in the wild and not reach out for one. As for less familiar seeds and berries, people often ask me, "Is it edible?" I always reply, "Yes, everything is edible — once!"

Sometimes people assume that because birds eat a certain fruit it is safe for humans to eat. Birds eat poison ivy, virginia creeper and European buckthorn (*Rhamnus cathartica*), too. *Cathartica* comes from the word *catharsis*, which in turn means "purging." I ate just one berry of *Rhamnus cathartica* in the fall of 2001, fully aware that it was a purgative but thinking, "Heck, what can one berry do?" I barely survived the two days of absolute elimination.

The Seed Collector's Botany Bag

When I head out on a day's hike I take my "botany bag" with me. It contains

- envelopes and paper bags to contain dry fruit and seeds
- plastic bags for moist seeds or cuttings
- a compact umbrella to turn upside down to catch seeds shaken or thrown into it
- pencils and tags
- a Swiss Army knife
- pruning shears
- a water bottle
- a field notebook
- a container to carry everything

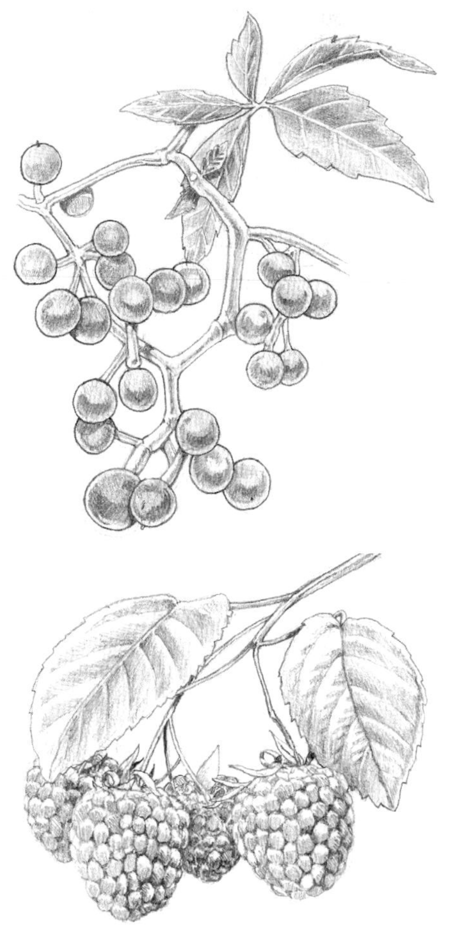

A few berries of the virginia creeper, Parthenocissus inserta *(above), definitely are enough, whereas it is rare to have enough of the delicious red raspberry,* Rubus idaeus *(below).*

It's a good idea to consult books on wild edible plants to guide your own indulgence. Some of our wild fruits are wonderful. I have gorged on juneberry (*Amelanchier stolonifera*), eaten pawpaw when the fruit was safe to eat and stuffed handfuls of my favorite, black raspberry (*Rubus occidentalis*), into my mouth. A note of caution, though. Sometimes information about the edibility of certain seeds and fruits is passed on through hearsay rather than personal experience. In this book, I have indicated edibility in Chapter 6 only if I have eaten the fruit myself.

*The showy petal-like bracts of the eastern flowering dogwood (*Cornus florida*) delight in spring.*

The Ethics of Picking Seeds

The temptation to reach out and pick some seeds from a low branch or from the ground is often irresistible. Seed gathering can become an almost addictive pastime. But it is unacceptable in arboreta, botanical gardens and parks. Seed collecting is incompatible with the aims and objectives of these special places. The ability of every visitor to see the flower and fruit of both common and unusual plants is important, including the amount and arrangement of fruit on the ground. Occasionally, the seed is required for research or fundraising. These institutions are simply off-limits for picking fruit without special permits. Also, because botanical gardens display plants from all over the world, a native plant could be pollinated by a closely related exotic species, thus making its seed unacceptable for adding to the natural landscape.

Trails are making many parks and conservation lands more accessible. Although these areas are off-limits for bulk seed harvesting, you are not likely to be admonished for gathering a few seeds. But rushed and careless picking can result in broken branches and trampled vegetation. You would be surprised how often this happens.

If you want to collect seed on privately owned land, obtain permission first. I once stepped over a fence to examine a massive serviceberry tree that stood in an abandoned pasture beside an old stone fence. A very upset owner soon appeared. He had had enough of people taking stones from his fence line and couldn't understand my interest in the tree. I bashfully apologized when he suggested that it would have taken very little time for me to knock on his door. End of discussion.

On another occasion, I noticed an eastern flowering dogwood in a villager's yard near Grand Bend, Ontario. I long ago realized that local rare species will often show up in someone's yard, dug from the wild nearby. I knocked at the door and a very friendly elderly man told me he had found it "up in the valley" and would be happy to show me a special old dogwood, the tree his dogwood had perhaps descended from, in a few days' time. He suggested that if I really wanted to know more about the trees in the area, I should go and see Frank in the next village.

I did meet Frank, and he has twice taken me to an amazing forest — one that has been respected by four generations of farmers. They harvested trees from the forest but always left the healthiest and best trees standing. Frank showed me a stand of flowering dogwood growing with witch hazel and maple-leaved viburnum under magnificent tulip trees.

A few days later, in the pouring rain, Frank's friend took me to the old dogwood. It was a lone tree, 16 inches in diameter and possibly 26 feet tall. Barely visible was an iron ring embedded in the trunk. My guide said that his grandfather had tied his horse to it. This tree was dead when I visited it again eight years later. The seedlings grown from the seeds I collected from this tree in 1995 are its only known legacy.

Take No More Than What Is Needed

I have been told that when native healers are gathering herbs, the plants themselves tell them to always walk past the first ones they find. It is easy to see how this practice would ensure that an area is not stripped of a locally rare species. The same philosophy applies to picking seeds.

Before reaching to gather any seeds, consider where you intend planting the seedlings. What is the intended site like? How does it compare to the habitat you are now in? Take into account factors such as the soil, drainage and exposure to sun and wind. Is it a reasonable match? Identify the species and determine its relative abundance. If it is rare, or one of only a few plants with a few seeds, leave it alone.

Determine how many seeds are in the fruit you pick, and take no more than you need. It is very easy to gather far more seed than required. Seed collectors are encouraged by the Society for Ecological Restoration (SER) to leave a large percentage of the seed crop for natural dispersal.

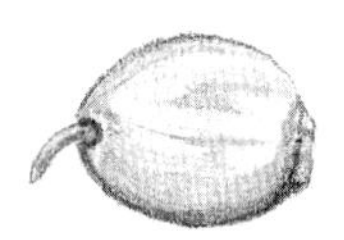

Germinating seeds of the chinquapin oak (Quercus muehlenbergii)

Preventing seed raids

When you have a bag of seeds here and seeds spread out to dry there, an overnight raid by mice or a chipmunk can completely devour your harvest. Always, always protect your seeds from scavenging rodents. They find seeds surprisingly quickly — they are eternally searching for food.

Chapter 3

Seeds to Seedlings

Don't judge each day by the harvest you reap but by the seeds that you plant.

— Robert Louis Stevenson —

The fascinating process of seed germination becomes visible when the root (radicle) emerges from the seed — the jewel of nature. I have witnessed over and over again a uniformity of germination that resembles an orchestrated ballet, synchronized as if the seeds were dancing together. To see the push of shoots through the mulch on top of a clay pot or seed frame is a most poignant experience.

Temporary Storage

During some of the rich seed-dispersal times of the year, it is possible to arrive home with seeds of several species and not have enough time to handle them all. I still manage to do this. As there is little reward in having your seeds overheat, rot or dry out too much, here are some insights on temporary storage.

Fresh-picked fruit must not be allowed to heat up in a closed container or a plastic bag — the seed will start to ferment and will heat up above a critical temperature, which will destroy the seed. As soon as you get home, spread out the fruit loosely in a cool place, away from direct sunlight.

The seed of some species is best extracted from the fruit or, in some cases, sown or treated soon after collection. The seed of other species can be dried and stored for sowing or treatment later. As a rule of thumb, if fruits are dry on the plant before dispersal, keep them dry; if they are fleshy, keep them moist. Here are some guidelines for temporary storage:

Make sure to properly label your seeds; include the species, the date collected and where they were collected. I use an HB pencil — it's even more permanent than permanent marker.

- Dry fruits of maple, elm, ash, tulip tree, ironwood and musclewood can be kept temporarily in closed and labelled paper bags (*see* illustration on opposite page).
- Dry capsules of ninebark, kalmia, potentilla, Labrador tea and shrubby St. John's wort can be placed directly in labeled paper envelopes.
- Dry seeds from the catkins of birch and alder are best held in labeled paper envelopes.
- Evergreen cones can be kept in labeled paper bags or spread out to dry immediately after collection. The extracted seeds are kept in labeled paper envelopes.
- Nuts of buckeye, hickory, hazelnut and walnut are kept dry, in open containers, for up to a week. Oak acorns and beech nuts are short-lived and must not be allowed to dry out for more than a few days.
- Fleshy fruits of juniper, dogwood, viburnum, cherry, serviceberry, holly, spicebush and hackberry, if they are mature and firm, are kept in partly open containers to ripen. Ripe fruits can be refrigerated for up to three or four weeks. Serviceberry can be spread out to dry before storage.
- Fleshy capsules of euonymus, bittersweet and magnolia are set out to dry for a week or so until they open. These are best stored as dried "berries."
- Dry pods of honey locust, redbud, Kentucky coffee tree and trumpet vine can be left out for weeks.

Some seeds can be stored for decades at very low temperatures in specialized facilities. Gardeners and nursery folk should generally avoid storing seeds for this long, but a refrigerator, not a home freezer, is good for a year or two. Seed should be air-dried for several days and placed in an envelope in a sealed jar and then in the fridge. When you take the jar from the refrigerator, bring it up to room temperature before opening it. This prevents condensation from destroying any seeds you later put back into cold storage.

CAUTION

Both beaked hazel, *Corylus cornuta*, and ironwood, *Ostrya virginiana*, have minute, sharp hairs covering the fruit. When the fruit is dry, the hairs easily embed into the skin and can be very irritating. Wear thick work gloves when handling these two species.

Ironwood (Ostrya virginiana)

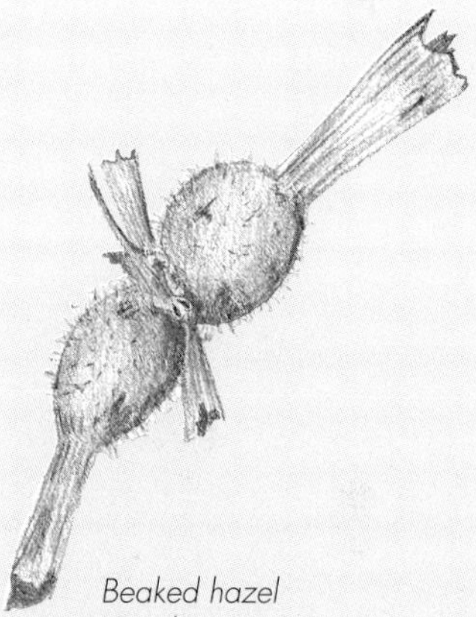

Beaked hazel (Corylus cornuta)

Seed Cleaning

When you first begin seed collecting, cut open a few fruits of several species to see the different ways they are organized, to identify the seed part, and to determine if the seeds are sound or empty. When you release the seed from the fruit you are looking at one of nature's most amazing exhibits.

Many seeds will not germinate if the fruit remains with the seed. The cleaning requirements for the seeds of each species are detailed in Chapter 6, but here are a few general guidelines:

- The winged seeds of elm, maple, ash, hop tree, birch and trumpet vine should not be separated from the wing. They land on the ground intact, and the radicle, the part that develops into the primary root, easily pushes through the seed's encasement.

- The dry seeds in bladders of ironwood and bladdernut and even musclewood seed, which is only partly enclosed at the base of a leaf-like bract, can be rubbed free of the fruits between very sturdy gloves (ironwood bladders are covered with minute, sharp hairs). The seeds are then easily separated from the shattered sacs by screening them on a sieve to remove the fine dust and hairs. The seeds are fairly heavy, so coarser material can be blown off with a fan.
- Willow and poplar capsules are collected just as they start to turn yellow or even after they have started to split and fluff out. They can be placed in a paper bag or open container in a dry room with no air movement. The capsules will split open in a day or two and release puffs of very light "cotton" that wafts away at the slightest breeze. The seeds are each attached to a cluster of tiny filaments and can be sown in that form.
- Squirrels remove the husks of hickory, buckeye, beech, hazels, sweet chestnut, butternut and walnut, and that is what we need to do, too. If the husk is not easily removed, it is likely that the seed is not alive, but always cut them in half to check. A 24-hour soak may be needed to make it easier to remove the husks of walnut and hickory. Buckeye, beech and sweet chestnut husks split wide open when they are dry. Use pruning shears to cut away hazel husks. Green walnut and butternut fruits can be whacked with a rubber mallet to pop off the skin. If the walnut and butternut have blackened or browned already, first soak them and then, wearing rubber boots, mash them in a wide bucket on the ground. Flush the hulls and inky pulp away with a hose. Wear your oldest clothes and gloves when you are working with walnut and butternut, because the hulls will stain everything dark brown — clothes, furniture and skin.
- The mature fleshy fruits of bittersweet, dogwood, sumac, viburnum and wild plum can be left in a mostly closed paper bag to over-ripen for a few days. There is a fine line between what is needed to soften the fruit and when decomposition sets in. Look at these fruits sooner rather than later. Once the fruit is fully ripe and softened, it is easily cleaned in a grit bag.

A simple glass jar is a great tool for separating seeds from grit. The grit falls to the bottom, the pulp and empty seeds rise to the top, and the seeds rest on the grit.

The Grit Bag

One year, an overwintering mockingbird got into the habit of perching on a branch in a sunny, sheltered spot outside my dining room window. I collected the snow underneath its perch for about a month, stratified the seeds from its droppings, and planted them. Roses, junipers, bittersweet, hawthorns, crab apples and nightshade germinated that spring. Afterward, I fantasized about keeping a few captive birds and feeding them the native fruit that I wanted cleaned.

Birds play an important role in cleaning seeds. The fruit is macerated by the grit in a bird's crop, with the added benefit that the seed coat is scratched, or scarified. The tearing action of grit on the tough outer proteins of the seed (called scarification) may help some of the fleshy-fruited species such as viburnum to germinate.

I don't keep birds, but I do make use of their technique by macerating fleshy fruit in a 1-quart (1-liter) milk bag containing coarse grit (sharp sand). I simply place seed and grit together in the bag, press the air out of the bag, and grind the mixture firmly between my knuckles. Seeds that are not hard enough to take the firm grinding in a grit bag are indicated in Chapter 6.

Once the fruit is well macerated, dump the contents into a tall, clear juice container or glass jar and flush with water. The pulp and empty seed will rise to the top, the grit will drop quickly to the bottom, and the filled seeds will settle on top of the grit. Separating the seeds from the grit takes practice. First, carefully decant the floating pulp and empty seeds into a sieve and then dump them into the compost. Add more water to the container, vigorously swirl it to suspend the seeds and then decant them again into the sieve to be saved. Repeat as needed. This should result in most of the grit remaining in the container and your cleaned seeds on the sieve.

Most ripe fleshy fruits, or those that were soaked first, will clean up nicely in the grit bag. The cleaned and scratched seeds can then be surface-dried and placed into stratification (which is discussed later on in this chapter) or sown directly. Some fruits are very greasy or oily, and adding a few drops of biodegradable detergent to the grit bag will dissipate the oils and make pulp separation easier. Spicebush, euonymus and magnolia fruits are very oily but the seeds are heavy enough to sink and separate from the pulp. Bittersweet fruits are also oily but the seeds are sometimes so light that they may still float. When a cut test indicates they are filled, the seeds and pulp of bittersweet need to be dried together. Once dried, the whole works can be hand-rubbed to detach the pulp from the seeds. Then the dry pulp can be fanned away or winnowed in the way that grain used to be separated from the chaff.

Once you have separated the seeds from the grit and thrown the empty seeds and chaff into the compost, use a sieve and catch-basin to decant the remaining seeds.

Understanding Dormancy

Seed dispersal occurs over the entire year. Paper birch seed is dispersed from July until March, elm seed during a week in early June, black cherry seed during a few weeks in August and witch hazel seed in October. Sumac retains its fruit for more than a year. There are almost as many dispersal patterns as there are families of plants. Depending on the time of year that the seed falls, seeds may need to wait for appropriate growing conditions before they germinate.

All temperate plants have hormones that prevent the leaf buds from leafing out during a warm spell between September and March. This "resting" stage is called dormancy. To break out of dormancy, the plants must go through a cold temperature period between 32°F and 41°F (0°–5°C). The amount of time required is regulated by the plant's genetic adaptation to the local climate. Cold temperatures trigger the enzyme-controlled slow breakdown of the growth-inhibiting hormones within the winter buds.

A similar chemical reaction occurs inside the seed of many species. Seed that is dispersed in early summer, such as poplar, elm and silver maple, germinates quickly in warm, moist conditions. But seed that is dispersed later in the summer and into the fall must remain dormant to prevent it from germinating during an autumn warm spell and then being destroyed by the first frost. These seeds have a dormant embryo that must experience cold, moist conditions before the shoot and radicle will push out of the seed.

Breaking Seed Dormancy at Home

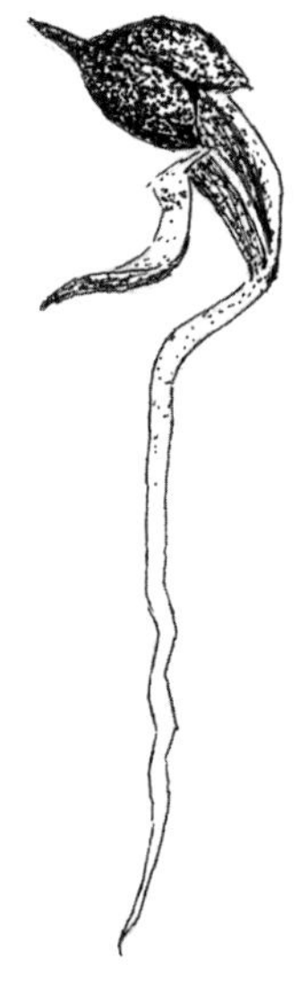

A sugar maple seed is easy to germinate but it first needs 90 to 120 days of cold, moist conditions.

In nature, cold, moist conditions occur before and after the ground freezes. But, more significantly, once the snow accumulates, the ground thaws beneath the insulating layers of leaves and snow. One year in late January, after almost a month of –4°F (–20°C) temperatures, I moved 16 inches of snow and the leaf litter aside to find a trillium rhizome for study. The surface soil was partly heaved, very damp, and had ice crystals in it, but I was able to dig the rhizome out of the soil just using my fingers. That same day, the city nearby had to thaw out frozen water lines under exposed driveways. The insulation provided by snow and leaves on the forest floor is truly remarkable.

In the natural world, seeds that fall in suitable habitat will break dormancy naturally and receive appropriate moisture and protection. Scientists have developed various methods to mimic nature's ways. Breaking seed dormancy by using stratification, a cold, moist treatment period, is one of them. The duration of stratification treatment is very specific, and seeds from the same species, collected from several locations, will all have very similar requirements. This makes it possible for us to use a guide to achieve uniform and consistently high germination — using fresh or properly stored seeds, of course.

Sugar maple and red oak are good examples of species that break dormancy with stratification. If you plant sugar maple or red oak seeds indoors after collecting them, they will not germinate. If you store them dry and cold over the winter, they will not germinate. The seeds of these two species must experience 90 to 120 days of cold, moist conditions before the radicle will emerge. These are among the easiest trees to germinate, and, because of their large seeds, their frequent occurrence in the wild and their importance in our forests, they make good candidates for education programs.

Above, a cluster of ripe seeds and a single seed of the hop tree (Ptelea trifoliata). This tree requires a warm period and a cold period to stimulate germination.

In contrast to sugar maple or red oak, the radicles of white and chinquapin oak emerge soon after the acorns fall to the ground. After a few days of rain in early October, I have even seen chinquapin oak with white root tips emerging from ripe acorns still attached to the tree. Yet, to grow, even these seeds need a cold period to develop into shoots. When I first started working at the University of Guelph Arboretum, several dozen pots full of chinquapin oak seeds collected the previous fall were stored in a warm greenhouse. I was told they had all germinated but had never grown a shoot. These "half-germinated" seeds required a cold period to break the shoot dormancy and could not possibly have grown when kept warm over winter.

Double Dormancy

The autumn seed crop of some species germinates unevenly, with a few seedlings emerging the first spring and the rest the following spring. Their ripe seeds have an immature embryo that must break out of a double (two-stage) dormancy. The embryo requires a warm period of one to three months either before winter or during the following summer to complete the first stage of germination. These species then require a period of cold, moist conditions to break down the inhibitors that prevent shoot elongation — the second stage of germination.

Species that tend to germinate unevenly are tulip tree, ash, hawthorn, burning bush, basswood, musclewood, hop-tree, some dogwoods, some viburnums and summer-sown red maple seed that dried too long before being sown. For these species, be sure to label and date your seed frames and pots clearly so you don't disturb the seed for two complete springs. Or be prepared to replant germinating seeds when lifting out the few seedlings that might germinate the first year.

Double-dormant seeds of ash, hop-tree, tulip tree and possibly basswood can be "pushed" to a higher germination rate in the first spring if the seed is sown when it is mature. Pick the fruit when it is changing color from green to yellow. If these yellowish seeds are planted immediately in September, in outdoor frames, optimum soil temperature conditions are still available for embryo development. This will increase the possibility that most of the seeds will germinate the first spring. If this isn't possible, the mature seeds of these plants and the ripe seeds of all other double-dormant species can be given an artificial warm, moist treatment followed by a cold, moist treatment to obtain similar results.

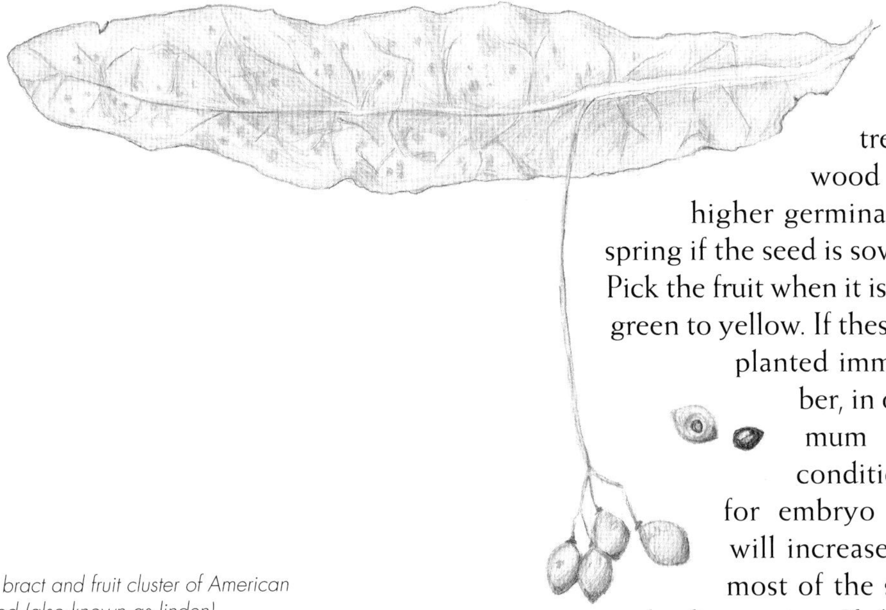

The fruit bract and fruit cluster of American basswood (also known as linden)

Seed Pre-treatment

Seeds that have been dried for storage need to be fully rehydrated by being soaked in water before treatment or planting. A 24-hour presoak draws out some of the germination-inhibiting chemicals in the seed coat, which partly mimics a good rainfall. Use about four times the volume of water to seed. Appendix III indicates whether or not a pre-soak is required for each species.

Fermentation

Gibberellic acid (Ga) is produced when fleshy fruits are allowed to ferment. (This is how alcohol was discovered.) Ga is known to improve germination in plants such as euonymus, honeysuckle, elderberry and viburnum, and some growers use commercially prepared Ga as a seed treatment for these plants. You can ferment the fruits of these plants yourself by placing them in a closed plastic bag for a few days; longer than that and you risk damaging the seeds.

Stratification

The term stratification refers to the procedure for treating seeds inside a plastic bag filled with a damp mix of sand and peat (see below), or in bags that are buried in a soil pit, to stimulate germination. Warm, moist soil conditions are re-created by placing the bag in a plastic storage container and keeping it at room temperature. This mimics early fall or summer conditions. Cool, moist soil conditions can be re-created by placing the bag in the fridge or a root cellar (not the freezer) to mimic winter conditions. Under the insulating blanket of snow and leaf litter, the soil stays a little above freezing for much of the winter. This is the condition we are trying to imitate — damp and cold, not frozen solid. Although some people may have success storing bagged nuts in an unheated garage, a refrigerator is ideal for the small-seeded species.

The stratifying mix is a damp mixture of about 50 percent washed and drained, but still wet, horticultural or concrete sand, and 50 percent dry peat moss. The peat moss seems to be important to re-create the slightly acid conditions that are common on the soil's surface, where organic decay takes place.

To stratify, mix the pre-soaked seed with about three times the volume of stratifying mix in a plastic freezer bag. (Sandwich bags are too thin and don't retain moisture well.) Close the bag tightly, but leave some air inside so that the seeds can breathe. Attach a label with the plant's name and the date. I find it best to place all the bags in a large, sealed container (the kind you use to store food in the fridge) to reduce moisture loss.

Check the moisture content of each bag monthly and re-dampen if necessary. Species that are in warm stratification need the moisture replenished more often. If the mix is too dry, the chemical reaction comes to a grinding halt; if it is too wet, the seeds will rot. Try to aim for the dampness of a wrung-out sponge.

Sometimes molds form on the pulp remaining on partially cleaned seeds or on dead seeds in the stratification bag. I have found that this doesn't affect the sound seeds. If you are concerned, the seed lot can be screened from the moldy mix, washed, checked for soundness and placed into a clean stratification mix to complete the treatment.

As the end of the cold stratification period approaches, check in case the seeds germinate sooner than expected. If roots emerge early, the germinating seeds are ready to be planted.

Before you start the stratification treatment, make sure that the duration of treatment takes you to springtime, when your

Bladdernut (Staphylea trifolia) *capsule and seeds*

soil can be easily worked and the seeds planted outside. Or, if the seed quantities are small, you might want to plant them earlier in a pot under grow lights, in a greenhouse or cold frame or on a window sill — whatever works best for you.

Some species will not germinate in the first spring if pregermination conditions were not optimal. As long as the seed is still alive and the soil moisture is managed, they will usually germinate the following spring. Patience is often well rewarded.

Hard Seed Coat Treatment

Basswood, bladdernut and hawthorn have a dormant embryo in combination with a very hard seed coat, which can result in delayed germination. In nature, these seeds may lie in the humus layer for years before the seed coat decomposes enough to take up water. These species should be well labeled and left undisturbed in the seed frame through two full spring seasons, and even a third. It is crucial to protect these seeds from mammals during the entire period.

Hard seed coats may be an evolutionary adaptation to help seeds survive the heat of brush fires. Redbud, honey locust and sumac all respond rapidly to a boiling water treatment that essentially mimics the steaming of soil during a ground fire. Store these seeds dry until springtime, when the day length is appropriate for growth. Place the seeds in a coffee mug and add about five times the amount of boiling water (it should be at a full rolling boil) as seeds. Let it cool and leave the seeds in the cold water for 24 hours. Seeds that receive enough heat to break down the hard seed coat proteins will take up water and swell to two or three times their original size. (Sulfuric acid is the classic treatment for denaturing hard seed coats, but I have always tried to avoid its use because it is dangerous to use and difficult for the public to obtain.)

Kentucky coffee tree is a species that does not respond well to boiling water or a scrubbing with grit. It germinates readily only after filing or grinding off a small patch of seed coat and soaking the seed for a day in cold water until it swells. Swollen seeds must be planted right away. Most legumes, and all sumacs, will germinate in about 10 days after exposure to the heat from the boiling water.

Don't Throw Out That Old Blender!

Used with caution, a blender can replace the grit bag for cleaning fleshy fruit or for scarifying hard seed coats. Old blenders with a detachable blade can be customized by grinding the sharp edge off the existing blade for use with hard seeds or by substituting a customized piece of polyethylene pipe for soft seeds. Or small pieces of tubing can be slipped over the blades. To nick the seed coats of very hard seeds such as hawthorn and juniper, start at the lowest speeds and keep checking to make sure the seeds aren't being cut open.

Sowing the Seeds

Other than rebuilding the surface organic layer with compost, you do not need to alter the soil of a seedbed for indigenous plants. If you sow your seeds in beds that have been thoroughly protected from rodents, the beds require little maintenance, just watering during a drought. However, a seedbed is not well suited to sowing very small-seeded species, especially those that need to be sown on the soil surface. They are best started in containers.

The seedbed should be well drained. To ensure this, don't locate it in an area that is flooded at any time of the year. The soil should be worked deeply with a spading fork or a shovel to provide good air circulation. Avoid overusing a Rototiller — it tends to pulverize and pack the soil instead of retaining it or building an open soil structure. When you first set up your seedbed, pry up the soil with a spading fork before making the first pass with a Rototiller to incorporate compost into the surface layer.

One of the most amazing aspects of seed germination is the amount of energy that a germinating seed has available to break through the soil and surface organic layer. In nature, seeds have adapted to germinate under a variety of conditions. Most require some level of site disturbance. Birch and ninebark seeds are small enough to germinate on a patch of soil where white-throated sparrows have been scratching for food. Pine and spruce will germinate through the loose needle litter. Dogwood and viburnum seeds are deposited in leafy humus by defecating animals. Cherry seeds are stashed by chipmunks, and hickory nuts are planted by squirrels.

Once again, horticultural practices attempt to mimic nature. Cover your seeds with soil to a depth of about twice the smallest diameter of the seed. If a thin mulch is also applied, the soil covering can be reduced a bit — it is not that critical, so please don't use a ruler! It is better to plant too shallow than too deep. Seeds that are planted too deep — more than three or four times their diameter — may fail to break through to sunlight.

Sowing in Tube Trays

Some people prefer to raise seedlings in tube trays under lights. Plastic tubes seem best suited for larger seeds that have been stratified and can be expected to germinate in a few weeks' time.

Tubes require frequent watering once the seedlings are underway, and a commercial soilless mixture is generally used to provide good drainage. The tubes are available in a variety of lengths. The best ones have ridges on the inside walls that guide the growing roots downward to an open bottom. When the tubes are arranged on an open wire-mesh surface, the root tips dry up when they emerge from the bottom of the tube — a root-control method called air pruning. This container design is particularly convenient for mass plantings of trees, shrubs and herbs because it produces compact roots on easy-to-plant seedlings that establish themselves very quickly. Watering and shading are crucial on hot, sunny days.

Weed Control in the Seed Bed

Weeds are part of a healthy soil, but vigilant control is required to prevent them from growing over your seedlings. Try to prepare your seedbed a week before you sow your seeds. This will allow a weed crop to germinate, which you can then slice off with a sharpened hoe. If you are able to prevent the weeds from going to seed, you will have a growing area that is almost free of weeds in a few years.

Direct Sowing in a Seed Bed

The larger nuts are best planted in a rodent-proof seed frame soon after harvest in the autumn. They should be permanently mulched with a thin layer of leaves. The leaves need to be thinned in early April before the seeds germinate, or the new seedlings will be elongated and deformed or may suffocate when they try to grow through the mulch.

Sow your stratified seeds in small patches in a rodent-proof seed frame in the spring, after their treatment time has been completed. Seeds that have been stratified in the peat-sand mix are usually sown with the mixture lightly pressed and covered with a single to double seed-depth of soil. Mulch the planted seed patch with a thin layer of coarse sawdust for small seeds, wood shavings for medium-sized seeds and wood chips for large seeds. Don't substitute a thick layer for a thin layer; its decomposition will remove too much nitrogen from the soil.

After sowing elm or red maple seeds in June, I have placed a piece of plywood or burlap over the patch of seeds during a heat wave, removed the covering at the first sign of germination one week later, and obtained excellent results. Mice love to tunnel under plywood, so a rodent-free wire frame is still required.

Growing Seeds in Clay Pots

For many years, I have used unwashed 4- to 6-inch clay pots filled with potting soil to grow many species at home. I prefer to use my own composted soil mixed with local, washed sand and roughly one-quarter peat moss to increase aeration and acidify the soil a bit. Most of our indigenous woody plants germinate best in slightly acidic soil — even those that grow in limestone ecosystems. The commercial soilless mixes will do, but their high peat content is quite acidic and may unduly stress seedlings that will ultimately grow in neutral to slightly alkaline soils. I doubt that anyone has tested this, but it has always been my gut feeling.

Clay pots are very convenient for small quantities (10 to 30 seeds) of all the medium-sized seeds such as the dogwoods, viburnums, spicebush and magnolia. Fresh seed can be sown and stratified in the pots in the fall or stratified seed can be sown in the spring. Pots of fresh seed can be buried, plunged up to three-quarters deep in the soil in the cold frame, mulched and covered with leaves for the winter. The leaves must be removed in springtime before germination. Be sure to set your labels deep into the pot so that they aren't pulled up when you remove the leaves.

Once the clay pot has been set into the soil, it is able to draw moisture from the ground and there is little risk of the pot drying out, even if you are away for a week. The pots are best suited to smaller-seeded species, and slugs need to be monitored and trapped.

Sow your fresh seed in a clay pot in the fall, add mulch, then plunge the pot up to three-quarters deep in a cold frame and cover it with leaves.

Very Small Seeds

Very small seeds need to be grown in a small pot or tray. To sow, scatter them on the damp soil surface and water with a fine-nozzle watering can or a coarse mister to avoid washing them away. Because they are not covered with soil, these seeds are best germinated by putting a plate of Plexiglas over the seed flat or pot to maintain 100 percent humidity until germination is evident. Once the seedling leaves begin to lift from the soil, raise the Plexiglas for a few days and then remove it entirely, preferably on a cloudy day to lessen the shock of the change in humidity. When the seedlings are up, they should be thinned out if they are too dense. If they are transplanted into a clay pot, it can be plunged into earth in a cold frame or placed in a protected seedbed frame.

The radicles of your seeds will emerge when the moisture, temperature and oxygen conditions are just right. There are several possible reasons why some never emerge: the seeds may have been dead when you gathered them; they dried out or rotted after you collected them; seed-treatment conditions may not have been optimal; the seeds were planted at the wrong depth; or the seeds were eaten by rodents.

Soon after the root begins to extend deeper into the soil to obtain moisture and nutrients, the shoot begins to elongate and push up through the soil surface. Germination is completed and your tree, shrub or vine is on its way.

Chapter 4

Beyond Germination

Keep on sowing your seed, for you never know which will grow — perhaps they all will.

— Albert Einstein —

I often find myself going out to examine seedlings simply because I've never stopped being fascinated by their near magical emergence from the earth. The first appearance of a carpet of red maple seedlings from two old roadside trees was a particular thrill, and I quickly fetched the students who had helped to pick up the seed from the roadside two weeks earlier.

New seedlings are at a crucial stage of growth due to their susceptibility to fungus, rodents and slugs. I still remember a day in June 1982 when I was checking through the arboretum's seedbed nursery. Of special interest to me was a recently germinated plot of chinquapin oak, because it was a species I had not known about before the previous autumn. To my horror, the germinated seedlings were all uprooted or chewed off. A chipmunk had feasted on the food remaining in the acorns. This was new to me — I had rarely grown tree seedlings before, and certainly not oaks.

I soon built seed frames to protect seedbeds and young plants. Still, from time to time, a chipmunk, mouse, squirrel, rabbit, groundhog or deer outwits me and gets into some part of the nursery. One year, I had hundreds of beech seedlings growing under a wire cage. They disappeared one night. This stuff is so delicious that a chipmunk had tunneled under the frame to reach them.

Seedling loss is significant in the wild — very few seeds even germinate, let alone survive. But seedling loss also is inevitable in cultivation. After your investment in

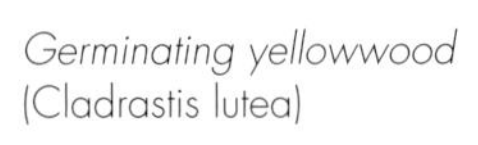

Germinating yellowwood (Cladrastis lutea)

time and nurturing, it is heartbreaking to lose protected seeds or seedlings, but it happens occasionally. You develop a healthy respect for chipmunks, sometimes bordering on hatred, but you have to admire their excavating skills.

There are two other major threats to young seedlings — slugs and a fungal disease called "damping-off." Both are dependent on constantly moist conditions. Slugs are more likely to be found in a cold frame and damping-off disease is more likely to occur in a container, just as the seed leaves are unfolding.

There is no ideal way to control slugs. Yes, chemical poisons will kill them, but the chemicals also remain in the soil and in the plants. Putting a toad in a cold frame to eat the slugs is an excellent method to combat them (*see* illustration p. 43), but be sure the toad always has water to drink. Another chemical-free method is to soak a newspaper in water and put it down where the slugs are dining on your plants. Slugs will hide beneath the wet paper late at night and can be killed or relocated the following morning. Lastly, slugs are attracted to a shallow dish of good beer (not "light" beer) — they drink too much and drown.

The damping-off fungus spreads out radially from the initial point of infection to attack the base of seedlings, causing them to wilt and fall over. Too-moist soil and stressed plants, usually due to a combination of under- and overwatering, promote its growth. Damping-off will wipe out an ever-widening ring of seedlings every day if the seedlings are closely spaced and the soil surface remains moist.

If damping-off invades your seedbed, you must stop watering and isolate the surviving seedlings in an "intensive care ward," separating them from the diseased seedlings to avoid fungal transfer. Use several clean spoons to scoop out the infected area and to scoop out the healthy seedlings at its edge, since the fungus has already spread beyond the dead and dying seedlings. Leave the hole open to allow it to dry out. Wait until the soil surface is well dried before watering again, but not so dry that your remaining seedlings begin to wilt. Although commercial fungicides are available for those inclined to use them, I prefer to use this natural method when possible.

To reduce the risk of fungal infections, your watering habits must change as soon as the seedlings begin to break ground. Before the seeds germinate, they must not be allowed to dry out. But, by the time the shoot breaks the surface, the root is relatively deep into the soil and the soil surface must be allowed to dry out between waterings. To counteract damping-off, choose a sunny day to water extensively. If you need to water on a cloudy day, it is best

Pricking Out

Pricking out is done when pot- or tray-grown seedlings grow three or four leaves above the seed leaves. Seedlings may look uniform, but usually there is a difference and the smallest ones can be discarded. At this stage, the seed leaves are still green and supporting the seedling with stored food reserves. This provides a two- to three-day period needed for the roots to reestablish themselves in the new soil.

Fill a clay pot or tray with soil mix and press it down lightly. With a knife or stick, make a hole that is large enough to accommodate the seedling root without bending it. This is especially important for tree species. If the root is too long for the hole, pinch or snip it shorter, insert the root, and then press the hole closed.

When I was about 14, my mother taught me how to "prick out" seedlings. We sat at a table with a wooden flat (a shallow wooden box) filled with potting soil between us. The tiny seedlings were plucked from a seedling pot and set about ¾ inch apart in small holes pricked in the soil. The seedlings responded well once free of the dense spacing needed for germination, and a month later they were transplanted into a nursery bed with little danger of damaging their roots.

How Woody Plants Grow

Woody plants begin by increasing in size exponentially once they have germinated. Every spring, their unique cambium cells between the wood and the bark are reactivated. They divide and add a layer of new cells to the water-conducting sapwood (the plumbing system) and a layer of new food-conducting cells to the inside of the bark. The stem growth pushes the bark out and causes it to fissure, flake or peel in distinctive ways. The new plumbing system each spring is perfectly designed to support the new shoots, leaves and flowers and a new crop of fruit. New roots in the soil join with mycorrhizal fungi that will assist in the uptake of phosphorus and other nutrients as the stored energy and food in the seed is used up.

The shallow roots spreading underground may be as long as the height of the plant above ground. They exist within the oxygen- and nutrient-rich top 12 inches of soil. The first seedling root — the tap root — will grow and thicken with age and may reach 3 feet or more deep into the soil before abruptly dividing into many short branches just above the water table. You can often see this on old windthrown trees.

done in the early morning so the moisture left on the plants and on the soil surface will dry off during the day.

Avoid the urge to water all containers uniformly by checking your seedlings first to determine which ones need water. Some species are thirstier or more developed than others and will require a drink more often. If a pot full of seedlings needs to be watered, water it completely so that it can be left alone for a few days. This gives the soil surface time to dry — the key to preventing damping-off. Once seedlings are a month past the "just emerged" stage, they toughen up and are progressively less prone to damping-off.

Watering plants at any stage of development should follow one rule: if a plant needs to be watered, make it a complete soaking, with as long an interval as possible between waterings, even to the point of the plant just starting to wilt. It has been my experience that many gardeners have a tendency to water lightly and frequently. This is a sure way to stress all but the toughest plants, ensuring that fungi have plenty of opportunities for infection. Even willow seedlings, a wetland species, will damp-off in a container.

Setting Up a Plant Nursery

My father maintained that a nursery was all about growing a root system that could be transplanted — anyone could grow the top of a plant. A good plant, according to him, was one that had been dug up, root-pruned and replanted to produce a root system that was compact and easily transplanted again.

A plant nursery is exactly what the word implies — a place to care for young plants. Your nursery can be anything from a single pot full of seedlings to a small bed in a garden to a full-scale commercial operation. The most critical consideration, regardless of size, is protection from wildlife. Fencing or enclosure is preferable to trapping or hunting, because it allows wildlife to continue to coexist with us as they have always.

Domestic cats and the automobile have only partly replaced the role of natural predators. Coyotes and wolves are important predators of the deer who daily dine on plants. A healthy population of coyotes or red foxes will control rabbit and groundhog populations. Fishers, red-shouldered hawks and occasionally owls are predators of gray squirrels. Coyotes, foxes, bobcats, hawks, owls and the endangered shrike help to keep the mice and chipmunks in balance.

Water availability is your next important consideration. Rain or pond water is preferable to any other water source. If you must use drinking water, let it sit overnight in a watering can or barrel so the chlorine evaporates. Well water usually has a high level of mineral salts that may deter seedling growth

or even kill newly germinated seedlings.

For the small nursery, a rainwater reservoir or a system of connected barrels is a great investment. Open barrels should be stocked with a goldfish or two to consume mosquito larvae, and reservoirs should be cleaned periodically. When a period of rain is in the forecast, I try to empty my rain barrels and rinse them out in advance.

A cold frame shelters young plants and is easy to build.

You can improve your rate of seedling survival by providing shelter from wind and sunlight. Avoid growing seedlings in hot areas beside driveways or against a south or west wall, or in low-lying ground that may be a frost pocket. Regular snow cover is fine, but if your property has a long fetch and you set up your nursery on the lee side of a windbreak or building, you could find a crushing 7 foot high snowdrift on top of your young plants. Generally, the east side of a building or windbreak provides ideal light levels, protection from drying winds and the option to collect rainwater.

Few woody plants can start their lives in full sun. About 50 percent shade can be provided in many ways: a burlap-covered frame raised above your plants on forked sticks, a walk-in style post-and-beam structure with a slat roof, or cedar beams covered with snow fencing. For temporary seedling shade, I use horizontal pieces of 2-inch angle iron welded to 2- to 3-foot-long sections of rebar. They hold up a set of 2 × 2 wooden rails on which I can place burlap, branches or snow fencing — whatever is handy. If you are shading a seedbed, you can place the shade material on top of the wire-mesh cage that is used to protect the seeds or young plants from animals. (*See* photo, p.255.)

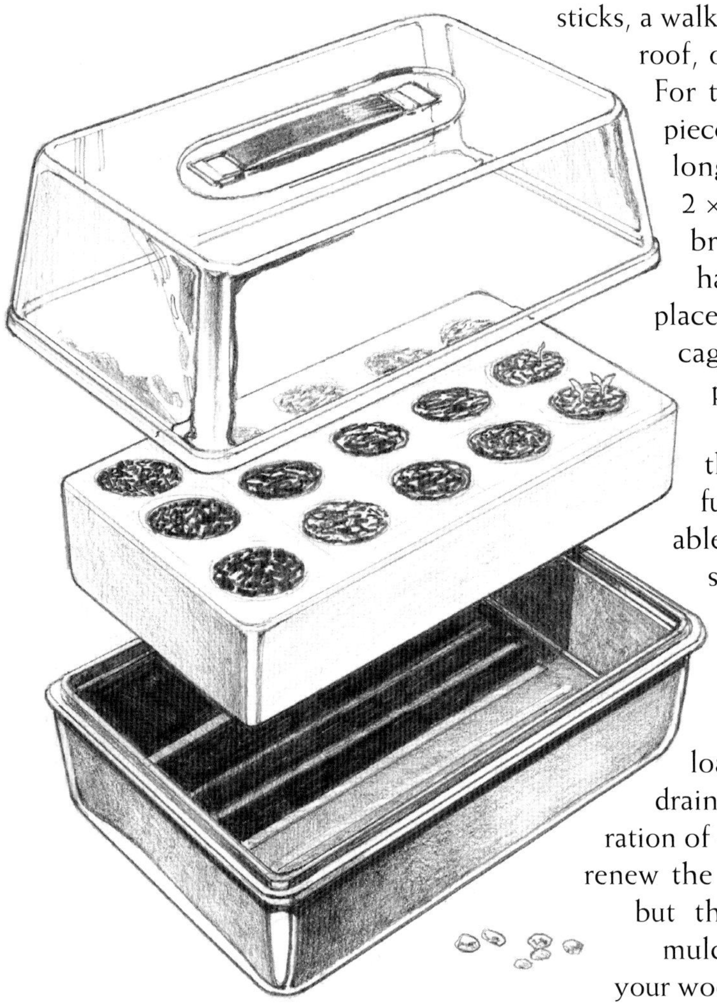

If you choose to grow your plants under the shade of trees to provide partial or even full shade at midday, make sure there is workable soil with no tree roots present to consume some of the available moisture and nutrients. Also, the germination- and growth-inhibiting toxin exuded from black walnut roots must be avoided.

Your soil should be well drained. Clay, loam, gravel or sand are all appropriate, if well drained. Most soils will require an initial incorporation of old manure or well-decomposed compost to renew the soil's health. Over the long term, regular but thin applications of wood chips and leaf mulches will enrich and condition the soil for your woody plants.

A commercial sweatbox

Growing areas may be associated with a downspout to provide additional water, or slightly sunken beds can be made with more acidic soil for plants having special requirements, such as winterberry holly, tupelo or swamp white oak, which are destined for a marsh edge. Plants destined for wet areas may be potted up for a growing season to plant during drier conditions associated with late summer.

An "old-fashioned" nursery raises trees and shrubs in the ground and limits outplanting to the spring, preferably, or fall dormant season. I still think this type of nursery is far superior to the nurseries that raise woody plants in containers filled with fertilized bark and peat moss from day one. When seedlings are freely rooted in unfertilized natural soil and naturally thinned, they are predisposed to thrive in the final planting site.

A cold frame is very useful for providing protection from wind, sun and climatic extremes. A tray of water should be provided if you temporarily employ a toad to control the slugs. The sloping cold frame should face north or east and can be put on the east side of a building to keep it even cooler in summertime. A hinged wire-mesh lid provides protection from foraging squirrels throughout the year. The cold frame can be covered with plywood for the winter when sunlight is not required. If sunlight is required during the winter months (i.e. there are green leaves), place several layers of clear plastic over a glass lid. Alternatively, cold frames or "hoop houses" can be purchased in many sizes.

A sweatbox (*see* illustration opposite) is indispensable for the germination of very small seeds that are sown on the soil surface, or for cuttings or poorly rooted divisions. It consists of a plastic-enclosed frame with a plastic liner on the bottom and a door or flap for access. The bottom is covered with a layer of wet peat moss. Premade plastic enclosures are available from garden centers, but most are too shallow to be useful for all applications. Many do have lovely dial vents to control temperature. If you don't want to go to the trouble of building or buying a sweatbox, surrounding the plants with a clear plastic bag and securing it over them will also serve.

Transplanting Outside

Transplanting seedlings outside is traditionally done on a rainy day when the air is humid and cool. These conditions ensure that there is very little stress on the newly planted seedlings. To reduce stress, the transplants may be planted under a temporary burlap shelter.

If your seedlings are in a clay pot or tray and they are growing slowly or have stopped growing altogether for the year, they can be unpotted as a whole and set into either a cold frame or a seedbed, depending on how tough the species is. One advantage of planting the seedlings as a group is that you don't have to water them as frequently, or at all if they are mulched. Left undisturbed in their original soil mix, they will quickly grow roots into the new soil. These plants can be lifted the following spring, teased apart and planted out at an appropriate spacing in a nursery.

At some point, everyone will experience a group of seedlings growing too large for their space in the second year and crowding others out. With the exception of a few slow-growing species, most seedlings should be transplanted after two seasons of growth, and fast-growing species like poplar, silver maple and sycamore, after only one season.

Many seedlings grow best when they are grown with their siblings. I like to sow densely. This lets seedlings thin themselves out through competition. The variation in size allows me to determine which seedlings are weaker and can be rogued by hand-thinning or transplant selection. Do not hesitate to compost your weakest plants, especially if you have

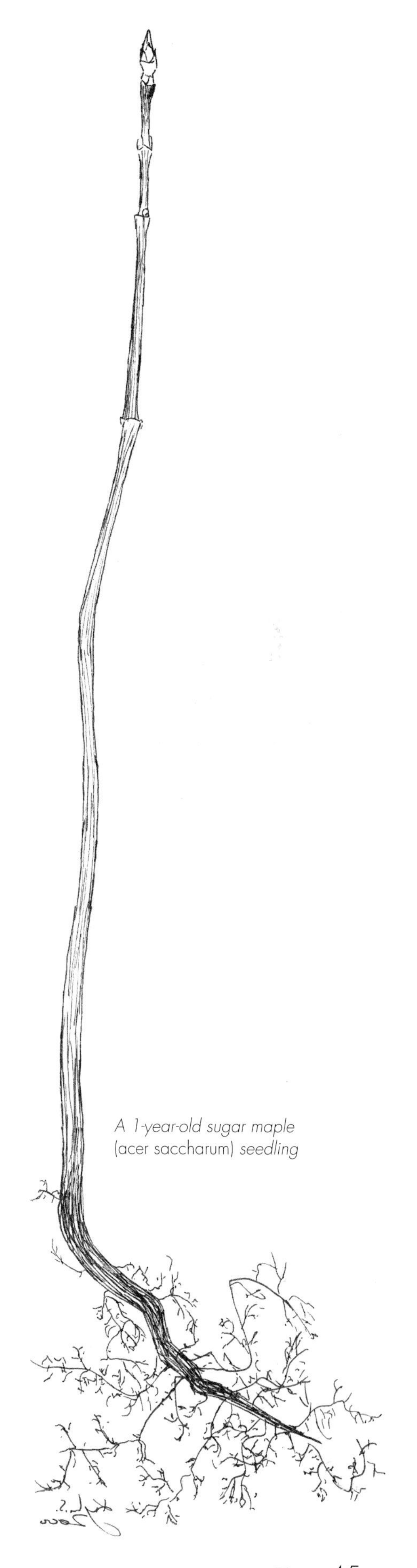

A 1-year-old sugar maple (acer saccharum) *seedling*

more than you need. There is little to be gained by planting trees, shrubs and vines that are not performing well at the seedling stage. Saving the weaker seedlings may seem kind-hearted, but, in nature, the weak do not survive.

Beyond the Seedling Stage

To plant seedlings in a nursery bed to grow for another year or two, first place a board on the bed to reduce soil compaction when you kneel down to plant. Loosen the soil with a spading fork, plunge your hand — fingertips first — about 6 inches into the soil, and pull back a little to leave a slot the width of your hand and 4 to 5 inches deep. The seedlings must be root-pruned to fit the hole without bending or twisting. To do this, after discarding the thin and very small seedlings, I bundle the healthiest ones together so that their root collars are all at the same height and then cut the roots so that they are a uniform 4 to 5 inches long. I then plant the seedlings about 3 to 4 inches apart in blocks, keeping in mind that I will lift a block of seedlings at a time when I dig them up a year or two later.

Digging up dormant plants grown in blocks in a seedbed or transplant bed is easy. It is important to prepare the new planting site in advance. Use a sharp spade to slice deeply all the way around a block of seedlings. With most fibrous-rooted species, you can reduce the possibility of tearing the roots by lightly prying the block up with the spade with one hand while simultaneously pulling several seedlings at a time with the other. If you are digging tap-rooted trees such as hickory, walnut and oak, first dig a trench on one side of the block about 4 to 6 inches away from the stems. Next, use the spade to slice as deeply as possible on the other side of the patch. Then tip the block of seedlings over toward the trench and pull on several seedlings together. This is best done in early spring when the soil is still moist, and you can often pull very deep tap roots out of the subsoil.

My father had another axiom: "A root out of the ground is like a fish out of water." After you have pulled up the group of seedlings, gently shake the soil off their roots until you can separate the plants. Place the roots immediately in a plastic bag and out of the sun and wind to keep the soil moist. Roots must not be allowed to dry out. I am amazed how many times I have seen plants with the roots exposed to the wind during planting operations. "Oh, they survive," say the workers, but I maintain that there is a huge difference between a plant surviving and a plant really thriving.

Move the seedlings to a shaded and sheltered location to take a close look at the roots. Use sharp pruning shears to hand-prune the roots to an appropriate length based on root

Plant Salvage

The misguided policies of both state and provincial governments continue to allow suburbs, new highways and mall sprawl to ruin natural ecosystems. The feverish "rescue" of plants and animals in advance of the bulldozers wrongly lets the decision-makers off the hook. Our aim should be to preserve wild nature, not to rescue it. However, the argument that "they are going to be destroyed anyway" tempts many plant collectors. Here are some tips to reduce the mortality rate of rescued plants:

– The removal of seedlings and young woody plants from ditches, new roads or destruction sites is best done when the plants are dormant. Transplanting trees more than 7 feet tall almost always ends in failure. Wild plants do not have the compact root system of transplanted nursery stock.

– All deciduous trees and shrubs have dormant basal buds that will sprout if the stems are severed by rabbits or mice. You can take advantage of this evolutionary adaptation by severing salvage plants a handswidth above the soil line before digging them up. If plants are lifted after they have leafed out, the foliage must be removed to reduce stress.

branching. Cut below the branching roots to increase the number of locations for new roots to develop.

The Planting Site

It is often said that native plants are better adapted to growing in our climate than nonnative plants. This statement is fundamentally wrong. Whether a plant thrives or not depends on the planter's ability to match the species to its ecological needs. There are many Eurasian species — French lilac, Norway spruce, European mountain ash, crack willow, European larch, mugo pine, dog rose, horse chestnut and a mega-list of invasive species — that are far better adapted than some indigenous plants to growing in much of our region. These exotic plants do well because they have adapted to the wide range of disturbed soils that are the result of human settlement. Our indigenous woody plants, on the other hand, have, for much of their history, thrived in rich, undisturbed soils.

To plant native plants successfully, you must pay close attention to the planting site's soil-moisture conditions and exposure to sunlight and wind, as well as the seed source. The existing soil and site characteristics determine which species will grow there — a fundamental principle of gardening that is all too often ignored when we choose a plant without listening to what the site has to say. Chapter 6 provides a general guide to the habitat requirements of each species, but you also need to become familiar with how the different species behave in your area.

Some Dirt on Soil

Natural soil develops in layers, and root systems have evolved to exploit those layers. Mixing the soil disturbs this natural process. Leaves and branches naturally lie on the earth's surface and, through the work of microorganisms, the organic matter is slowly incorporated into the soil. The longer it takes for decomposition to take place on the soil surface, the more acidic it tends to be.

The microorganisms (the numbers of these wee beasts in a teaspoon of soil staggers the mind) and the organic acids that are released during decomposition are important for maintaining the structure of clay and organic soil colloids. Incorporating well-rotted manure or compost into the soil will partly improve the soil structure, but it does little for the health of the soil if decomposition on the soil surface is absent. It is the decomposition process at the soil surface that is important, not the final product.

A Plant in a Pot

Many nurseries grow plants in pots. Most herbaceous plants and shrubs can tolerate being container-grown with their roots wrapping around and around inside the bottom of the pot. However, container-grown trees are likely to experience a serious structural problem called root girdling — a condition where the wrapping roots will gradually strangle the tree to death. This may occur a few years or decades after outplanting.

The Truth about Earthworms

Recent studies of earthworms reveal that none are native to the glaciated regions of North America, and, contrary to popular gardening belief, they do a huge amount of damage by dragging undecomposed organic material into the soil, where it does not belong. Even worse, where they have invaded natural forests (people who go fishing have regularly dumped their leftover bait), earthworms diminish the leaf litter and increase the rate at which forest soils dry up.

Protection from Browsing Animals

You may need to set up a wire or netting fence around your plants to protect them from browsing deer, rabbits or groundhogs. Avoid commercial tree tubes. The trees produce elongated shoots in the tubes that have poorly developed fiber cells and often bend over once the tube breaks down or is removed. Saplings must sway in the wind to strengthen their trunks.

The soil of the planting site does not need to be altered beyond enrichment of the surface with a thin layer (2 inches) of wood chips and an annual contribution of leaves, especially on soils that have been compacted or cultivated for decades. A thin layer of mulch does not impede air circulation and restricts invasive weeds for the first several years until the woody plants are well established.

Soil compaction is one of the greatest threats to soil and root-system health. The decline of sugar maples in schoolyards is due to compaction. Species such as bur oak, on the other hand, show a tremendous tolerance to it. Highly compacted soil should be repaired by pulling a chisel plow or subsoiler through it. Then a thin layer of compost and wood-chip mulch should be spread over the area.

Planting Method

During the process of digging up your woody plants from the nursery bed, you might have broken a few roots. Use your pruning shears and cut off any damaged roots. If your plant is a tree species, remove any roots that cross back on themselves or wrap around themselves. Shorten any roots that are really long so that outplanting will be easier. Generally, for every ⅓ inch of stem width you should have about 4 to 8 inches of root length. Place the roots back in the plastic bag right away or plant them in a temporary trench, heeled in, for planting later.

"Heeling in" is a way to store plants so that their roots don't dry out. A trench is dug in a sheltered and shaded location and the roots are set into the trench closely together, covered with soil, shaken to filter the soil into the roots, firmed a bit and watered. Most gardening books suggest that you lay the plants on an angle, but this is not necessary. In fact, if the plants are set upright and you run out of planting time, they can even be left in the trench for the rest of the growing season and planted the following spring. Species such as oaks and hickories that are difficult to transplant should be dug up, heeled in and left for a week to provide an unstressed time for new roots to be initiated before planting out.

A fine day for planting is one that is overcast, misty and calm and the soil is not wet and rain is in the forecast. For large-scale plantings, try to time your work to precede a period of cooler, rainy weather. Pre-organized community events often don't (or can't) take the weather into account. Planting in soils that are wet from recent rain makes it difficult to handle the soil without damaging its structure and may result in poorly planted stock. Planting during hot, windy weather is also likely to reduce survival rates.

When the soil and weather conditions are suitable for planting, dig up a few plants or pull the plants you want from where they are heeled in, moisten the roots and place them in a plastic bag that is kept out of direct sunlight. Don't store them in a bucket of water — plant cells need oxygen for respiration and the roots may be damaged or drown. It is better to simply dip bare-root plants in water before planting.

Gardening book diagrams that show a planting hole twice as wide as the roots and packed with manure, peat moss and bonemeal (dubbed the "five-dollar planting hole") do not make sense. These artificial "container gardens" tend to encourage the roots to stay within the nutrient-rich planting zone. This is counterproductive. Woody plant root systems must spread beyond the planting hole as rapidly as possible to obtain an adequate supply of moisture and nutrients.

When I dig planting holes, I first lift the herbaceous or humus surface layer and set it on one end of a 3-foot-square tarpaulin. Then I put the 4 to 6 inches of topsoil in another pile, the deeper soil in another pile and finally the subsoil (if I have to go that deep) in another pile.

To start with, place the root collar (the place where the root and stem join) about an inch below the soil surface. Then pour the loosened soil (subsoil first, followed by mineral soil and organic soil) over the roots. Shake the plant upward to allow the loosened soil to filter into the roots. Stop pulling when the root collar is level with the soil surface. A light press on the soil around the outer edge of the hole will secure your plant in place, and a thorough watering, not a heavy stomping, will set the soil into the root system. Replace the humus layer and add a thin mulch. New root development will take only a few days, and the thin mulch of wood chips helps to retain moisture and reduce weed competition.

Transplant Stress

Transplanted trees and shrubs may be killed by overwatering (drowning the roots) just as easily as by drought. An initial soil wetting is adequate for dormant bare-root plants. Once the leaves begin to unfold, the moisture loss from the plant increases. However, well-established roots are able to meet all but a hot, windy day's demands for moisture from a moist soil. If the soil is moist, adding more water doesn't remove heat stress. Instead, you need to reduce the plant's need to cool itself by providing shelter or by giving the leaves a cooling shower at midday, when desiccation rates are highest. The effort you make to cool down your plants during periods of heat stress will pay off: it will dramatically increase their chances of

The Value of Good Tools

Over the years I have found that a cheaply priced tool is not worth owning. I am a minimalist, and the few tools I have are easy to clean and maintain in good condition. A sturdy digging knife is a great weeding tool and can be used for digging up seedlings, too. A sharp spade will cut cleanly through roots. A good file is needed to keep the spade sharp, and a wire brush to keep it clean. A spading fork is indispensable for working soil and a steel rake is useful for seed- and nursery bed preparation. Pruning shears are essential. I swear by the Felco 2, but I have a large hand. Get one that fits your hand size. Last, for those inclined, a very sharp knife is needed for chip budding.

Buy your specialized tools from fine horticultural suppliers. My father maintained that he was never so wealthy that he could afford poor-quality tools.

survival. Of course, this is more easily done on private property, in close proximity to a watering can, than in the larger landscape.

Hickory, oak and other species that are considered difficult to transplant can be protected out in the landscape by a temporary shelter. A few branches of pine, or some sticks stuck into the ground and topped with burlap, will protect your plants while they are becoming established.

Some plants just do not grow well, often because the soil doesn't suit the species. Soil that is too wet or dry for a particular plant will damage its roots, resulting in small leaves, weak growth or death. Plants that are grown in soil that is too alkaline for them may have an iron and manganese deficiency, resulting in chlorotic leaves — leaves that are yellowish with green veins.

Alkaline soils tend to stress most seed sources of black oak, sassafras, holly and witch hazel, and severely stress striped maple and hobble-bush. Beech seedlings become chlorotic in the absence of their specific mycorrhizal fungi. Horticulturists like to practice intensive care on chlorotic plants, using manganese and iron foliage sprays or adding peat moss to the soil in order to turn their plants green, but to what end? If acidic-preference plants are not in the right soil, they may never really thrive. You will be tempted (we all are at first) to use these horticultural life-support methods to help them, but, if time is limited, try to focus your efforts on the plants that are most suited to the planting site.

English novelist and gardener Vita Sackville-West (1892–1962) advised, "Never retain for a second year, a plant which displeases you." I have heard occasional stories of plants that were nurtured along and have grown into healthy trees or shrubs, but most weak plants belong on the brush pile. If the soil characteristics are not optimal, my experience has been that weak plants usually continue to decline.

When growing plants with special requirements on inappropriate soil, such as many of the oaks, red maple and tupelo, a few dark green healthy seedlings may flourish among dozens of chlorotic seedlings. The healthy ones may have a genetic trait for adaptation or tolerance and continue to do well. It is worth spending extra time on them.

Propagation of Woody Plants by Division

Growing plants from seed is not your only option. The horticultural industry strongly encourages the public's interest in unusual forms of plants, and most of these are cloned by division, cuttings or grafting. These nursery-grown cultivated forms (cultivars) now make up the bulk of the planted landscape in urban areas. Every individual of a particular clone is identical, which guarantees getting what you want in flavor, form or color.

Many native woody plant species are self-cloning. They spread by underground stems or root sprouts and most of them can be easily propagated by division. This involves dividing an existing plant into pieces, with each division having both roots and a growing part that is or will become the stem.

It is often difficult to obtain seeds of hazelnuts, sassafras, juneberries and native honeysuckles because the fruit is consumed so rapidly by birds and other animals. Seeds of raspberries, currants and gooseberries are easily obtained, but may not produce the precise flavor that you are after, if that is your goal. In these cases, division may be the answer. All of these species and many more are self-cloning plants.

If you only want to propagate a few plants, division permits the occasional collection from an established plant, while retaining the original plant as part of the landscape. This method is particularly useful at salvage sites, where divisions can be taken from several mother plants to help retain the genetic diversity in the landscape you are building.

Divisions are best made in the dormant season. Woody plant stems have dormant basal buds that are often not visible at ground level. When a shoot is nibbled off by rabbits or burned by fire, some species have evolved to recover very rapidly when these dormant buds sprout. Division is rarely successful if you do not take advantage of this natural tendency and cut the stems back to just above ground level. When first attempting this, it will be very difficult to find the courage to cut the shoot off. Knowing the result makes it easier — basal shoot growth is inevitably more vigorous than stem bud growth. Removing the stem forces the dormant basal buds to sprout, but with a sufficient delay to allow new root growth in the divided plant to start first.

"Ground layering" is a common occurrence in the wild and in gardens. Roots will often form where stems bend over and touch the ground. In Chapter 6 the plants that are self-cloning are identified and information is given as to whether they are colony forming, suckering or layering plants.

For sassafras, wild plum, pawpaw and other plants that sprout from roots, use a sharp spade to slice down about 4 inches out from the stem. Slice down at least the depth of the blade and all around the stem before trying to lift it with a section of the spreading root. For roses and hazelnuts and other species that spread by underground stems, dig deeply enough to determine the presence of roots and sever a section of rooted stem. Plants that naturally layer themselves, such as dogwoods, are severed from the parent plant once a well-rooted branch is found.

If you are taking a division from a wild plant on a friend's property, remove your division from the soil and put the soil back in place — leaf litter and all. Cut the stem of the divided plant close to its base and plant it right away, or keep it damp and cool in a plastic bag by placing it in the fridge for no more than a day.

Easy to Difficult Divisions

Dividing the species that are difficult to propagate should only be attempted in the dormant season. Plant them in a cold frame or protected location for at least the first year and pay close attention to matching soil characteristics, especially for the acidic species. Moderately easy to propagate species are best planted in a nursery area first, while easy to propagate species can be planted directly into the planting site. You can increase the survival of divisions that have few roots on them by rooting them in a sweatbox.

In the following species, root shoots rise directly from the spreading roots of the parent plant, and the division must include a section of the spreading root. There is no point in trying to grow beech trees from root shoots, because the shoots do not rise from roots that are small enough to sever.

- **easy** — bittersweet, nannyberry, bladdernut, elderberry, prickly ash, gray dogwood, red raspberry, blackberry, swamp cottonwood, sandbar willow, virgin's bower, burning bush, rock elm
- **moderately easy** — staghorn and shining sumac, largetooth and trembling aspen, maple-leaved viburnum, wild crabapple, round-leaved dogwood, bristly sarsaparilla, devil's-club, black locust, speckled alder, greenbrier
- **difficult** — Canada and American plum, pawpaw, hawthorns, sassafras, pin cherry

The species listed below slowly spread by rhizomes. Rhizomes are underground stems with roots and buds. They grow between an inch and 3 feet each year, depending on the species, and can be propagated by division. The fewer the roots on the division you sever, the more important it is to plant it in a well-protected location.

- **easy** — poison ivy, bush honeysuckle, meadowsweet, purple-flowered raspberry, rose (swamp, carolina and smooth), holly, snowberry, chokeberry, bayberry
- **moderately easy** — American and beaked hazel, mountain holly, wild raisin, sweet gale, alder-leaved buckthorn, sheep laurel

- **difficult** juneberries (*A. humilis*, *sanguinea* and *stolonifera*), blueberries, sweet fern

Some shrubs have many upright stems from a common base. Occasionally it is possible to slice a rooted stem off.

- **easy** woolly-headed, upland, slender, sage-leaved and heart-leaved willows, arrowwood
- **moderately easy** steeple bush
- **difficult** downy arrowwood

Ground-layered branches are easily severed from the parent plant. In the unusual case of black raspberry, long, arching canes grow into the ground and produce roots at the shoot tip. The new plant grows "backward" from basal buds at the shoot tip.

- **easy** black raspberry, red osier dogwood, Virginia creeper, currants and gooseberries, highbush cranberry, dwarf raspberry, buttonbush, wintergreen, horizontal juniper, trumpet vine
- **moderately easy** running strawberry, flowering dogwood, all honeysuckles, all grapes, bearberry, leatherleaf, white cedar, Canada yew, prairie rose
- **difficult** Labrador tea, hobble-bush viburnum, sand cherry, trailing arbutus, black spruce

Stem Cuttings

Black willow will grow naturally from cuttings — some stems break off, take root and produce a new plant. But horticultural techniques that accelerate the rooting of cuttings are now so reliable that most shrubby plants in commercial nurseries are grown from summer (softwood) cuttings. However, cuttings from most tree species will not produce roots. Avoid wasting time trying to root maple, birch, coffee tree, pawpaw, tulip tree, hop tree, pines, leatherwood, redbud, oaks, hickory, walnut and witch hazel. Shrubs are a much better bet.

The art of initiating roots on cuttings depends on the ability to provide a humid environment, the time of year and the degree of difficulty of the species. Although willows are easy to root in a glass of water, most other plants require a sweatbox and the use of rooting hormones.

Roots are produced from the cambium tissue in the stem, in response to the wound made by severing the stem. Cuttings are generally taken from the current season's growth but may include a "heel" of the previous season's growth. The healthy shoots from the most vigorous parts of the plant are easiest to root, while the weaker growth from old, slow-growing plants is very difficult to root. Softwood cuttings should be 2 to 4 inches long, consisting of three to four nodes, with the leaves stripped off the bottom ¾ inch.

The cuttings are planted in containers filled with a rooting medium. I have used pure washed sand, a 50/50 peat moss and perlite mix, and a sand and peat mix. As long as the medium drains easily and holds the cuttings up, it will be fine. Wet the mix thoroughly before sticking your cuttings in. When the cuttings are in, flood the mix to settle it about the stems.

An all-purpose liquid rooting hormone or powdered hormone to stimulate root initiation can be purchased in three strengths: #1 for early June cuttings, #2 for midsummer cuttings and #3 for hardwood (winter) cuttings. Woody plant species respond differently from each other and from week to week as the stems mature. The best time to collect most softwood cuttings is the first half of June. Novices should use a range of cutting dates at first to build their plant-propagating skills. The development of a white, swollen growth at the base of the stem is called callus tissue (undifferentiated parenchyma cells) and indicates insufficient hormone strength or that the cutting was taken too late. Black and rotting stems indicate excessive hormone strength or a cutting taken too young. Ideally, the stem produces several roots in two to five weeks, depending on the species and the quality of the cuttings.

It takes two weeks to three months to root different species. Once softwood cuttings have rooted, they need to be weaned from the sweatbox by venting. Try to make the transition to dry air last for a few days, and take advantage of cloudy days for the transition. Once weaned from the sweatbox, you can transplant the cuttings into a clay pot. Once again, always endeavor to ease the transition by placing the pot in the vented sweatbox for a few days.

Hardwood stem cuttings of willow, poplar (not aspens), grapes and red osier dogwood are taken from November to early December. They are dipped in #3 hormone before being plunged into a sandy mix in the cold frame. They produce root initials by late April. They can then be set out in the nursery at wider spacing, watered thoroughly and shaded. These will be large enough to set out into the landscape the following spring. I have found it interesting that hardwood cuttings of the native pussy willow, *Salix discolor*, are consistently very difficult to root.

Root cuttings can be made from most species that spread by stems rising from roots. Dig out several 6-to-8-inch-sections of sturdy roots, up to ¾ inch thick. Lay them horizontally 1 to 2 inches deep in the cold-frame soil. Before laying out the root sections, cut back any shoots to within

1/3 inch of the root. The roots should be left undisturbed for at least one growing season while the shoots develop. Then they can be transplanted into your nursery.

Grafting

Grafting occurs in nature when two branches of the same plant or, rarely, two plants of the same species join together at a place where they cross. Natural grafts are more commonly seen in the root mats of wind-thrown trees, where two or more roots of a tree have grown together. Horticultural grafting mimics this natural process by uniting the cambial tissue of two closely related plants. It usually involves grafting a bud or the stem of one plant onto the stem of a rooted plant.

Grafting does not change the genetics of the new plant. The grafted shoot reproduces a clone of a plant with desirable qualities or characteristics, using the root of another. The rootstock may influence the rate of growth, but it does not change the genetic characteristics of the bud or shoot grafted onto it. Grafting is time-consuming and is performed only on plants that do not produce roots easily by division or cuttings — a distinctive maple, oak or apple, for example.

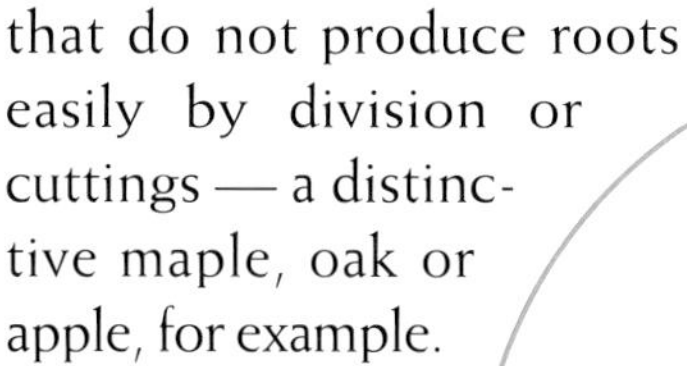

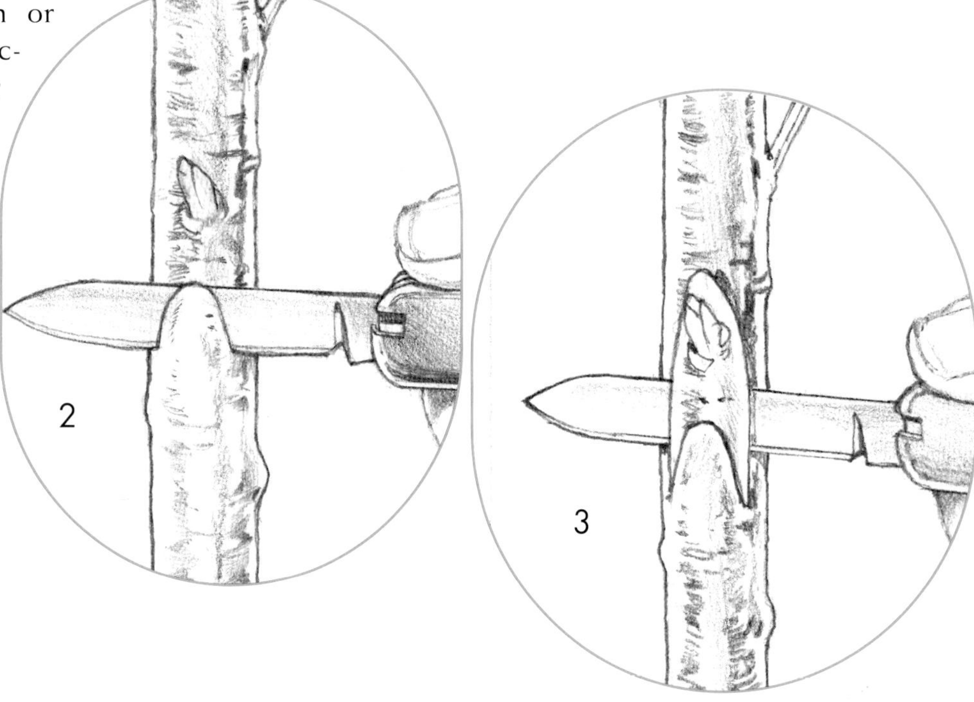

Weed Prevention

I enjoy weeding to this day, even though I spent many days of my youth filling the farm trailer full of weeds. Weeding is a meditative time but also an exploratory time. A journey into the planted landscape provides the time to witness and ponder the changes that nature's hand is forcing, and the new habitats that are being produced as your plants grow.

Be careful not to introduce weeds into your nursery from purchased container-grown plants. I scrape the surface soil into the bottom of the hole that I have dug for the plant in order to bury and decompose any potential weed seeds.

Several grafting techniques exist, but chip budding is by far the easiest for the layperson. The bud of one plant is inserted into the seedling stem (the rootstock) of another plant (preferably the same species) and held in place with a plastic strip. A bud "takes" in about three weeks, when the union begins to swell (from cell regeneration), after which the plastic is removed. The grafted chip bud remains dormant through the fall. The stem of the rootstock is cut off just above the grafted bud the following spring to force the new bud to grow strongly. You may need to tie the emerging shoot loosely to a stick to aim it upward at first.

Grow the seedlings to serve as rootstock a year or two in advance of collecting the bud wood from a desired plant. The rootstocks are healthy seedlings of about pencil thickness. The bud wood is obtained in early August. Cut off a shoot of the current season's growth and remove the top and bottom quarter (the buds are best in the middle of the shoot). Then pull the leaves off. These "bud sticks" must be kept cool and moist and can be stored damp in the fridge for up to a week.

Chip budding requires a very sharp, straight knife. A horticultural grafting knife or a Swiss Army knife is best. The cuts are made with a single, smooth draw stroke, producing very flat surfaces that easily fit when bound together. Make the first cut slanting downward beginning about ⅓ inch below a bud. Make a similar cut starting about ¾ inch above the bud, and then draw the knife down beneath the bud to remove a chip of wood containing the bud *see* illustrations 1, 2 and 3, on the previous page. Repeat this procedure on the shoot of the rootstock where you want to place the chip bud.

Chip buds are best placed on the north side of the rootstock, within a foot or less of the ground. Wipe any soil off the stem first. After you cut the bud from your bud stick, avoid touching the cut surface by using your thumb to hold the chip against the knife and then using the knife to set the bud into the rootstock. The wedge left on the rootstock by the angled cutout will hold the chip in place (*see* illustration 4, left). Try to match the cambial line between bark and wood on both sides of the chip. Most professional grafters bind the chip in place with precut plastic strips, but ⅓ inch wide strips can be cut from a sturdy plastic bag for the few buds that a gardener will do.

Wrap the chip tightly in place by overlapping the strip, and leave the bud exposed (*see* illustration 5, opposite page). Three weeks later, you can remove the plastic strips and look for white callus growth at the bud chip edges. You should now have a completed graft that will remain dormant until spring. Buds are attractive to rodents, so you will need a tree guard for protection.

Rootstocks have a tendency to sprout from below the bud union or even from underground. Your challenge is to continually remove the sprouts from their point of origin (dig underground if needed) until the "sucker" sprouts stop forming.

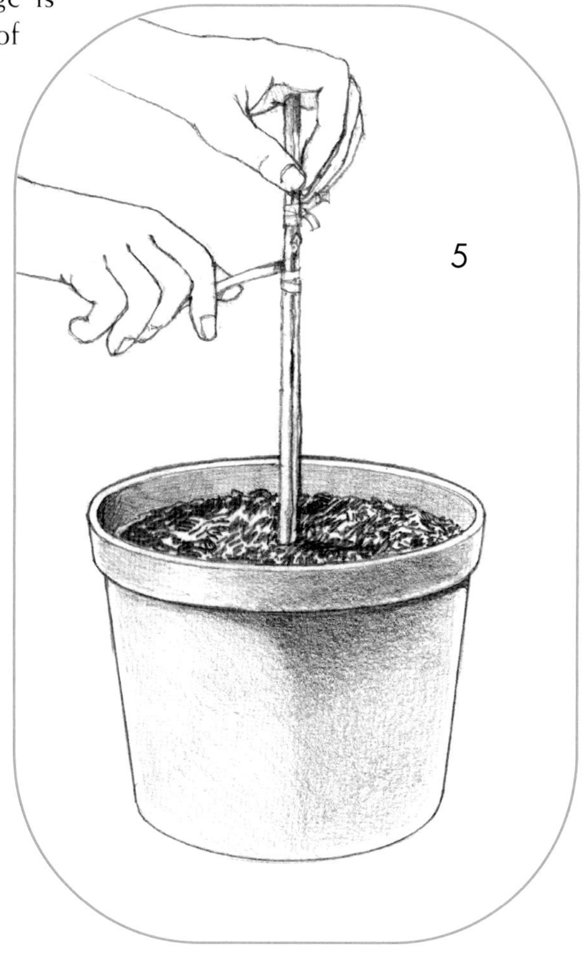

Such is the amateur's nursery. As your skills develop, you will discover that it is relatively easy to grow far more seedlings than you can handle. Try to focus energy on those that have a planting destination; otherwise, do what I have always done and create a brush pile. Avoid donating inferior-quality plants or the cast-offs — these must go on the brush pile.

Work with your local community to establish native woody plants in your area, especially on urban brownfield sites. You will obtain your greatest success by planting native plants appropriate to your area while respecting their specific soil and site requirements. The better the match, the more likely the plants will thrive and help to sustain the precious natural diversity of the region.

Chapter 5

Restoring the Landscape

Two forces shape the world — nature and human nature

— Paul Leet Aird —

There is an innate tendency in all of us to attach sentimental value to trees. I know several oak enthusiasts and many others who think the tulip tree is the greatest tree that ever lived.

The desire to pick up and plant the seeds from a tree that is sacred to us creates the parallel desire to look after it and to remain nearby. Those who call a place home tend to be interested in protecting its trees and woodlands from the onslaughts of planning gone wrong.

It is important to plant a single tree — where only a single tree makes sense for shade or beauty — if nothing else. Some homeowners may want to establish primary trees that are well matched to the soil and moisture, reach the largest size for the space available and are native species from local seed sources. Trees located so that they prevent summer sunlight from directly entering windows will reduce and may eliminate future air-conditioning costs. To reduce heat buildup, trees can also be placed so that they shade a dark roof and asphalt or concrete surfaces.

Woody Plants and Climate Change

A woody plant must stand its ground. No other living organism has the ability to stand through blistering heat or violent winter storms without the option of running for cover. Nevertheless, whole populations of plants do move through the landscape in response to climate change or disease. At a rate that is almost imperceptible within a human lifespan, the natural movement of seeds and migration of species is unending.

The advance and retreat of glaciers through the Great Lakes basin forced species to retreat to tiny refuges deep in the southern states, and with each interglacial warm period, a northward recolonizing occurred. Based on fossil evidence,

Chickadees, shown on the opposite page, are excellent natural seed dispersers. Henry Thoreau writes of a mistaken belief in Pliny's time and in his own time that seeds spontaneously developed in the soil in order to produce the trees, which appeared as if from nowhere.

Shown above, the fruit of the sycamore. The sycamore habitat is threatened by drainage ditches, where its seeds often fail to germinate because the soil is much too dry; it is very easy to grow in the right conditions.

many species became extinct or were extirpated from North America during the glacial period.

The current natural global warming trend is being enhanced by our contributions of heat and greenhouse gases from the burning of fossil fuels. Generally, vegetation responds to a warming period with a northward shift in life zones. However, extensive clearing and settlement of land, especially in the Great Lakes region, virtually prevents any possibility of natural species migration through this region. Additionally, ecosystem fragmentation increases the possibility of extirpation of local species. Hemlock, *Tsuga canadensis*, for example, at the southern edge of the mixed-forest life zone, cannot tolerate hotter, drier conditions. With global warming, we can expect a natural decline of hemlock and many other species in the southern extremities of their range, along with an impeded northward advance.

Should we then consider which species or seed sources are better suited to the changed climate rather than focus on preserving a species on the fringe of its receding adaptive range? Because of constraints in natural seed movement, should we consider obtaining seed from more southerly or drier areas? One way to help us answer these questions in the future is to begin now to maintain a record of the seed sources of our planted trees and shrubs. By doing this, you and those who follow will have a valid basis for evaluating the success or failure of trees and shrubs planted 20, 50 and even 100 years earlier. These data might best be kept with the property deed.

Lessons from the Sycamores

As settlers cleared and leveled the land, the deeper swamps, sugar-bush woodlots and steeper grades were avoided. Ultimately, though, many wet woodlands and swamps were drained, and dams and reservoirs had to be built to reduce the flooding caused by destruction of the natural vegetation cover that used to retain the water in the soil. The remaining swamps and wetlands in the lower Great Lakes region cover about 30 percent of their pre-settlement area.

The sycamore is at the extreme north end of its natural range in the Great Lakes region. It is usually found in a riparian environment, but also grows in former swamps that are now drained. Sycamore is a moist-soil specialist, and when its seeds fall on drained land, they do not germinate because the soil is too dry. In these conditions, relic trees are at risk of being locally extirpated. Other swamp species with widely scattered populations at risk include shellbark hickory, black gum, tulip tree, Shumard oak and even "common" species like winterberry, buttonbush, cottonwood and black ash.

If you find old relics of a species in your area and notice virtually no natural regeneration, make the effort to gather and grow some of the seed produced, even from overmature, dying trees and shrubs. Plant the new saplings on the site to restore youth to the population. A declining population will also benefit if you establish seedlings on the site from other, nearby populations.

Preserving Stress-Tolerant Trees and Shrubs

Our indigenous trees and shrubs are in the early stages of developing a seed bank of mature, stress-tolerant individuals on disturbed sites. In many cases, natural regeneration is thwarted by our heavy footprints (grazing, mowing, spraying, weed whacking, cultivating, asphalt, concrete and fire suppression) and requires our assistance. This calls for the preservation of the last great natural areas with healthy populations of stress-tolerant individuals for seed collection and the restocking of degraded sites.

Perhaps strongly stated, but poignant, William Least Heat-Moon writes in *Heart of the Land*: "Even to speak of 'Last Great Places' reveals how pressing the need, an urgency resulting from a most uncharacteristic national slowness, to respond to immediate threats to our abundant land. If a foreign power tried to take from us a mere square yard of American soil, we would answer with arms. Yet, when our own outdated policies and attitudes work for an even greater loss, our response too often is a despairing, 'what can I do?' Here are a few ideas.

Lessons from the Maples

The sugar maple is native to cool, enclosed forest ecosystems. The species holds a significant place in upland wooded communities, and after the clearing of those lands by the mid-1800s, it regenerated from cut stumps and saplings. By the early 1900s, the maple regrowth had reached flowering age, and seedling regeneration became the principal source of stock for trees planted along roads and streets. An Ontario provincial program that started in 1883 was designed to offset what had become a bleak and exposed landscape. Farmers were instructed to dig saplings from their own woodlots to plant the local field lines and were even paid 25 cents for every tree that survived the first year. Similar programs were undertaken across eastern North America.

Seventy years later, in the 1980s, sugar maples began mysteriously dying, primarily on roadsides but also in some

The Nine Lives of a Woody Plant

When we do the things on this list, we must expect to see some woody plants decline and die at an earlier age than necessary, and, in some cases, see a significant collapse:

1. Plant in the wrong soil type or moisture regime.
2. Place in the wrong light and temperature conditions.
3. Cut, injure or bury major roots during construction.
4. Feed with nitrogen fertilizer or blood meal so that the plant grows foliage in excess of its ability to take up moisture.
5. Water too frequently so that the surface soil is saturated and prevents air from circulating to the roots.
6. Rake the leaves away and thus prevent them from decomposing on the soil surface to maintain a healthy soil.
7. Select out disease resistance in favor of an aesthetic trait such as colored leaves.
8. Remove the dynamic habitat that sustains the diverse parasites and predators needed to fight an imported pest.
9. Import a disease for which the plant has a poorly developed resistance.

woodlots. Acidic precipitation was first to take the blame, then road salt, then over-tapping and poorly timed or sloppy pruning, then verticillium fungus wilt — none of which could explain all of the deaths. "Sugar maple decline," as it came to be called, was in fact due to multiple stresses.

Not understood at the time, but chief among these stresses, was the stress associated with the loss of the American elm that began a few decades earlier. The elm, by all accounts, embraced many roads and fields and was almost eliminated from the landscape by the introduced elm disease. With few elms left to reduce wind velocity, desiccation had a significant effect on the sugar maples with their massive mature canopies. Most of the roadside maples — now at 70 years old — fared poorly under conditions they had been able to tolerate in their youth.

Older sugar maples that are surviving along road corridors are genetically predisposed to handle a significant level of disturbance and drought and are likely the most appropriate seed source for planting in our disturbed sites After surviving the droughts of 2001–02, the remnant sugar maples (even those in decline now) are even more significant as a seed source.

Like these sugar maple survivors, other species have populations that have adapted to drier conditions. Alternate-leaved dogwood, bitternut hickory, basswood, bur oak, musclewood, witch hazel, spicebush and many others, while often found in the interior of forests, are also found at the forest edge, out in the open or along exposed fence rows in significantly drier conditions than found in the woods. These plants tend to have low branches that seed collectors can easily reach, and could be collected from for that reason alone, but they may also be the most appropriate seed sources.

Pests and Diseases

Of all the inquiries I received by mail and all the questions I have been asked at lectures, the most frequent were about pests and diseases: "How do I control x?" "Does it spread to other plants?" "Will it kill my plant?"

Woody plants are very resilient and it takes significant stress to kill them. Valiant life-support efforts can often extend the life of a woody plant but are rarely successful over the long term if the conditions causing the stress are not changed. Climatic extremes and pollutants may create stresses that predispose woody plants to decline from historical pests and diseases that the plant could have tolerated in a less stressful environment. This is impossible to prove, but it makes intuitive sense. An introduced pest or disease, on the other hand, may have a devastating impact on a species because coevolutionary selection or natural predators or both are absent. It is not uncommon to want to save a dying tree or shrub, but pest management involves habitat and predator management to deal with the cause and not just the symptoms.

The larvae of the morning cloak butterfly feed on trees such as Chinese elm (an invasive species) and willow.

The powerful concern for the health and well-being of the woody plant world is appropriate. It is, after all, an essential part of our own life-support system. Standing by and watching trees decline and not doing anything about it seems irresponsible — and it is if we are not learning why they are declining. In a natural setting, trees that are dying from stress must be allowed to drop out of the reproductive pool for the health of the species. A dead plant is part of a healthy ecosystem where replacement trees and shrubs are always present. But in our yards, parks and roadsides we usually eliminate the seedlings that would have otherwise replaced a dying tree or shrub.

Most insects, whether boring, chewing or sucking, tend to be host specific. Insects are part of the ecology of a species, and efforts to control what we call "pests" will directly or indirectly reduce the natural predators as well. This results in a cycle of dependence on control measures and pest resistance to the pesticide. It resembles the way bacteria can build up resistance to overprescribed antibiotics. But some effort is required in new plantings to minimize defoliation events for a couple of years while the roots get established.

Diseases (bacterial, fungal and viral) tend to be even more closely host specific, although the new disease called sudden oak death that has shown up in the western United States, England and Holland has an unusually broad range of hosts. Diseases, too, must be seen as part of the biology of a plant. Plants coevolve with diseases, and their immune systems are strengthened in the process. Any effort to treat diseased plants with pesticides will serve only to keep the weak plants in the breeding population. Any effort to clone disease-tolerant individuals will result in every individual's being equally susceptible to a new race of the disease. And any effort to cut trees in advance of a migrating disease will risk losing the rare individuals with a strong immune system that could cope with the disease. Nature plays a game over and over again. It's called survival of the fittest. It is not a very gentle game, but one we should regard with profound respect.

Pests and diseases have been ravaging the plant world since the beginning of time, yet there is an innate fear of allowing new ones to run their course. We seem to think that plants (trees in particular) need our help. It is true that we have so severely fragmented and stressed our settled landscape that we have left it with few resources to cope with new pests. The predators are not numerous enough to cope with the initial outbreak. But even in the highly fragmented landscape, parasites and predators do rise to the bait and diseases are thwarted by sustaining genetic diversity.

How Plants Cope Naturally with Pests and Diseases

I remember my reaction — a deep sense of awe — on seeing a 2.5-billion-year-old fossil of algae cells in the Royal Ontario Museum in Toronto. In comparison, the woody plants that are discussed in this book are relatively recent — in the 40-to-90-million-year range — but they are linked by genetic threads to all of the plants that came before.

Like all organisms, not every individual comes with a guarantee that it will live a long life. Attrition of individuals due to intolerance of drought, flooding, pests and disease is natural and important for the species — such are the rules of natural selection.

Every tree and shrub species is a member of a community of mammals, birds, insects, fungi and bacteria. Most are beneficial organisms that make life possible. Look at our own bodies, coated inside and out with beneficial bacteria and occasionally invaded by harmful organisms.

Plants may control pests by depositing various toxic chemicals in their leaves and inner bark or sapwood to discourage feeding. It has been suggested that this chemical deposition can rapidly increase in response to a pest feeding event.

According to Dr. Gard Otis, an entomologist at the University of Guelph, some pests feed on the leaves of a great number of species, while others have a narrow host range, sometimes a single species. Genetic diversity in the host species creates a spectrum from good to poor hosts for pest insects. The community of insect "associates" — various parasites and predators — is always in a state of flux in relation to the pests. These checks and balances have worked for millennia and are often called the "balance of nature."

Disease control in plants is managed through selection. The weak die out and the disease-tolerant live. The oldest plants then tend to drop the seeds that are most fit for the local area and most likely to tolerate disease in the future. Healthy plants in a healthy vegetation community are generally not affected by pests, diseases and even drought. They have cell walls that are tough enough to resist fungal invasion or entry by juvenile insects, resulting in a high juvenile-pest mortality. If you feed the plant with nitrogen (including the acidic precipitation from nitrous oxide auto emissions), water it too frequently, or breed and select for fast growth, the cell walls will be softer. Soft cell walls are more likely to be invaded by diseases or pierced by the tiny mouthparts of juvenile insects.

Some scientists want us to believe that genetic engineering is the answer. They suggest that trees can be engineered to be resistant to a particular pest or disease. In reality, the clonal release of genetically engineered organisms flies in the face of nature's guiding and sustaining principle of life — genetic diversity. Genetically engineered plants, in my humble opinion, are unacceptable in the landscape. Evidence abounds that pests and diseases quickly build up resistance to the engineered traits. Engineered plants do not and cannot work in the long run.

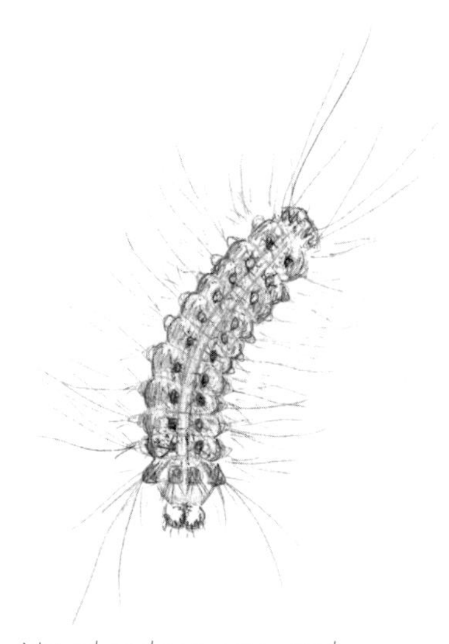

Natural predators, not pesticides, have held the destructive gypsy moth (above) in check.

Lessons from the Gypsy Moth

Foresters have always cried foul at threats to their economic resources, be they environmentalists or pests. The introduced gypsy moth escaped confinement in a greenhouse near Boston around 1870, and, by 1890, the U.S. government was convinced that it would wipe out the forests of eastern North America. A pesticide rain drenched the forests for years, yet the gypsy moth continued to spread. Trees were defoliated and leafed out again, albeit a little weaker. The weakest trees or those located on unfavorable habitat for their species died.

The gypsy moth now covers extensive parts of eastern North America and the forests are still relatively healthy — if one ignores the stresses associated with climate change and other industrial activities. Dozens of predators and parasites now attack the gypsy moth. *The Gypsy Moth Handbook* states: "Fungi, bacteria and viruses play a role in holding the gypsy moth in balance but a nucleopolyhedrosis virus is consistently capable of bringing an outbreak to total collapse — leaving the larvae hanging wilted from twigs."

Examples of the balance of nature being played out again and again in the woods are easy to find. Viburnum leaf beetle swept through the Great Lakes region in the mid-1980s. Most of the information available on the beetle will tell you that it can be controlled with insecticides and suggests planting species that are not eaten by the beetle or its larvae. However, although many viburnums on dry and compromised sites were killed, predators did establish themselves, and now the beetle numbers have been reduced to near insignificance. What caused the decline? Cornell University web notes recommended encouraging beneficial insects such as stinkbugs, which feed on the adults, and lacewings and ladybugs, which eat the larvae. They suggest avoiding the use of broad-spectrum insecticides, maintaining instead a diversity of plant species to provide habitats favorable to beneficial insects.

In the 1980s, an explosive outbreak of forest tent caterpillars turned Ontario forests into leafless trees in May. The trees resprouted in June. After two or three years of defoliation events, the forests recovered without a spray program being implemented. The tachinid fly, the principal parasitic predator of the caterpillar, took a few years to build up its own numbers in response to the rampaging host — a classic example of predator–prey balance.

A New Invader: The Asian Longhorned Beetle

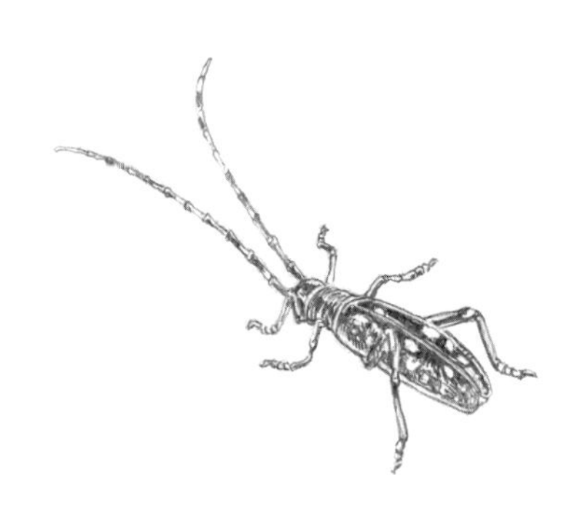

Asian longhorned beetle (shown approximately half-size)

The Asian longhorned beetle was discovered in New York in 1993, in Chicago in 1996 and in Toronto in 2003. Once again, the alert went out that there are no known predators. This beetle is a cambium-zone feeder; it consumes the inner bark and outer heartwood of hardwoods. Maple, horse chestnut, elm, poplar, willow, birch and alder are among the known hosts.

In the infected trees I examined, the beetles had dug tunnels deep into the trunks. There were also palm-sized patches of adjoining tunnels right under the still-visible egg-laying patches. The larvae spend a significant amount of time feeding in tunnels just underneath the thin bark — exactly where insect-loving woodpeckers would search for food. Unfortunately, there are no local forest habitats in this industrial area to support woodpeckers, let alone any of the other predators and parasites of wood-boring insects. It is a biological desert. In the longhorned beetle habitat in China, deforestation, extensive settlement and monoculture plantations have reduced the number of its natural predators as well.

New import rules for wooden crates will substantially lower the risk of another release of Asian longhorned beetles. A program to search for infected areas and destroy all host trees in Chicago appears to be very close to exterminating the beetle there. In New York City, more than 6,000 trees have been removed since 1996. A search-and-destroy program is underway in Toronto and, since the invasion zone is small, the beetle may be eradicated. But if the beetle breaches the present zone and keeps spreading, how many thousands of trees will be cut before we decide to give predation a chance? We may find that eradication campaigns will fail because they do not allow predator populations to build up in response to invasions.

It is an extreme measure to cut down every host tree, infected or not, for a distance of up to half a mile, a distance a female beetle will rarely travel. I think a little discretion and improved monitoring are needed. Cutting down high-quality, uninfected trees — particularly sugar and silver maples — as a precaution has a huge cost, too. Another option is to send a climber in to inspect each tree, develop a tree stethoscope to listen for the munching sounds, and apply a systemic insecticide to the very best trees that are at risk.

If the beetle escapes our ecologically fragmented cities and enters natural forests, should we raise our hands in despair? I don't think so. We need to be observant — to watch for and encourage natural predators. Even more important is the search for trees that naturally tolerate the new pest.

Another New Invader: The Emerald Ash Borer

The emerald ash borer from China was first discovered in 2003 in the Detroit-Windsor area. It was reported to be killing weak and healthy ash trees alike. But who is to say that the "healthy" ash were not weakened already by drought, nitrous oxide–rich rainfall and other destructive forces. Ash is an early-succession species that seeded into the landscape in great numbers when the elms were cleared out. It is not predisposed to live long under any but the best conditions.

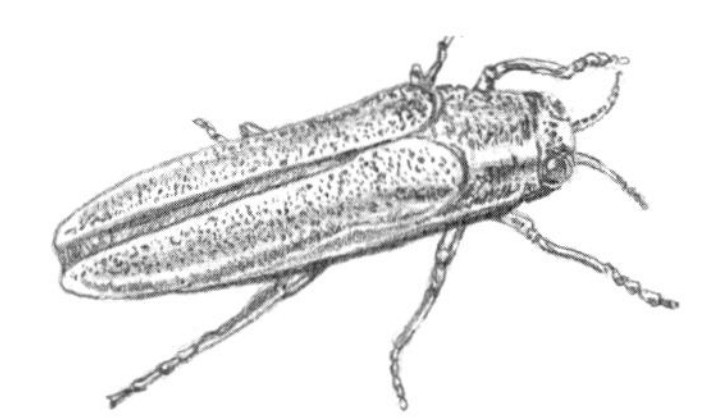

Emerald ash borer (shown at about twice life-size)

This ash borer is an *Agrillus* beetle, in the same family as two native flat-headed borers: the bronze birch borer, which attacks stressed birch trees, and the two-lined chestnut borer, which attacks stressed oak trees. Downy and hairy woodpeckers will take to this new pest once they learn how to find it, and many parasites and predators of the native *Agrillus* will inevitably cross over and consume or parasitize the new invader.

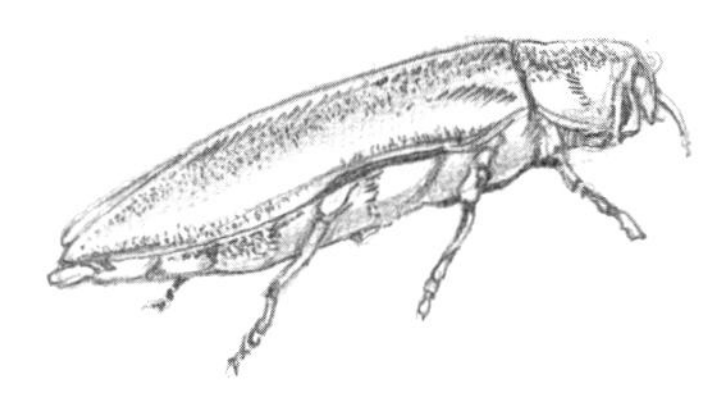

Bronze birch borer (shown at about twice life-size)

By late 2003, this insect had spread into several counties in Michigan, much of Essex County in Ontario, and Ohio. Canada, in an effort to isolate and eradicate the beetle, subsequently cut down a wide swath of trees. Since then, the insect has escaped confinement and this effort to save trees has failed. We will now have to learn to leave it alone long enough for predator numbers to build up.

The emerald ash borer has its origins in China, where it coexists with ash trees. Removing ash trees in Ontario in advance of the insect's spread risks losing the rare genetic traits that will produce a native tree that can stand up against this introduced borer.

Nature is amazingly resilient but it requires both genetic and biological diversity. A human hand and sufficient funds are needed to restore the habitat diversity in our most highly fragmented landscapes. We all need to plant a greater diversity of native trees on our own properties and parks. We should also protect places where dead trees can be left standing or be simply felled and left intact to help sustain natural predators and parasites. Since the arrival of the chestnut blight in eastern North America in the early 1900s and Dutch elm disease (*see* photos pp. 260 and 261) in the 1930s, there has been

The Way to Run a Forestry Industry

Forests are still being trashed by irresponsible forestry operators or high-graded by unscrupulous log buyers arriving on rural doorsteps with checkbook in hand.

Ecological forestry, as practiced in its best form, takes pride in carefully selecting only trees that are starting to decline and even leaving some of those for predator habitat. It leaves the healthiest trees standing to produce the best seeds for renewing the forest. Future forestry, practiced in this way, should involve no tree planting at all.

widespread and somewhat frantic concern about tree survival as imported pests or diseases continue to spread across the continent. One significant problem is the focus on trees that die rather than the survivors. Although gene pools of many species have been affected by forest high-grading and land clearing, there remains a vast genetic wealth. Survivors are inevitable and it is important to accept that we must not destroy healthy trees, *especially* when a pest or disease is challenging a species.

Planting a Wind Filter

Historically, windbreaks generally disregarded the huge difference between the northwestern winter winds and the southwestern summer winds. A dense wind barrier of conifers, running in a generally southwest to northeast orientation, will deflect the prevailing winter winds over or around farm structures but won't appreciably reduce their velocity. It creates a zone of relative calm on both sides of the windbreak and downwind. This reduces energy costs in the building but increases crop losses by increasing moisture evaporation.

Crops are better protected by wind barriers that promote wind speed reduction rather than diversion. Wind speed reduction is best achieved with deciduous trees planted in a generally southeast to northwest orientation to meet the prevailing hot, dry southwesterly summer winds head-on. Instead of the wind being deflected, the open nature of deciduous trees allows the air to flow through the trees. The transpiration of water from the leaves humidifies the air, and the foliage and branches reduce the wind speed, thus reducing the rate at which the wind can suck moisture from the surface of the earth and the growing crop.

Creating a Healthy Landscape

With effort, we can restore landscapes. Conservation and preservation of the fragmented woodlands, wetlands and grasslands is the first step in creating a healthy landscape — "saving the pieces," as botanist John Ambrose calls it. Expansion of the pieces, particularly the larger, healthier fragments, is the second step in restoring ecological function. Linking the fragments to provide pollen and seed corridors and avenues of travel for animals becomes the critical third phase.

Community tree planting is an honorable activity. But often trees are needlessly planted on ecologically important old fields in conservation lands that already have significant tree cover, rather than along roads or as replacements for lawns and parking lots, where hardy native trees are most needed. When we blindly fill every empty urban space with trees to suck up tailpipe exhaust, we inadvertently destroy riparian scrublands and succession edge communities that support a greater diversity of life — meadowlark, bluebird, savannah sparrow, red-tailed hawk, screech owl and fox.

In urban areas, we need to persuade governments to fund the protection of our urban forests and their natural edge communities. More emphasis should be placed on selecting appropriate indigenous species to plant along our streets to avert the continuing tragedy of street monocultures. And we need to remove exotic species from the many brownfield sites in industrial areas and either leave them as ecologically important open areas or replace them with native trees and shrubs.

Woody Plants and Climate Moderation

The sound of the wind in trees (the proverbial "whispering pines") is the sound of branches "lifting" the wind well above the surface of the earth. Trees also release water into the air by transpiration (the vegetable world's version of sweating) at up to 132 gallons (500 l) per day for a large tree. This moisture increases the density of the air, which reduces the rate at which the near-earth atmosphere changes temperature.

The climate-moderating effect of tree communities can readily be experienced by entering an area of mature tree cover from an open space. It is not just the shade but also the reduced wind speed and the moisture in the air that make a forest feel more pleasant.

Without the mature tree community, the land heats up, generating strong thermals that tend to pull low cloud cover into thunderheads. The result is more violent storms with more localized and heavier precipitation. Although average rainfall may not change appreciably from decade to decade, the amount that soaks into the ground is considerably less and the amount that flows off the land is much more, and can be measured by flood-control efforts. The droughts, floods and landslides worldwide are ample evidence of the consequences of forest destruction.

More important than filling urban parks with trees is the planting of woody plants in rural areas to restore climate moderation in the region. Locally, tree planting for climate moderation is the costly and almost thankless task of private landowners. Because this activity is benefiting all residents, we should all be paying for the planting of trees through a carbon tax.

Reforestation of Fields

Desertification is a natural response to the extensive removal of vegetation cover. After the clear-cutting of the Great Lakes watershed, deserts with dunes were forming in three southern Ontario counties by the mid to late 1800s. Under the direction of pioneering forester Dr. E. J. Zavitz, pines were planted in the early 1900s, and the desert formation was halted and reversed. The sterile pine plantations ultimately functioned as a nurse crop, eventually becoming diversified as indigenous flora slowly seeded back in from fence-line remnants and adjacent woodlots. On very exposed and eroded sites, the planting of a nurse crop is one way to rebuild the organic content of surface soils so that the site once again can support a diverse plant community.

Recent observations by ecologist Mathis Natvik suggest that there are significant advantages to be gained by closing the field tiles and unleveling the land before reforestation begins. Mathis manages the reforestation efforts in the field gaps surrounding Clear Creek Forest, a new Nature Conservancy site in the municipality of Chatham-Kent, Ontario.

After the trees were cleared for farming, the original rough topography was gradually flattened by plows, harrows and land levelers. Mathis made the brilliant observation that before he could replant the forest, he had to first restore the naturally rough topography and associated hydrology of the site. He showed an equipment operator the natural pit-and-mound (knob-and-kettle) microtopography of the old-growth forest adjacent to the site. The operator quickly grasped the idea and proceeded to unlevel the first field in 2001.

The pits filled with water in the spring and frogs arrived. Maple, elm, willow and poplar seeds landed on the water in the pits and blew over to the "shore," where they germinated. Squirrels planted nuts on the mounds hundreds of feet from the forest edge and birds perched on the mounds to drop their loads of dogwood, *Amelanchier* and viburnum seeds. Sedges, grasses, asters and goldenrods all took hold in short order. Although seedlings were planted at Clear Creek, Mathis soon realized that without planting a single seedling, the regraded site would naturally and rapidly regenerate from the local seed bank. Planting needs to be reserved for sites without an adjacent seed source.

On our own properties, we can create downspout gardens to capture and hold more of the rain that falls on rooftops. By determining the dimensions of the pits and mounds in remnant old-growth forests in your area, you can determine the dimensions suitable for your own property. Mathis spoke of a man in Windsor, Ontario, the noted environmentalist Bruno Sfalcin, who decided to mimic the upturned roots of a giant tree by burying branches in mounds of surface soil that he built by digging pits on his property — one every year. A pit can be partly filled with compost to create a mini marsh garden. Pits will dry up within a few days to a week, depending on soil drainage characteristics, and create dynamic sites for growing a range of dry to wetland species.

Plant identification skills will help you recognize the diversity and composition of plant communities and increase your ability to plan and develop a personal or community naturalization project. Weeding out nonnative species in naturalization areas requires a critical ability to distinguish plants when they are young. Even the nursery and landscaping trades confuse many exotic species with their native counterparts. This means that many naturalization projects will have exotic species unintentionally planted in them that will need to be edited out!

The Goal of Ecological Restoration

A healthy landscape should appear from above as a green web, with wind-filtering tree lines along roads and field edges and diverse vegetation corridors along streams and through rural and urban properties. All these green web strands should be connected to the remnant forest fragments.

I read somewhere that the largest recorded songbird migration at Point Pelee, Ontario, was on May 3, 1952. This was more than half a century ago. The numbers are much reduced now. Seasoned birders moan that they no longer see the incredible "waves" of songbirds. Pesticide spraying against insects and to destroy weeds since the 1950s has led to birds lying dead in urban yards and rural fields across the Americas.

I watch birds. I watch their behavior and what they are eating. We owe much of the health of our forests to our feathered friends. In my humble opinion, the feeding of birds does not respect these animals; it degrades them to be partly domesticated and it degrades us because we have created landscapes that do not support them naturally.

Gerry Waldron asks poignant questions in his book *Trees of the Carolinian Forest*: "What span of time is necessary for a dynamic woodland community to arise from a barren field? A property along Big Creek in Haldimand-Norfolk is certain to evolve more quickly than an isolated patch (of bare earth) on the (drained) lake plains of Chatham-Kent. In the latter place it might take hundreds of years to achieve the species diversity of the virgin forest. Where would species come from? It could take thousands of years to recreate the micro-topography (pits and mounds). Can we accelerate the process? This is the challenge of ecological restoration."

Ecological restoration is a wonderful goal, but the idea of restoring the landscape to pre-settlement conditions needs to be put to rest. It is simply not possible. None of the pre-settlement conditions exist anymore. The humus layer is gone, the soil is compacted and depleted, rainfall is more intense and runs off quickly because the land was graded, and the climate is changing.

Terminology that suggests "invasive aliens are all bad" is unacceptable. Exotic plants are now part of the landscape, as is the international human community. Indigenous people included, we humans were all invasive exotics in the Great Lakes region from when the glacial ice melted back 12,000 years ago, and many brought food plants with them.

Although visiting a natural area should be part of a learning landscape related to the flora and fauna of the local bioregion, schools might also establish some of the amazing diversity of plants that arrived in both historical and present times. These could be set into formal and food gardens associated with the school buildings' "footprints." The combination of natural and cultural landscapes lengthens the threads of learning opportunities that help students and the community understand the fabric of a large landscape.

Learning about landscapes evokes a more complete understanding of the natural human relationship to plants that provide food, fiber, medicine, shelter, climate moderation and ecological function. Teaching the basics of the natural world in schools benefits everyone. Creating an understanding of how to repair the fragmented landscape is the real goal. The goal of restoring the fragmented landscape of the Great Lakes region may seem lofty, but it has immense biological and moral worth.

A New Culture of Nature

Alex Wilson, in *The Culture of Nature: North American Landscape from Disney to the Exon Valdez*, suggests: "We must build landscapes that heal, connect, and empower, that make intelligible our relations with each other and with the natural world: places that welcome and enclose, whose breaks and edges are never without meaning. Nature parks cannot do this work. We urgently need people living on the land, caring for it, working out an idea that includes human culture and human livelihood. All of this calls for a new culture of nature, and it cannot come soon enough."

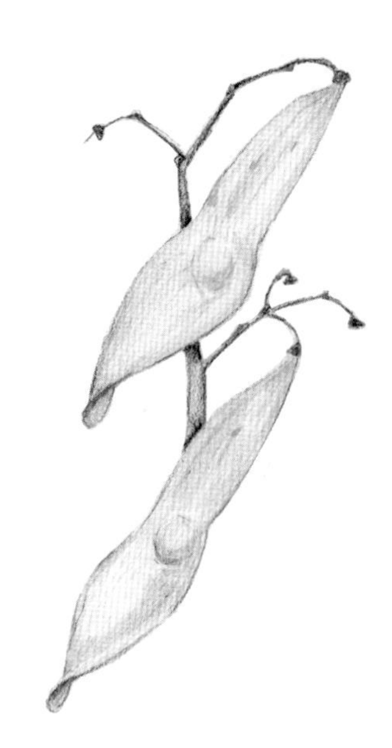

The fruit of the exotic tree of heaven (Ailanthus altissima)

Chapter 6

The Woody Plants

In all things of nature there is something of the marvelous.

— Aristotle —

This section is arranged in alphabetical order by genus, the botanical grouping of species. Much repetition is spared this way, since in most genera, the seeds of each species are handled in a similar if not exactly the same way. Both the botanical and common names are indexed.

For the botanical naming of plants, *Flora North America* and the *Ontario Plant List* are followed, with the exception of the genus *Amelanchier* and the omission (for sanity) of most hawthorns, raspberries, currants and gooseberries. Botanical "splitters" (those who like lots of species) love this sort of stuff. I am a "lumper," tending to group plants that frequently mate or intergrade with each other. The important stuff, their ecology, is often equally inexact and also requires interpretation; I provide my experiences.

Overall Intent

The information for each genus is organized as running text rather than titled sections is to reduce the temptation to skip the journey — a more complete understanding of the plant — in order to get to the seed. All indigenous species described here are listed in the Seed Treatment Guide (Appendix III) for those who do wish to go straight to the seed treatment.

My association with these species is mostly in Ontario. All of them have a range that extends beyond the Great Lakes watershed, and the ecology of a species and its site tolerance may be quite different in other life zones. The specific ecological variations of species in your area only can be determined by examining and noting the site characteristics where you find them growing.

Illustrations showing distinctive aspects of the fruit and seeds of the different species accompany each genus, where called for. They are depicted at the size you will see them unless otherwise noted. Color photographs on pp. 233–264 provide further reference.

The imperial measuring system is used throughout. A handy chart on p. 269 provides conversion factors to convert imperial measurements into the metric system.

The species information provided is what I believe is the minimum needed to understand the ecology and identity of a species — enough to not only germinate it, but also to provide it with an appropriate place to grow.

Abies, the FIRS

The firs have smooth bark with rounded swellings that contain a fragrant, sticky resin, which can be released by pressing the lumps with your thumbnail. Seed cones are erect; the scales fall off after the seeds are dispersed, leaving a persistent bare axis. There are about 40 species of fir of the north temperate world, with some species reaching south to the mountains of Guatemala and north Africa.

Abies balsamea, BALSAM FIR, is usually abundant in dense groves in conifer swamp forests or as scattered, stunted trees in the mixed-forest understory. The few individuals found on drier, open sites may be significant seed sources due to their apparent tolerance of greater exposure. Balsam fir has a considerable range within the mixed and boreal forest zones, from Newfoundland to Alberta, south to Minnesota and Wisconsin and across to Pennsylvania and New York.

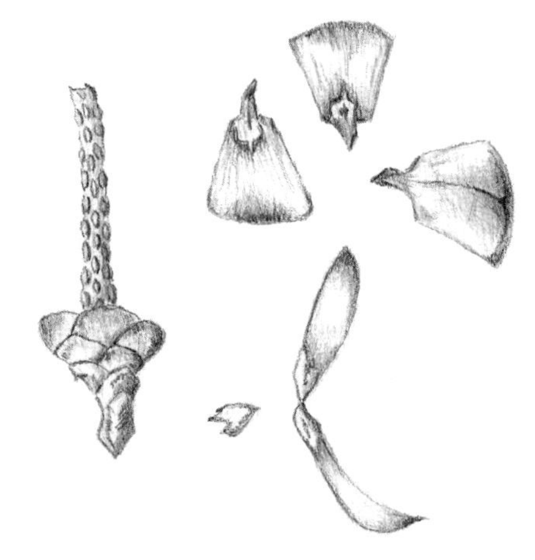

Cone skeleton, seeds and scales of Abies balsamea *(balsam fir)*

Natural seed dispersal occurs from September to January, when finches, red squirrels and wind shatter the upright cones on the tree. The winged seeds may be carried great distances by strong winds, but most land downwind, in close proximity to the parent. Those that aren't consumed by grouse and mice germinate in the first spring while soil temperatures are still quite cool. Female cones are produced on the previous year's growth. Because they appear only at the very top of the tree, and the male cones appear on lower branches, cross-pollination is promoted. Wind carries pollen in updrafts to fertilize cones of downwind trees in May, and the cones mature within one growing season.

Seeds are difficult to obtain because the mature cones are very sticky and often inaccessible, and shatter soon after maturity. Foresters obtain seed by scheduling a cut in the seed collecting area so that whole, mature cones can easily be picked from felled trees. If cones are fully ripe and starting to shatter, a tarpaulin spread out just downwind of a seed tree will capture a portion of the seed that rains down. Once a tarp is in place, you can also attempt to throw a stick into the cone crop to dislodge some seeds into the breeze.

Balsam fir seed is stored in a cold, dry place until ready for sowing. The seeds require a cold, moist stratification for three weeks and will germinate only in a cool location. This is accomplished most easily by planting the seeds close together in a clay pot in March or April and plunging the whole pot into

The seeds and distinctive cone of the exotic Pseudotsuga menziesii *(Douglas fir)*

the soil of a cold frame. They typically will grow only about ¾ inch in the first year. In early summer, the seedling cluster can be tipped out of the pot and set directly into the cold-frame soil. After two to three years of undisturbed growth at close spacing, the seedlings should reach 4 inches. They must then be separated, root-pruned and planted 3 to 4 inches apart in a nursery. At 8 to 12 inches tall, they are ready to plant in a partially shaded backyard or a river-edge or swamp-edge environment. Porcupines feed on the bark, while deer, rabbits and moose will browse on the foliage. Protect seedlings according to which browsers are in your region. (*See* photo p. 234.)

Abies fraseri, FRASER FIR, is an eastern fir limited to high elevations of the Appalachian Mountains from Virginia to North Carolina. It is commonly grown for the Christmas tree industry in the north. Other than having a darker and more deeply notched needle and cones with projecting bracts, it is so similar to balsam fir that it will undoubtedly be inappropriately substituted in native tree planting projects out of its range.

Exotic Alert

The west coast Douglas fir, *Pseudotsuga menziesii*, has naturalized in close proximity to older planted trees in some parts of the region, especially on moist, sandy soils with thick needle cover in pine/oak woodlands. Douglas fir has pointed buds and distinctive pendulous cones (with protruding, three-pronged bracts) that are shed after the seeds are dispersed. (*See* photo p. 234.)

Acer, the MAPLES

The maples are a complex and dynamic genus of more than 100 temperate species, concentrated primarily in Asia. Individual species are relatively habitat specific, but the many different species of this region occupy virtually every habitat, from deep shade to full sun, and from swamps to sand dunes and rock outcrops. Maples flower in subtle though, once noticed, rather stunning displays in April and May. The fruit is a winged key or samara with a single seed, typically borne in pairs. Our climate produces fall colors in the maples that are unsurpassed in much of the temperate world.

One is rarely far from a native maple in the natural areas of this region. There are six native species, including two natural hybrid complexes. The maples break into two groups based on the timing of fruit maturation: the "soft maples" (the red and silver maples), with fruit maturing in early summer, and the "hard maples," with fruit maturing in the fall.

Most maples start to bloom before they are 15 years old, but sugar and black maple take 30 years. With the exception of sugar, black and striped maple, which flower in three- to five-year cycles, all other native maples flower and fruit every year.

Maples are wind pollinated and produce a paired fruit (the maple keys) either in clusters or chains, depending on the species. Once fully ripe, the keys separate and are disseminated by wind, employing the classic helicopter spin. Although a strong windstorm will carry fruits for several hundred feet, most fruits fall downwind in the local area and may be further dispersed by small mammals. I have found germinated stashes that contained over 50 maple seedlings within a 4-inch-radius. Much of the seed-rain or "mast" of all maples is consumed quite rapidly by many animals, including chipmunks, mice, turkey and grouse. Seeds of red and silver maple germinate soon after dispersal in late spring, while all other maples are dispersed in the fall and germinate readily in cool soil the following spring.

Each fruit contains a single seed and has an attached wing. A cut test will reveal the folded green cotyledons, indicating a healthy seed. (With one other exception — pawpaw — all other genera have seeds that are white inside.) Germination is improved if the seed wings are detached from stored seeds prior to the one-day presoak (neither are required when direct-planting fresh seed). The wings can be removed from dry stored seeds by rubbing the fruit in gloved hands and then removing the broken wings by fanning.

In all maple species, there is a significant premature drop of thin, green, aborted keys. Fruits are mature only when they shift in color from green to yellowish or reddish brown. At this time, squirrel feeding may be observed, and seeds can be hand picked from trees with low branches. Fruits are fully ripe when they turn tan-brown and, if the natural seed fall has started, branches can be shaken to dislodge keys onto a tarpaulin. Keys can also be gathered where they have fallen on the ground, but close inspection will probably reveal that most of the seeds have been carefully extracted by rodents. Temporary storage is best in a paper bag.

Although fruits of all maples can be harvested at the mature stage, allow them to ripen fully by drying the seed for a few days prior to sowing. If this is not done, they will rot in the ground. Seeds can be sown close enough together that the wings overlap. Outdoor seed frames should be mulched with leaves over the winter. Be sure to remove most of the mulch in April to prevent the seedlings from smothering as they sprout in early May. Completely enclose seed frames until all seeds have germinated. The seeds in unprotected frames will not survive chipmunk and mouse entry (keeping in mind that with the summer sowing of soft maples, some seeds will lie

Maple Tar Spot

Tar spot is a native fungus disease (*Rhytisma acerinum*) that infects most maple species but primarily the silver and Freeman's maple and, increasingly, the Norway maple. As might be expected, the worst infections occur on stressed trees. The early summer symptoms of yellow-green spots are rarely noticed, but in the late summer the spots turn black as the fungus produces fruiting bodies on the underside of the leaves. Tar spot is not serious enough to kill maples because it doesn't defoliate the tree. Zero-tolerance gardeners try to rake and dispose of the leaves or spray their trees — time better spent reading a good book or enjoying a walk in the woods.

dormant until the following spring). Seedlings require shade protection for the first two years.

For greenhouse culture or sowing in a cold frame, take note of the stratification time requirements listed in Appendix III. If you begin stratification in September for sugar maple, the radicles will emerge in January, when light levels are too low and day length is too short to support growth. Radicle emergence in all maples takes place at cold temperatures near the end of the required cold stratification period; it is not possible to keep seed dormant for two or three more months until the ground thaws. Species with a short (up to 90-day) stratification period should be dried at room temperature for several days prior to dry-cold storage.

Growing maples requires considerable attention to protection, especially outside the urban core. All maples are important food for rodents and large herbivores. Moose and deer browse on the foliage and twigs in summer, while mice, voles, rabbits, beaver and moose eat the high-sugar-content twigs and bark in winter. I have seen trees 8 inches in diameter completely girdled by meadow voles. Unprotected seed frames will not survive chipmunk and mouse entry.

Many horticultural cultivars have been selected from *Acer rubrum*, *A. saccharinum* and *A. saccharum* for use in the landscape trades. These are unacceptable for naturalization projects.

The Spring-Fruiting "Soft Maples"

Acer rubrum, RED MAPLE, is a tree of higher ground on sandy soil and rocky granitic outcrops, but also swamps. It is usually found as a forest canopy tree, widely scattered in rich upland woods or on well-drained embankments above cascading streams. Red maple is found in the deciduous and mixed-forest life zone from Newfoundland west to Minnesota and south all the way to southern Florida!

Red maple is relatively slow-growing and particular to well-drained, rich, undisturbed soil. Only seeds from trees growing at a similar latitude to your site will thrive, and that is why some cultivars (garden selections) are difficult to establish. Despite any difficulty finding fruit of red maple, your efforts will be well worthwhile. It is a singularly stunning small to medium-sized tree. Fruit size and shape are the only reliable ways to distinguish species in this group. (*See* photo p. 235.)

Acer saccharinum, SILVER MAPLE, is a lowland species that may reach massive proportions. Pure stands occur in swamps that are inundated for long periods of time, and individual trees can be seen in rich bottom lands along the edges of quiet streams and small lakes. It is found chiefly in the deciduous and mixed forests from New Brunswick and Maine across southern Quebec, Ontario and Michigan to Minnesota and south to Louisiana and northern Florida.

Acer x *freemanii*, FREEMAN'S MAPLE or HYBRID SOFT MAPLE, has likely been in the landscape for millennia, but it has only recently been widely recognized. This self-perpetuating hybrid is a natural cross of silver and red maple, and forms nearly pure stands in swamps that are dry for a longer part of the year than those in which silver maple is found.

Both silver and Freeman's maple seem to have a broad adaptive range. They grow surprisingly well on roadsides and urban side streets, often producing some of the largest urban trees. Because they are a swamp species, their roots can tolerate low oxygen levels underwater, similar to conditions that exist under pavement and compacted soils. Interestingly, other swamp species tend to thrive along urban streets as well. These maples are very adaptable to cultivation but require pruning to prevent narrow-angled branching, which can lead to trunk splitting later in life.

Silver maple and Freeman's maple bloom in late March and early April, one to two weeks earlier than red maple. At this time of year, the red-flowering female trees can be easily located in the landscape of bare branches. Female trees are well worth noting, since, at the height of mosquito season, you won't want to spend more time than necessary searching through the woods to find a seed tree.

Comparison of fruit size in the red/silver maples is the best way to distinguish Freeman's maple from the two species. After pollination, the developing fruit of red/silver maples are brilliant red, then green, then partly reddish again before finally turning tan as natural dispersal begins.

Annual fruit crops of the soft maples are dispersed in late May and early June. Look for squirrels eating seeds in the trees. This indicates the fruit are mature and whole clusters will be dropped to the ground, an opportune time to obtain a few handfuls. The very small fruit of red maple, often high in the canopy, make it difficult to obtain fruits otherwise.

Once the fruit are ripe, a warm southwestern wind will disperse much of the crop within a few days. A huge percentage of the fruit are consumed on the ground by chipmunks, turkeys, grouse and mice. If moisture conditions are right, the surviving seeds germinate in about 10 days.

Mature (red-tinged) seeds that are gathered or picked should be dried for a day or two until they begin to turn tan, then sown immediately. Seeds of the red/silver maples should not be allowed to dry for more than a week. The seed of red maple, in particular, will often become dormant when dry and only germinate the following spring. If only some of the seeds you sow germinate, rodents will uproot

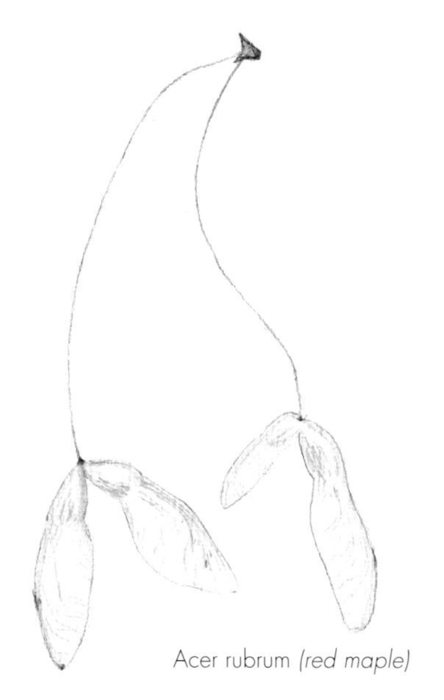

Acer rubrum *(red maple)*

Acer saccharinum *(silver maple)*

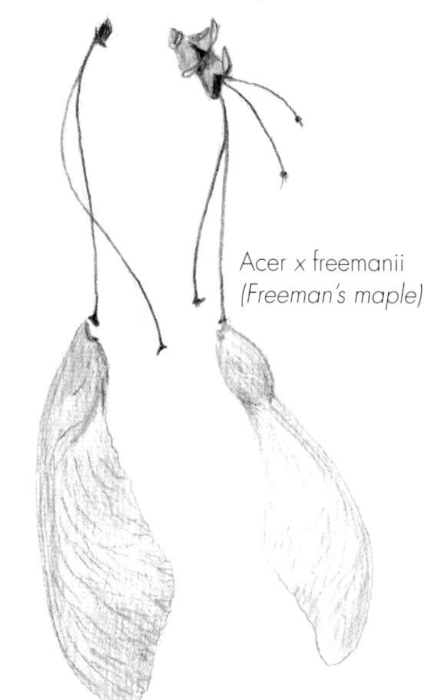

Acer x freemanii *(Freeman's maple)*

Fruit size and shape are the only reliable ways to distinguish species in the spring-fruiting group.

Sex Change in the Red/Silver Maples

Although the fruit-size breakout indicates that the red, silver and Freeman's maples are fairly stable, the possible continuous crossing would suggest that they really should be called a red-silver maple complex or hybrid swarm.

Trees in this group have a remarkable tendency to go through a sex change. Young trees start flowering with all male (yellowish, pollen-bearing) flowers, and most slowly switch to female (bright red, pistillate) flowers as they age. Individual trees may go through several years with both sexes of flowers produced on the same tree, either within individual flower clusters or on entirely separate branches.

the entire seed frame, including the first-year seedlings, in search of every last ungerminated seed. Be sure your seed frame is enclosed for an entire year.

The Fall-Fruiting "Hard" Maples

Exotic Alert

Norway maple, *Acer platanoides*, has become synonymous with settlement and exotic invasion in eastern North America. It is the most prominent exotic maple, causing considerable concern throughout the deciduous forest zone. *Acer pseudoplatanus*, the sycamore maple, is beginning to invade natural areas south of the Great Lakes.

Norway maple has been relied on heavily for roadside and street planting for a half a century, especially since the near demise of the American elm in the sixties and sugar maple decline in the eighties. The dark red–leaved and round-headed selections belong to this species, too. They are all host to the tar spot fungus disease, round black dots on the leaves in late summer. Norway maple escapes cultivation when seeds from its abundant annual seed crop move through storm sewers into ravine forests or blow directly into adjacent natural areas. This maple is filling urban and highway woodlands, casting a very heavy shade, which is causing a serious decline in biodiversity as well as significant erosion on forest slopes. Of course, Norway maple can always be girdled and left standing, ultimately to become the coarse woody debris so integral to woodland health.

Acer saccharum, SUGAR MAPLE, thrives on moist, well-drained soils. It occasionally dominates the forest community, reproducing in the shade of mature forests but rarely in the open. Sugar maple is widespread and abundant in the mixed forest from Nova Scotia through southern Quebec and Ontario to Minnesota and south through the deciduous forest to northern Mississippi and Georgia. (*See* photo p. 235.)

Acer nigrum, BLACK MAPLE, a close relative of sugar maple, is generally found in soils that are heavier and moister than those in which sugar maple occurs, such as floodplains and bottom lands. It is abundant in widely scattered pockets, with a more southerly distribution than sugar maple, occurring mainly from northern New York State through southern Ontario and Michigan to southern Wisconsin and Minnesota, south to Tennessee. (*See* photo p. 235.)

Black maple has leaves with a very fuzzy undersurface and petiole, whereas sugar maple's are entirely smooth. Intermediates are commonly found with smooth petioles and less fuzzy leaf undersides. These two hard maples also hybridize with *A. barbatum*, the southern sugar maple, where their distributions overlap.

It is important to maintain the local genetic diversity by growing the range of sugar and black maple races prevalent in your area. My own observations indicate that, side by side with sugar maple, the black maple is often healthier and longer-lived in urban conditions. The sugar maple is an exceptional yard tree, but where browsers are present, it requires protection when young. In the end, observations suggest that we pay more attention to the tree's ecosystem than its name.

The flowers of mature sugar and black maples are yellowish when in full bloom in early May. Both male and female flowers occur in a pendulous cluster, with male flowers far outnumbering female flowers. After wind pollination, the male flowers with their hair-like stalks drop to the ground. Seed crops are at three- to five-year intervals almost uniformly across a considerable part of the range, with stressed trees

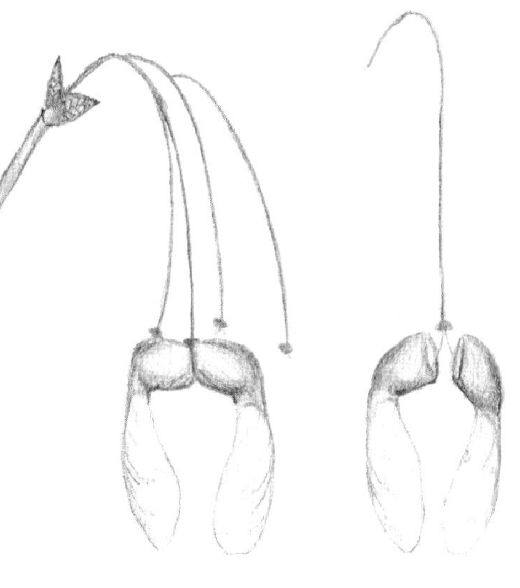

Acer saccharum *(sugar maple)*

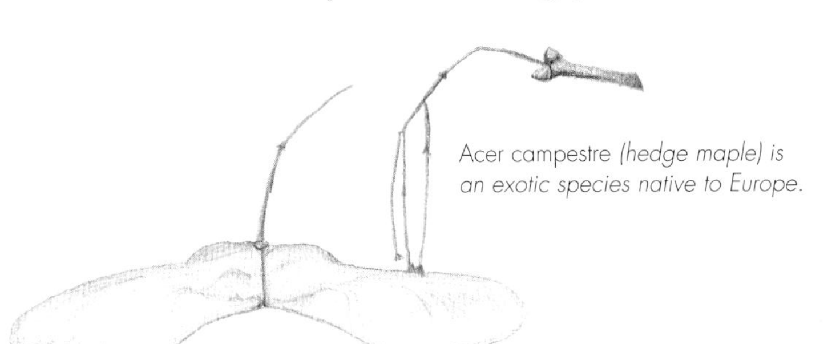

Acer campestre *(hedge maple) is an exotic species native to Europe.*

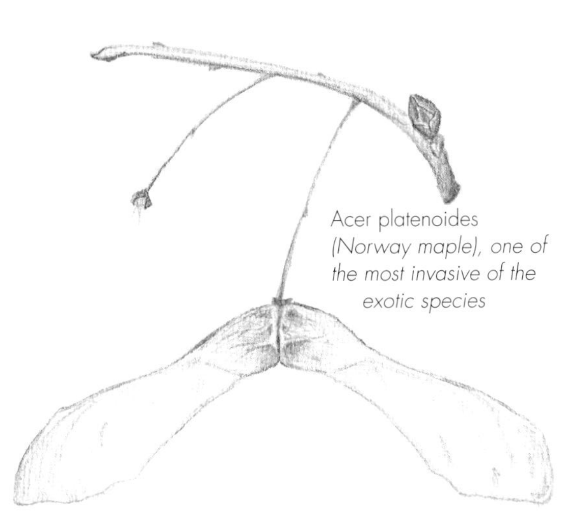

Acer platenoides *(Norway maple), one of the most invasive of the exotic species*

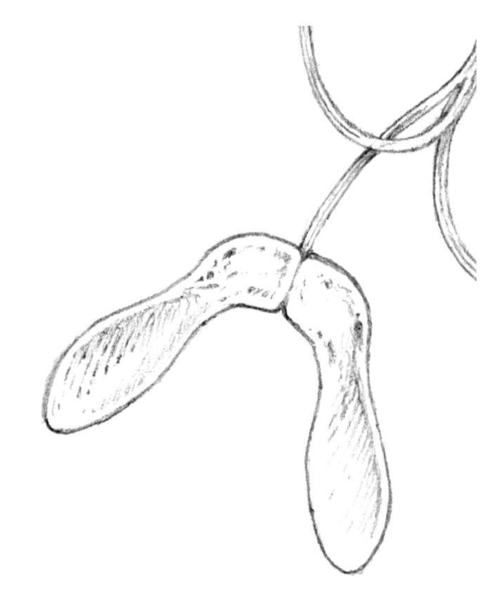

Acer nigrum *(black maple) is a native species.*

Acer pseudoplatanus *(sycamore maple), an exotic species native to Europe.*

Acer ginnala *(Amur maple) is an exotic species native to northern Asia.*

Seeds of native and exotic fall-fruiting species, shown above and on the next two pages, show many differences.

occasionally fruiting off cycle. Only one seed of each pair is filled, making these fruits unique among the maples (at least in my experience). Fruits are mature when the wing has turned brown and the seeds are still partly green. If harvested at this stage, they must be dried prior to sowing or storage. Ripe keys are dispersed in September, with rare individual trees holding fruit into January.

Natural germination is in the first spring and, after heavy seed crop years, the seedlings will carpet the forest floor and extend out into open fields in great numbers. As sugar and black maple seedlings are some of the most desired foods for browsers, natural and somewhat indiscriminate thinning or even complete elimination of new growth can be accomplished; saplings are not safe until they are above browse height. This also makes planting efforts in natural areas fruitless without adequate protection. Seeds should be sown outdoors when ripe, or dried for storage until early January, when they can be soaked and cold-stratified for 90 days. Like most maples, radicle emergence will begin in the cold stratification period.

Acer spicatum
(mountain maple)

Acer spicatum, MOUNTAIN MAPLE, is a beautiful species of the cool, moist forest understory, occasionally growing out of rotting stumps or logs. It is often found along streams but occasionally grows on surprisingly exposed, drier sites. Trees on exposed sites should be considered for their adaptive possibilities. It ranges from Newfoundland across southern James Bay to central Saskatchewan, through the Great Lakes region east to Pennsylvania and south through the Appalachian Mountains to northern Georgia. (*See* photo p. 234.)

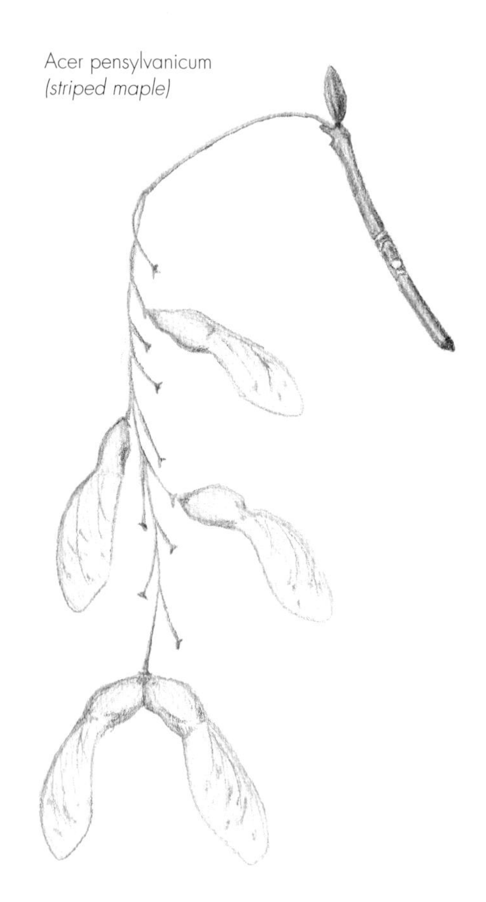
Acer pensylvanicum
(striped maple)

Mountain maples are noted for their exquisite orange-red autumn display and their striking upright spikes of white flowers in early June. Although most of the flowers of this species are perfect (with both male and female parts), separate male and female flowers do occur on the same tree. The annual crop of fruit is dispersed in October. Seeds are readily germinated after a 150-day cold stratification. Seedlings reach 12 to 15 inches after two years of growth and are easily transplanted. This is a lovely species, appropriate for setting into established restoration sites and backyard gardens, especially sites with a stream.

Acer pensylvanicum, STRIPED MAPLE. Anyone who sees this distinctive tree falls immediately in love with it. The outward arching branches, handsome bark and fascinating leaves are compelling to the gardener in each of us, but its strong association with deep, acidic humus soils in old-growth woodlands means that it is rarely seen in gardens or parks. It grows primarily on the Canadian Shield and Appalachian slopes; however, isolated populations grow on the limestone bedrock of the Bruce Peninsula and on the sand and limestone of Presqu'ile Provincial Park in Ontario. Seeds from these latter two sources are worthy candidates for trial in areas of limestone based soils. It ranges from Nova Scotia across southern Quebec to northern Lake Huron and southward through New England, from eastern Ontario to Pennsylvania and south through the Appalachian Mountains to northern Georgia. (*See* photo p. 235.)

Separate male and female flowers, in long, pendulous clusters after leaves emerge, can exist on one tree or different trees, sometimes changing from year to year on one tree. Fruits are dispersed in October, with seed crops at two- to three-year intervals. Seeds germinate readily after 150 days of cold, moist stratification. Seedlings require a fairly acidic soil to grow in or the leaves will become yellowish with dark veins, indicating an iron and manganese deficiency. This deficiency might be corrected with foliar nutrient applications, but if the soil is unsuitable, to what end? Unless seed sources can be found that will thrive on neutral soils, this species is not worth trying to grow outside its natural acidic soil range.

Exotic Alert

The nursery industry occasionally distributes *Acer rufinerve*, the Asian striped maple, which is able to grow in alkaline soils. It is sometimes inadvertently labeled as the native species.

Acer negundo, MANITOBA MAPLE or BOX ELDER, is an important pioneer species that readily establishes itself in open, highly disturbed, moist sites. Its natural range spans a few life zones, extending across most of central North America, from central Manitoba and Saskatchewan south to Texas and northern Florida, but absent in much of the eastern coastal plain and the shores of the northern Great Lakes. (*See* photo p. 234.)

Male and female flowers are on separate trees and are wind pollinated. An annual seed crop is shed gradually through the winter and provides a great food source for squirrels and seed-eating birds. This long dissemination period allows some seeds to occasionally be blown great distances over crusted snow. Seeds are mature in early October and require a cold stratification of only 60 days to germinate. Seedlings grow very rapidly, and one-year-old plants can be set into the landscape.

Trees are prone to upset and major branch breakage, but with proper pruning and management, they can be fairly long-lived. As drought conditions become more prevalent, *A. negundo* will play an important role as an interim wind-control species on farm lands and perhaps even as an urban tree in areas of highly-degraded soils.

In Praise of the Manitoba Maple

This tree is frowned on, but that should only be in relation to the statement that this shade-intolerant species makes on the degradation of the natural landscape, and the abuse it takes in the urban landscape. It is so detested in the east that many try to argue that it was never native here. I think it was always present along isolated stream banks and clearings, reaching out into the light. Once the forests were harvested for high-value trees, it spread rapidly. Native or not, this pioneering tree is here to stay. It ultimately gives way to other species but not before it has made a valuable contribution, stabilizing and renewing a humus layer on the degraded sites where it thrives.

Frederick W. Schueler of the Eastern Ontario Biodiversity Museum, Kemptville, poetically encourages an even greater appreciation of this much-maligned species. He documents the contribution, extending by way of food and habitat, to a vast array of organisms, from the *Fusarium* fungus that produces the red stain in the wood and *Cecropia* moth larvae that feed on the foliage to the evening grosbeak flocks that feed on the annual fruit crop in winter. The box elder bug that is spreading eastward is described as a nuisance when it gets into houses to overwinter in great numbers, but I suspect that the sanitization of modern landscapes has simply diminished its predators.

Acer negundo
(Manitoba maple)

Aesculus hippocastanum *(horse chestnut)* husk

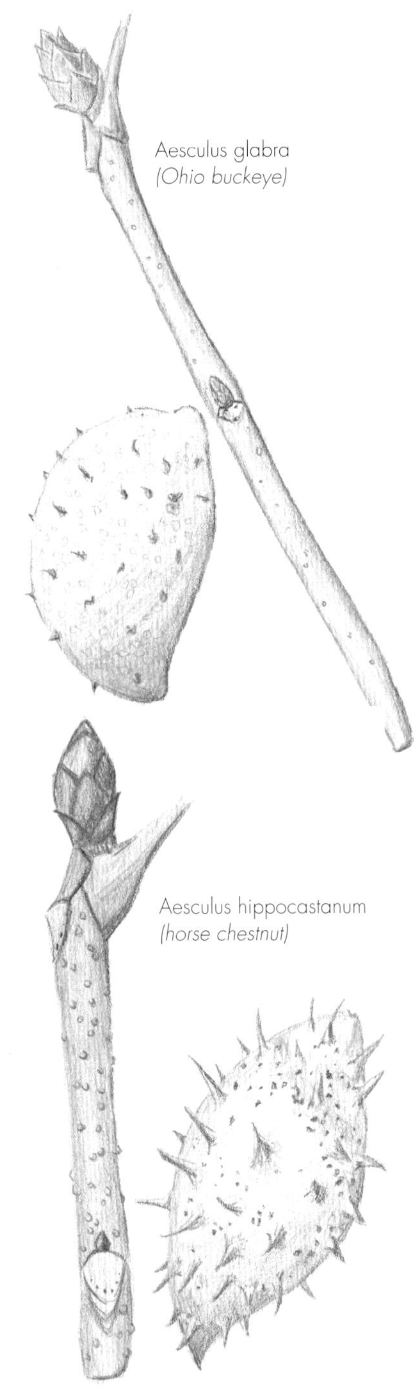

Aesculus glabra, *the Ohio buckeye, is readily distinguished from the European horse chestnut (also a buckeye) by dry buds and near-spineless fruit husks.*

Aesculus, the BUCKEYES

The seeds of all 15 species of buckeyes in the northern hemisphere are considered inedible for humans but are consumed by squirrels, and were possibly dispersed by the mastodon in millennia gone by. They are fed to pigs and horses in Europe, hence the name horse chestnut.

> **Exotic Alert**
>
> Most of us are very familiar with the European horse chestnut, *Aesculus hippocastanum*, with its bright white flowers. This tree is actually a buckeye, not a true chestnut of the genus *Castanea*, as the name might imply. It is increasingly found naturalized adjacent to settlement and is susceptible to leaf scorch disease. The leaf-mining larvae of a moth are presently causing extensive defoliation and tree decline in parts of Europe.

Aesculus glabra, OHIO BUCKEYE, is early to leaf out and early to drop its richly colored autumn leaves. Ohio buckeye is a small tree of bottom lands and stream banks in rich, moist woods. It is found in open woods principally in the deciduous forest of the Mississippi watershed, from Pennsylvania to Iowa and south to Texas, with outlying populations in southern Michigan and adjacent Ontario and northern Mississippi to Georgia. One natural stand occurs in southern Ontario across the river from Michigan, on Walpole Island First Nation lands. It is highly probable that the stand was introduced by First Nations travelers, for it is easy to imagine the beautiful seeds being shared as a gift or token by visiting tribes.

The greenish yellow flowers are produced annually in May and are pollinated by insects. Isolated trees in cultivation appear barren, the perfect flowers obviously requiring pollen from another tree. Each fruit contains two and occasionally three seeds, and squirrels disperse the annual seed crop. The seed germinates in the first spring and remains a highly prized food for squirrels well into the summer until the endosperm (kernel) has been used up by the seedling. Enclosure in a seed frame is thus required from planting date until midsummer to ensure seedling survival.

The fruit can be gathered from the ground or picked from trees in early September, when the husk shifts color from green to yellowish tan. Within a few days the husks will dry and split open, exposing the shiny chestnut-brown seeds. Buckeye seeds will dry out beyond recovery in a few short weeks of lying on the kitchen counter. Seeds should be planted outdoors within the week or held in a plastic bag in the refrigerator until ready to plant. Be sure to get them in the ground before freeze-up, since seeds will germinate in the fridge after only four months. A few may be grown in pots indoors so that you can enjoy their green leaves in midwinter, and then be placed outside in May after the danger of heavy frost is past.

Seedlings will reach 1 foot in two years and should be transplanted before the buds swell in the third spring, either to a nursery space or to their final destination. Groundhogs seem to favor the slow-growing buckeye and may keep it short for many years. Ohio buckeye is very adaptable to moist soils in urban conditions, but it will not thrive for long on dry sites. It is valued for its fall color display and its resistance to leaf scorch disease.

Aesculus flava, YELLOW BUCKEYE, has smooth, pear-shaped fruit. It occurs in the Ohio River valley (southwestern Pennsylvania to southern Illinois) and south through the Appalachian Mountains to northern Georgia. Two southeastern species, red buckeye, *Aesculus pavia*, and the shrubby bottle-brush buckeye, *A. parviflora*, are occasionally found in cultivation in the southern Great Lakes region. (*See* photo p. 236.)

Alnus, the ALDERS

> **Exotic Alert**
>
> *Alnus viridis*, green alder, is a circumboreal species that extends south only to the shores of the northern Great Lakes, and is distinguished by its overwintering catkins within a scaled bud and seeds with a conspicuous wing. The much planted *A. glutinosa*, European alder, grows 33 to 50 feet tall, usually with a single trunk, and has invaded many riverbanks and wet places, spreading from sites where it was once planted to stabilize stream banks. Both green alder and the European alder are now showing up in naturalization sites due to confusion in the nursery trade and improper seed-parent identification. Gray alder, *A. incana* ssp. *incana*, native across Eurasia and a close relative of speckled alder, may also be part of the array of alders found planted in the Great Lakes region.

Alnus incana ssp. *rugosa*, SPECKLED ALDER, a shrub usually 7 to 10 feet high but occasionally reaching 15 feet, is found in wet meadows and swamps and on stream banks. Alder swamps are virtually impenetrable except in the height of winter by snowshoe. It ranges from Newfoundland to Saskatchewan, southward through Ontario to Iowa, West Virginia and New York.

Alnus serrulata, HAZEL ALDER, a large shrub to 30 feet, found along stream banks, ditches, swampy fields, bogs and lakeshores. It is primarily an Atlantic coastal species in Nova Scotia and Maine south to Florida and along the Gulf Coast to Texas, also along the St. Lawrence River valley and the lower Great Lakes to the dunes of southern Lake Michigan.

Alnus incana *(speckled elder)*

Reproduction and Propagation

Natural seed dispersal takes place during late winter and early spring as the hard catkins dry out and the scales separate. The seeds will float for days if they fall on open water. Seeds are reportedly consumed by grouse and woodcock, and are definitely consumed by ducks and swamp sparrows on the northward migration. The northern distribution of the speckled alder may be explained by some seeds inevitably passing intact through these birds' digestive systems. Seeds germinate in the early spring when the ground is still cold and saturated.

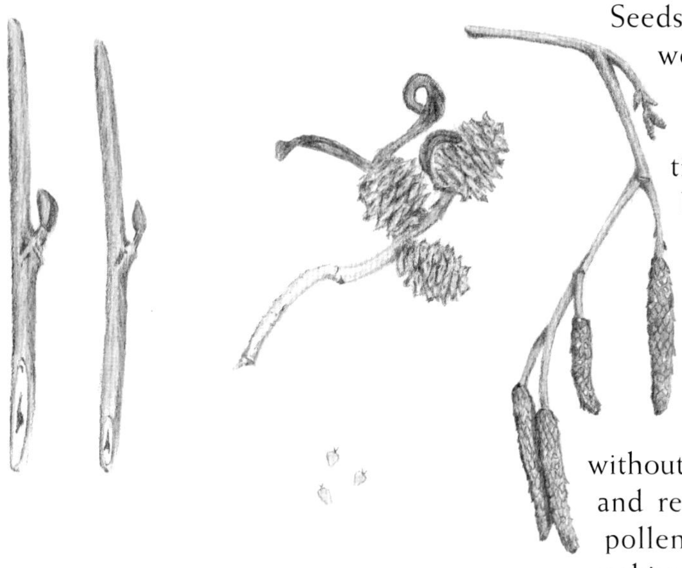

Like all members of the birch family, alders have preformed male and female flower catkins that go through the winter without protection. Male catkins elongate in April and release yellow pollen into the wind. The pollen lands on the minute upturned female catkins, which will mature during that growing season. The mature, cone-like fruit can be picked from late fall to early spring and dried at room temperature to open the scales. The seed is a small nutlet with a thin, ridge-like wing.

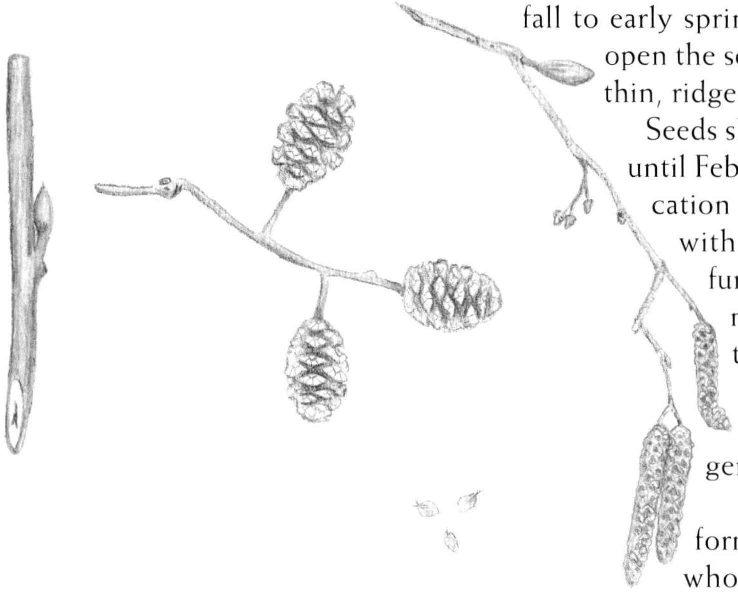

Seeds should be stored dry in the refrigerator until February and then placed in cold stratification for 60 days. The tiny seeds are sown with the stratifying mix and not covered further. Germination requires constant moist conditions. A small container of the sown seeds can be placed in a plastic bag or under a clear cover to maintain a saturated soil surface until germination.

Seedlings can be transplanted as they form their third and fourth leaf, or the whole pot of seedlings, after they have been thinned, can be plunged into the soil of your cold frame to grow for the first season. Planting out into your nursery can be done the next spring. Alders need no protection from rodents but should be shaded during their first year. Dense thickets of alders stabilize stream banks and are an important habitat for songbirds.

The fruits, seeds and buds of Alnus rugosa, *the native alder (above) and* Alnus glutinosa, *the European alder (below). The native alder,* Alnus rugosa, *has stalked winter buds and smaller catkins than* Alnus glutinosa.

Amelanchier, JUNEBERRY and SERVICEBERRY

The genus *Amelanchier*, with some 20 species in the northern hemisphere, reaches its greatest diversity in the Great Lakes region and the northeastern U.S., yet most of the species are not available in the horticultural trade. The species are specialists for the most part, each occupying a range from dry, open meadows to moist, well-drained open woodlands. The perfect flowers are pollinated by native bees as well as European honeybees. Annual crops of fruit ripen from late June to late July and most are quickly devoured by robins, catbirds and cedar waxwings. Natural germination takes place the following spring in relatively cool soils. Significant numbers of seedlings may be found under large old trees where the seeds have passed through the digestive systems of birds that have spent several days feeding in the same tree.

The flowers and the mature form of several serviceberry species are relatively distinct for each region, and every author seems to devise his or her own species characterization based on these differences. Confusion occurs because the genus is beset with hybridization, species complexes and changing botanical and common names. The common names for *Amelanchier* read like a phone book: serviceberry, juneberry, shadbush, Saskatoon, mountain juneberry, smooth serviceberry, Allegheny serviceberry, sarvis, sugar plum and more. The amelanchiers bloom in mid-May as the leaves are unfolding; this is an excellent time to try to identify this difficult group. The white flowers are so showy that this is also the best time to locate plants in your area for later seed collecting.

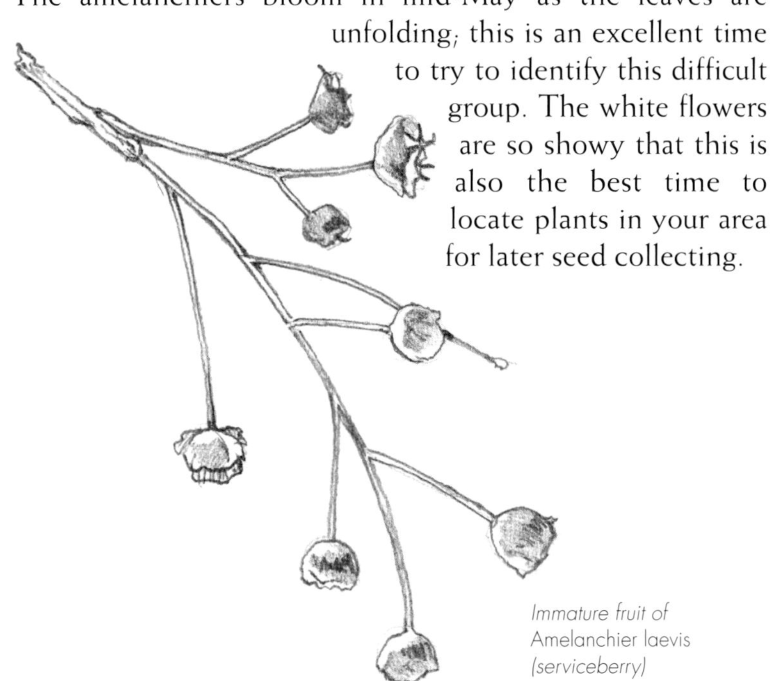

Immature fruit of Amelanchier laevis (serviceberry)

Fruit differences help to identify the species.

Exotic Alert

Saskatoonberry, *A. alnifolia*, a bushy, suckering midwestern species (rarely eastward, into northwestern Ontario and Minnesota), is commonly sold in garden centers throughout our area. A few cultivars have been selected from the hybrid *A.* x *grandiflora*, and several larger-fruited selections of *A. alnifolia* have been named (e.g., 'Smokey', 'Regent'). These cultivars are also readily available at nurseries. By collecting fruits from mature specimens of the indigenous species in the wild and using this seed source for planting stock, it is possible to reduce the impact of foreign pollen on natural stands.

Amelanchier canadensis, SHADBLOW SERVICEBERRY (or, inappropriately, Canada serviceberry), as presently understood, is an upright, tightly multistemmed, suckering shrub species of the east coast from Newfoundland to the Carolinas and more rarely inland to Alabama, Tennessee and western New York State up to the Niagara River. It has shorter, more rounded petals than the tree-sized serviceberries but the same sweet, juicy fruit. It can grow to 20 feet in a range of soil types. It is most frequently found in bogs, swamps and other wet ground.

Amelanchier stolonifera, RUNNING JUNEBERRY, blooms latest of all the amelanchiers, and has tight, upright clusters of short-petaled flowers that peak when the leaves are still folded and downy. This dwarf serviceberry is a shade-intolerant, suckering, colony-forming shrub to 3 feet high. It grows in dry clearings from Newfoundland to Minnesota and south to Virginia.

Amelanchier sanguinea, LARGE-FLOWERED JUNEBERRY, has loose, drooping flower clusters that are at their peak when the leaves are open and very downy. This species is a tightly branched, upright shrub rarely reaching 10 feet. It is found in sunny, dry clearings and rocky soils from southwestern Quebec to western Ontario and south to Iowa and North Carolina.

Amelanchier humilis (considered synonymous with the western species *A. alnifolia*), SHADBUSH, has dense, arching flower clusters that are at their peak when the leaves are unfolding and densely white and woolly. This species is an open, suckering, colony-forming shrub in rocky soils, growing 3 feet tall in dry sites and up to 16 feet tall in the partial shade of moist, open woodlands. It is found from Quebec, throughout Ontario to Manitoba and southward to South Dakota to Pennsylvania.

Amelanchier bartrameana, BARTRAM'S JUNEBERRY, has upright clusters of two to five flowers that are produced with expanding glabrous, reddish leaves. This rounded shrub with few trunks grows to 7 feet tall in the partial sun of open woodlands and forest edges. It is more abundant in the east, growing in sandy and acidic moist soils from Newfoundland to western Ontario and south from Minnesota to Massachusetts.

Amelanchier arborea, DOWNY SERVICEBERRY, has upright to nodding racemes of long-petaled flowers that open when the leaves are still folded and grayish and downy. This is a fence-line and forest-edge tree, often with multiple trunks bent toward the sunlight. It thrives in dry locations, occurring from Maine and New Brunswick to Minnesota and south to Oklahoma and northern Florida.

Amelanchier laevis, ALLEGHENY SERVICEBERRY. The signature of this large, single- to multi-trunked tree is its pendulous racemes of long-petaled flowers. They open at the same time as its smooth, coppery red leaves unfold. It grows to 33 feet or more along the forest edge or in interior woodland clearings and thrives in moist but well-drained soils from Newfoundland to central Ontario into Minnesota and south to Iowa and Georgia. Occasionally (twice in my experience) a pink-flowered form is found in the wild. *A.* x *grandiflora* is a natural hybrid of this and the previous species; garden selections of it are widely available at nurseries.

Seed Collecting and Propagation

The berries of most *Amelanchier* species are very tasty, despite the large seeds and persistent sepals. (*A. arborea* and *A. sanguinea* berries are a bit dry.) The fruit is often difficult to obtain because it is devoured by cedar waxwings and other birds just as it turns red and begins to soften — well before it is fully ripe. Fruit maturation is progressive through each cluster, starting at or near the base and a week or two later in the berries at the end of the cluster. There are usually one to four seeds in each berry but some have as many as ten.

Handpick only the fruits that have become red and swollen, or fully ripe and purplish. Mature red berries will ripen over the next few days if left at room temperature in an open container. Fresh fruit can be cleaned in the grit bag, but keep in mind that the seeds are not very tough. The very plump, dark brown healthy seeds are readily distinguished from the thin, light brown aborted seeds. Occasionally both filled and empty seeds will float to the top of the container. Stir a few drops of liquid detergent into the pulp to break the surface tension and allow the filled seeds to sink. You can sow cleaned seeds right away or dry and store the seeds, but dried berries store better.

Dry the fruits by spreading them out on newspaper until they are no longer fleshy and then place them in dry, cold storage for stratification starting in January. Seeds from dried berries can be extracted in two ways: crush the dried berries with a rubber mallet and hand-sort the seeds, or soak them for 24 hours and then clean them in the grit bag. Cleaned seeds are placed in cold stratification for about four months. As in most members of the rose family, radicle emergence takes place in the cold, so after three months you should check for germination weekly. Once radicle emergence has occurred or the four months have elapsed, the seeds should be planted.

Seeds are planted in a clay pot or seed frame ⅛ to ⅓ inches apart, depending on the number of seeds you have. Summer-sown seeds should be covered with a leaf mulch through the winter, but remember to remove it in mid-April. Spring-planted stratified seed only need to be kept moist. Germination is rapid in warm conditions, producing 3- to 4-inch seedlings in the first year. In two years they will be up to 10 inches tall and can be teased apart and planted into a nursery the following spring. They survive transplanting well as long as the buds have not yet started to swell. Plants that are several years old are best moved in the fall. The suckering species can be divided but divisions must be cut down to just above ground level to be successful.

Rabbits, mice and deer really go for *Amelanchier*. You will need to protect your seedlings until they are well established. Plants that have been severely debarked in the winter should be cut down to sprout from the base again. The amelanchiers are well suited to gardens but difficult to establish in natural areas where herbivores may destroy young plants.

Amorpha fruticosa, SHRUBBY FALSE INDIGO

Shrubby false indigo is a low shrub occasionally to 13 feet high found along riverbanks and in moist thickets. It occurs from Pennsylvania and New York to Wisconsin and south from northern Florida to Texas. Rare in southern Ontario and Manitoba and occasional in many western states, introduced in others to the north and west.

To germinate seeds, presoak for 12 hours in warm water, then sow in early spring in a greenhouse. Seeds typically germinate in 30 to 60 days.

Aralia spinosa, ANGELICA TREE

Angelica tree is a spiny, upright shrub to 33 feet occurring in moist woods and wooded slopes. Its range is from Maine to Illinois and Iowa and south from Florida to Texas.

Collect seeds when mature and sow them directly into a cold frame or protected seedbed. Dried and stored seeds will require 90 days or more of stratification for germination. Keep in a protected propagation area until they are 10 inches or more tall.

Aristolochia macrophylla, DUTCHMAN'S PIPE

Dutchman's pipe is a deciduous woody climbing vine that grows to 33 feet or more. It occurs in rich, moist woods and along streams, primarily in the Appalachian Mountains, from Pennsylvania and south from Georgia to Alabama.

Collect ripe seeds in the fall and sow into a seedbed or give a 90-day stratification and then sow in a greenhouse.

Aronia melanocarpa, BLACK CHOKEBERRY

In the spring, cloudlike mounds of white bloom make this shrub easy to find in the slightly acidic soils along marsh and bog edges. It grows in the mixed and deciduous forest regions from Newfoundland to Minnesota and south to Georgia.

Chokeberry spreads slowly by underground stems, occasionally forming dense thickets over 3 feet high. In late May, clusters of perfect flowers are pollinated by bees. The fruits turn black and are ripe in mid-September to mid-October. Seeds are dispersed by thrushes and other birds during fall migration. A portion of the crop persists, freeze-dried and shriveled, through the winter. The fruit is softened by spring rains and a few are consumed during spring migration, partially explaining the northward distribution of the species, but mammals such as foxes and wolves are more likely to move the seeds in the fall.

The berries are easily cleaned using the grit bag, yielding seeds almost identical to those of *Sorbus* and *Amelanchier*. However, because the seeds have a longer cold-stratification period (five months before radicle emergence), they can be stratified or planted in a clay pot or seed frame after harvest as soon as they are cleaned. Seedling growth is similar to, but not as vigorous as, *Amelanchier*.

Chokeberry is an important wildlife plant that provides much-needed late-season energy for the southbound migration of a number of bird species. It is well suited to urban water gardens and the wetland edge of natural areas. (*See* photo p. 236.)

Aronia melanocarpa *(black chokeberry) fruit, dried berries and seed.*

Red or Black?

Aronia arbutifolia, red chokeberry, is a thicket-forming clonal species growing 3 to 10 feet tall in bogs and wet woods primarily in the east from Newfoundland to Florida. Its dazzling red berries in October persist as a dried brown fruit through the winter if not consumed by wildlife. This species is commonly sold throughout the Great Lakes region, often substituted for black chokeberry. Both species are readily propagated by division of rhizomes from the parent plant.

Rare Hybrids

Aronia, Sorbus, Amelanchier and *Pyrus* all have the same number of chromosomes and once in a while produce exciting inter-generic hybrids. They are named by combining the generic names, as in *Sorbaronia, Amelosorbus* and *Pyronia.*

Asimina triloba, PAWPAW

When I first heard of pawpaw, I thought it was a mythical tree in a childhood song. The long, luxuriant, overlapping leaves hang downward, giving it a very tropical appearance, and, in fact, it is the northernmost member of a huge tropical and semitropical family. The custard-apple family has over 120 genera. One of them, *Annona*, contains some fine commercial fruits, such as the custard apple of Australia and the soursop of the Caribbean.

Our pawpaw grows in natural black walnut woodlands, on floodplains and in other moist but well-drained sites. The stems of this colony-forming tree are well spaced out, arising from a horizontally spreading root system. Found from New Jersey through extreme southern Ontario to southern Michigan and southeast Nebraska and south to Florida and eastern Texas, the pawpaw is a signature tree of the deciduous forest life zone. At least three other pawpaw species are found in Florida. Zebra swallowtail larvae feed on the foliage.

Pawpaw can form extensive colonies from a single tree, up to a hectare in size. If a colony is isolated, it produces very few fruits and often none at all, due to its self-incompatibility (its own pollen will not bring about fertilization necessary for fruit production). The fruit, when produced, is consumed by raccoons and opossums, and perhaps long ago by bears or other large mammals. First Nations peoples and early settlers are also implicated, as has been the case with people moving most any of the nutritious fruits worldwide. Pawpaw fruit drops from the treetops in mid-October and decomposes rapidly in the leaf litter, leaving the seeds lying in the same arrangement as they were in the fruit. Natural germination takes place in early summer of the following year. Because the seeds require considerable warmth to germinate, natural regeneration by seed is only occasionally observed in the northern end of its range.

Flower buds are produced in late summer along the length of the current season's growth and are visible throughout the winter. These plump, velvety buds swell to form stunning (one author suggests lurid) maroon flowers in May. The flowers have a slightly rotten smell and are pollinated by swarms of flies and beetles.

Collect the fruits when they can be shaken from the tree (and are still green), or ripe ones from the ground. The green fruit may need a few days to soften so that you can easily extract the 3 to 12 smooth, dark brown, flattened, oval-shaped seeds. The seeds are ½ to 1 inch long and all are viable. If you cut one open to examine it, you will find the pawpaw seed streaked with dark lines through the whitish endosperm, unlike the pure white seeds of most other plants in this region.

*Up to 20 seeds are present in the larger papaw (*Asimina triloba*) fruits.*

Beware of Eating Unripe Pawpaws

The pawpaw fruit is delicious. It has a smooth, custard-like texture and its flavor is often described as a combination of various tropical fruits. The green fruit can be stored in the fridge for up to a month. At room temperature, it soon turns from green to yellow-brown and is soft enough to eat. Either the pawpaw is peeled and the large seeds removed or it is eaten like an olive. Be warned, though: consuming unripe fruit causes stomach upset within a day. The entire staff at the University of Guelph Arboretum can attest to this. We were all off work the day after eating a cake made with unripe pawpaws baked by a well-intentioned co-worker. No legal action was taken!

Fall sowing outdoors is not very successful — at least in the northern part of its range. These seeds are best cold-stratified for four to six months and then planted in a cold frame in April. They are best planted on their sides ¾ inch apart and ½ to ¾ inch deep. It takes two to three months before the shoots emerge. Shoot emergence is slow at first, and some seedlings have difficulty shedding the seed-coat if it is pulled above the soil (not planting the seeds deeply enough will result in a large percentage of seedlings with this condition). If the seeds are germinated in clay pots, do not separate the young seedlings when you are ready to transplant them outdoors. The whole mass should be plunged into a shaded bed, being careful not to disturb the roots. After two years, when the plants have grown about a foot, the seedlings can be separated and planted in partial shade. Herbivores seem to leave this tree alone, and extensive colony formation is halted only by mowing.

Shoots taken from an existing colony may be planted in early spring. Select a shoot one to two years old, dig a foot deep and a foot wide around it, and carefully separate the roots from the soil. This should ensure there is enough root attached to the shoot. The shoot will survive best if cut down to two buds above the soil line. These can be set into the ground in an appropriately shaded location and protected by a burlap enclosure for the first year.

The cultivated varieties of pawpaw are propagated by grafting, and fruit growers in several regions are commercializing these tasty fruits.

Betula, the BIRCHES

The birches are usually found in great abundance in early successional forests but rarely in more mature forests. It is a very diverse and complex genus, with over 50 species in the northern hemisphere, mostly in Asia. The paper birch of our region is more closely related to the white birches of Eurasia than to the native yellow birch that it might be growing adjacent to. Hybrids of our native birches are found occasionally and might cause considerable head scratching.

Exotic Alert

European white Birch, *Betula pendula*, was undoubtedly preferred by the horticulture industry because the bark turns white so soon, making it easier to sell than a 10-foot paper birch still sheathed in coppery red bark. For some reason, even reforestation efforts of a few decades back focused almost entirely on planting the European white birch, possibly due to the convenience of an annual seed crop at a very early age, or to mistaken identification. The species has naturalized extensively in many areas. When found in the wild in southern Ontario and Michigan, it is often mistaken for gray birch or paper birch by those who think a birch is a birch is a birch. Pay attention to the base of the trunk, which differs from species to species. (*See* photo p. 236.)

Betula alleghaniensis *(yellow birch) has hairy scales.*

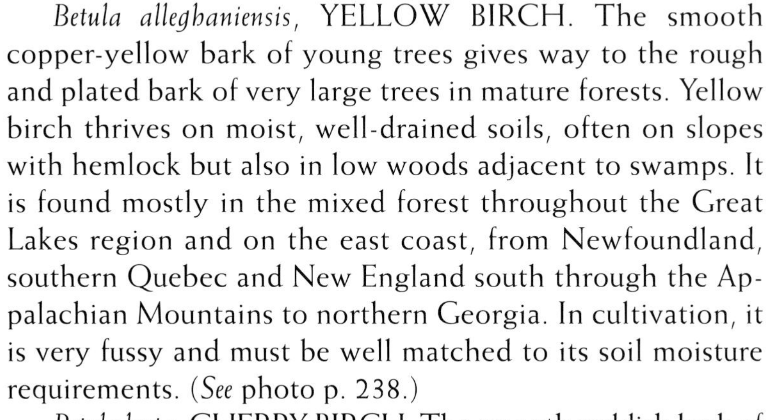

Betula lenta *(cherry birch) has smooth, hairless scales.*

Betula alleghaniensis, YELLOW BIRCH. The smooth copper-yellow bark of young trees gives way to the rough and plated bark of very large trees in mature forests. Yellow birch thrives on moist, well-drained soils, often on slopes with hemlock but also in low woods adjacent to swamps. It is found mostly in the mixed forest throughout the Great Lakes region and on the east coast, from Newfoundland, southern Quebec and New England south through the Appalachian Mountains to northern Georgia. In cultivation, it is very fussy and must be well matched to its soil moisture requirements. (*See* photo p. 238.)

Betula lenta, CHERRY BIRCH. The smooth reddish bark of cherry birch loses its shine and becomes furrowed as the tree ages. It grows naturally in moist upland woods and on river embankments throughout the eastern portion of the deciduous forest, from Maine south through New England and in the Appalachian Mountains to northern Georgia. A few populations are known on either side of the Niagara River, one in Ontario and several in adjacent western New York. This is a handsome, long-lived yard tree. (*See* photo p. 236.)

Betula papyrifera, PAPER BIRCH, is an early-succession species of moist woodlands, but it is also found on sand dunes (sand is often damp below the dry surface), exposed granitic rock outcrops and limestone cliff edges. It can reach considerable size, up to 82 feet under ideal conditions such as near the base of slopes and in other moist conditions. Its natural range is primarily in mixed forests from Labrador across Canada into Alaska and from the northeast states through Michigan and Minnesota, as well as down the Rockies into Montana. Paper birch is an irresistible yard tree; it is indispensable for providing quick shade in woodland gardens. (*See* photo p. 238.)

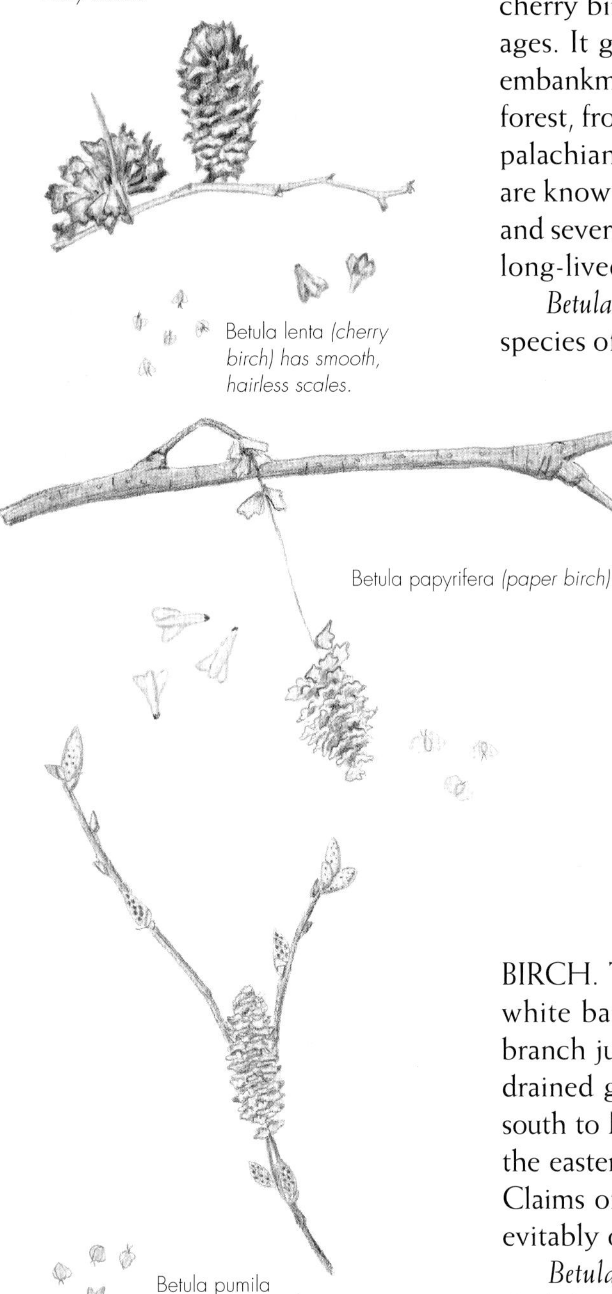

Betula papyrifera *(paper birch)*

Betula pumila *(swamp birch)*

Betula populifolia, GRAY BIRCH. This small birch has a trunk with unbroken, chalky white bark and distinctive black ridges angling down from branch junctions. It is usually very abundant on acidic, well-drained gravelly soils from Nova Scotia to southern Quebec, south to New Jersey and Pennsylvania, extending west only to the eastern end of Lake Ontario in the mixed-forest life zone. Claims of sightings farther west (even into Michigan) are inevitably of European birch. (*See* photo p. 239.)

Betula pumila, SWAMP BIRCH. This upright, shrubby, dark-barked species is adapted to thrive in bogs and swamps and along wet shores. It occurs from Newfoundland and Quebec to Yukon, south to Oregon, Indiana, southern Ontario and Michigan to New Jersey. The species is a great fit for urban wetland gardens, due to its small leaves and slow growth. Two other shrubby birches occur in North America, *B. glandulosa* and *B. occidentalis*, but well north of the Great Lakes.

Betula nigra, RIVER BIRCH, has distinctive tan to pinkish or reddish brown bark, curling back similar to paper birch. It is a tree of lowlands and riverbanks, occurring mainly from southeastern New York and Pennsylvania to eastern Iowa, and south from Texas to northern Florida. River birch makes a stunning, disease-free landscape tree. (*See* photo p. 237.)

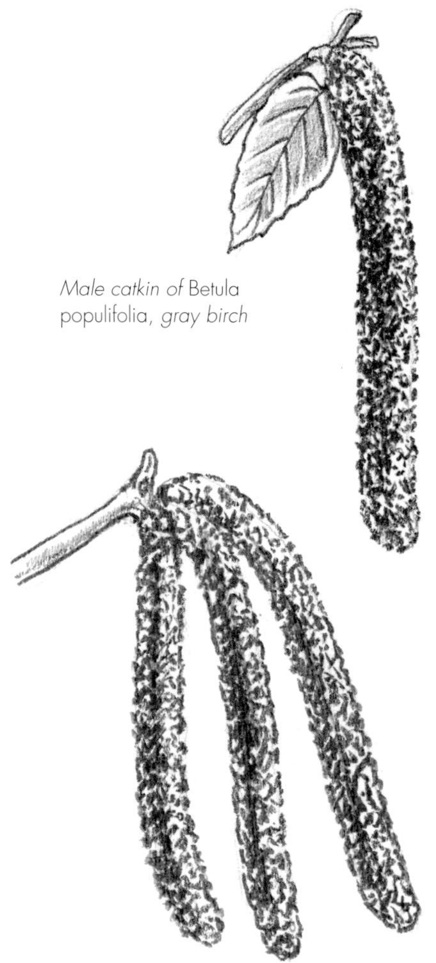

Male catkin of Betula populifolia, *gray birch*

Male catkins of Betula nigra, *river birch, usually come in threes.*

Reproduction and Propagation

Birch seed falls gradually from late August to early April as the catkins slowly dry and release portions of the seed crop. The peak dispersion occurs in November during windstorms. Intensive feeding by finches and chickadees causes the catkins to shatter, raining scales and seeds to the ground below. If you take a late winter walk through a stand of birch, you will see the snow surface speckled with both scales and seeds and perhaps evidence of ground feeding by juncos and grouse. With the exception of swamp birch, which produces annual seed crops, the native birches tend to produce a seed crop every other year. Birch seeds are blown considerable distances across smooth snow or ice surfaces during winter storms.

Male flowers are preformed in August and are visible in winter as small catkins at the ends of branches. The catkins elongate in mid-April as the leaves expand, turning bright yellow when the pollen sacs are exposed. On the same tree, reddish female catkins expand from the opening buds in spring, but at a slightly different time so as to receive windblown pollen from neighboring trees.

The fruit is a cone-like catkin with each scale protecting two seeds. When a mature catkin is shattered over white paper, the scales are readily distinguished from the minute seeds, which have a thin wing on either side.

Since the fruit naturally dries before dispersal, never subject the seeds to damp, airtight conditions. Always use a paper envelope for collection. Each catkin holds upward of 200 sound seeds — as much as anyone needs. The catkins can be picked from branches at the mature stage or you can shake the seeds from the catkins once they have started to shatter naturally. When collecting *B. pumila* seeds in the early spring, I've found snowshoes indispensable for making my way along frozen swamp edges.

Cherry Birch and the Pioneers

We nearly lost the cherry birch two centuries ago. The trees were cut and the twigs boiled for the valuable wintergreen oil that floated to the surface.

Is Cherry Birch Native to Canada?

The one stand on the shores of Lake Ontario near St. Catharines, Ontario, of less than 20 trees, is thought to be natural, but alternatively it could have been established by the early settlers in the late 1700s or early 1800s. Evidence on the natural-stand side includes documented records for the current site going back to the late 1800s; the largest tree (lost about a decade ago), estimated to be about 200 years old by its size and the growth rate of cored trees in the area; descriptive records from the mid-1900s of the stand as a natural forest; and the occurrence of stands on similar sites on the U.S. side of the Niagara River.

Beware the Bronze Birch Borer

The white birches in a planted, storm-sewered, dry urban landscape are particularly prone to destruction by the bronze birch borer (*see* illustration p. 67). This flat-headed borer's larvae tunnel along the cambium layer, consuming both inner bark and outer heartwood. Several circuits around the trunk will ultimately prevent water from moving up the stem and cause the crown to die. Evidence of the insect can be seen on the trunk, where wound-response growth forms a ridge over the borer's tunnel. It is related to the exotic emerald ash borer (*see* p. 130), which is now devastating ash trees and ash-dominated forests in the southeastern Michigan and southwestern Ontario area.

The native paper birch long ago evolved to tolerate or resist the bronze birch borer, which in turn is kept in check by natural predators. The European white birch was never exposed to this North American insect in Europe. Because it is often planted on dry sites, it is thus doubly compromised, and will rarely survive the borer. Nevertheless, some do, and large specimens are occasionally seen.

The catkins, whole or shattered, should be dried for several days at room temperature before long-term dry storage. Birch seed is said to need only moist conditions in the presence of light to trigger germination, but I have found that cold, moist conditions for about two weeks will enhance germination, at least for yellow and cherry birch.

The seeds can usually be separated from the heavier catkin scales by fanning. Scatter the seeds on the soil surface so that they receive light. To keep the soil moist (necessary for successful germination), use a clear cover or place the container in a plastic bag. Alternatively, seeds can be covered with small grit and cold-stratified for two weeks. Germination takes two to three weeks; the covering should be removed as soon as seed leaf lifting is observed.

Seedlings develop slowly in their first year. With the exception of swamp birch, which is much slower, the birches reach at least 20 inches in two years. At this time, they can be separated and planted. The birches are well adapted to moist conditions, with little tolerance for dry sites. They can grow surprisingly fast in appropriate conditions. Transplant success is reduced dramatically if seedlings are dug after the buds have started to swell in late April. Transplants need to be protected if deer and rabbits live in the area, but meadow voles seem to avoid the genus. Larger trees are an important food for beavers.

Campsis radicans, TRUMPET CREEPER

I was stunned when I first noticed the massive 6-inch wide trunks of trumpet creeper that adhered to the oaks and poplars in cottage front yards in Rondeau Provincial Park on the north shore of Lake Erie. *Campsis* grows naturally in well-drained, sandy and gravelly soils. It forms a widespread, scrambling, low shrub on open rocky shores, mounding over any structure. In open woods, it ascends tree trunks, holding fast by aerial roots. It is found in extreme southwestern

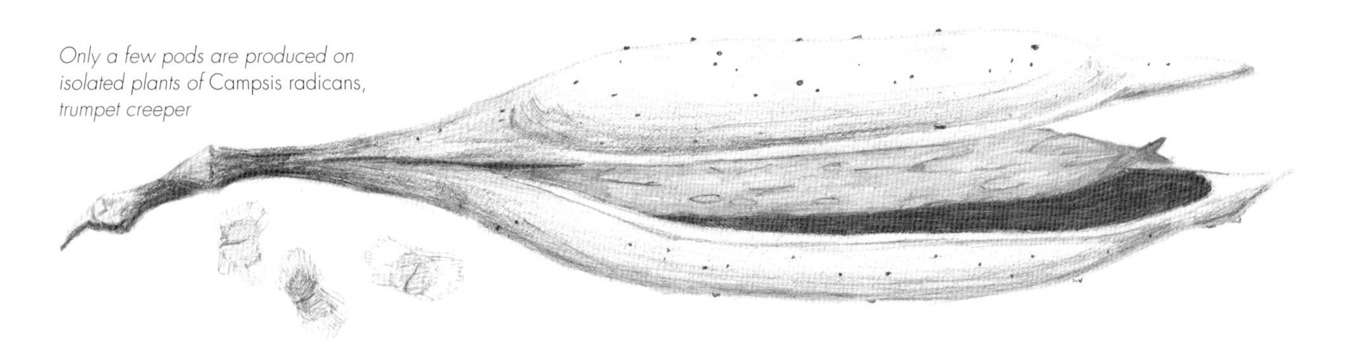

Only a few pods are produced on isolated plants of Campsis radicans, *trumpet creeper*

Ontario, especially the Lake Erie islands, and is widespread throughout the eastern U.S. from New Jersey to Ohio and Illinois and south from Florida to Texas. *Campsis* has naturalized from cultivated plants in a number of locations, particularly around old homesteads, where it covers embankments and old structures.

Natural spread is clonal, both by root sprouts and by layering of ground-hugging stems, as well as seed dispersal. When the pod-shaped fruits split open in late winter, the seeds can be carried by the wind for hundreds of yards. They germinate on warm, damp soil in the first spring. Fruit is rarely produced on cultivated plants in isolation, indicating a partial requirement for pollen from neighboring plants. The flowers are pollinated by hummingbirds and probably moths in July and August.

The fruits mature in late fall. The short, thick, woody pod encloses hundreds of very thin seeds, each with a paper-like wing attached to one side. The seeds are paired, with the wings extending in opposite directions, and are easily extracted by splitting the pod open. Fruits and seeds must be stored dry. No pre-germination treatment is required. Seeds sown in the spring germinate in several days and produce 6- to 8-inch high seedlings after one year. These can be planted out the following year. Chewing by mice or voles around the base of stems has been observed, and deer may browse on the foliage.

This species is often available from garden centers as cloned plants propagated by both root and stem cuttings. Plants grown from seed are uncommon but more desirable, because, if a few are planted, they will produce the attractive pods.

Carpinus caroliniana, MUSCLEWOOD or BLUE BEECH

Exotic Alert

There are about 25 species of *Carpinus* in the northern hemisphere, mostly in Asia. The European hornbeam, *Carpinus betulus*, is now found in some naturalization areas due to substitution or seed-source confusion in the nurseries. The European species is very similar to the native tree, but the former has larger fruit and seeds. In addition, *C. betulus* grows more tightly upright than the open, more spreading native species. Its leaves are generally thicker, with sunken veins, and they turn brown in the fall rather than the delicate shades of red, orange and purple of the indigenous species.

Whenever I pass a musclewood on a trail, I have an irresistible urge to feel the muscle-like ripples of the trunk that give this tree its most appropriate common name. One of its other common names, blue beech, should be discontinued, because it is not a beech at all. Occasionally it is called an ironwood, but this name rightly belongs to the closely related *Ostrya virginiana*, which has tan-brown bark in longitudinal strips and clusters of dry, sac-like fruits.

Musclewood is a small understory tree primarily in the deciduous-forest and southern mixed-forest life zones. It is common in cool, moist, rich soils but is found occasionally in drier, warmer, exposed or forest-edge conditions, which should be sought out as seed sources. *Carpinus caroliniana* is common in the southern Great Lakes region to New England and south from northern Florida to eastern Texas. This same species is also found in the mountains of Mexico and Central America, possibly as remnants from its glacial refuge.

Carpinus caroliniana *(musclewood)*

Musclewood seed is dispersed in October as the dry fruit clusters shatter in the wind. Windblown fruits don't travel far in the woodland, where wind speeds are not high. It is likely that woodland mice and chipmunks play a role in some dispersion, but natural seedlings are often found in close proximity to the parent plants, indicating little movement of fallen seeds. Natural germination is delayed until the second spring because the embryo is not fully developed at seed dispersal time.

Like all members of the birch family, musclewood produces male catkins during bud development in late summer. Visible in winter, these catkins elongate in early spring, turning golden yellow as the pollen sacs are exposed. Female flowers are wind pollinated when the styles protrude from the partially expanded lateral buds in early May. The fruit is a terminal cluster of nutlets, each held firmly by a partially enclosing leafy bract.

A good crop of fruits is produced at two- to five-year intervals and can be gathered from the low branches of forest-edge specimens any time after the fruit begins to shift color from reddish tan to brown. Immediately dry the fruit for a few days until the bracts are easily crushed. Rubbing the dried fruits between leather-gloved hands will dislodge the seeds from the bracts. The powdered bracts can then be screened out or separated from the seeds by winnowing. Seeds can be kept dry and refrigerated for several years.

High germination rates are possible the first spring if you soak the seeds for 24 hours (discarding the floating empty seeds), then give the seeds a 60-day warm, moist stratification followed by an 80- to 90-day cold stratification period. The seeds are planted about ¼ inch apart, covered with soil, and placed in a cool location (extra shade will help). Because this species is generally found in cool, damp woodland environments, it does not germinate well in a warm location.

Seedlings, which are 8 to 12 inches tall after two years, have a fibrous root system that makes them easy to transplant. They should be spaced 3 to 4 inches apart and must be shaded to grow well. Four- to five-year-old plants can be established in woodlands and shade gardens, provided you protect them from rabbits, which will reduce them to sticks where predators are absent.

Carya, the HICKORIES

The relatively fast-growing hickories are only slow-growing for a few years after transplanting. The delight of seeing the shaggy bark can be attributed to only two species; the other hickories often go unnoticed until the nuts are on the ground. The hickories are a small genus of some 15 species in eastern North America and Asia, with the several local species occupying the full range of soil-moisture regimes.

High squirrel populations and a hesitation to plant hickory have limited the number of these remarkable trees in the settled landscape. After the seeds mature in October, squirrels are the recognized dispersal agent, planting one seed at a time. The long-distance northward migration of these heavy nuts in postglacial forests, and how they crossed the rivers of the Great Lakes region, remains a mystery to me.

Carya illinoenensis, PECAN, is a hickory that occurs naturally from just south of the Great Lakes watershed to the lower Mississippi valley. It can be found in wet floodplain forests but adapts to well-drained soils. Hardy selections are in cultivation well north into southern Ontario. Pecan is highly valued for the commercial nut crop. (*See* photo p. 238.)

Carya cordiformis, BITTERNUT, is perhaps the most common hickory. This unusually smooth-barked tree with its yellowish winter buds grows in moist, well-drained woods. The distinctive thin-shelled nut has a very thin husk and, true to its name, is unpalatable. It is found from southern Maine and southern Quebec and across to Minnesota and south along the east coast and Mississipi Valley and as far west as Nebraska, and south from Texas to northern Florida. (*See* photo p. 238.)

Carya glabra, PIGNUT HICKORY, which has tight, furrowed bark and smooth, short-pointed buds, is found most commonly in gravelly or sandy upland woods. Its flattened, elongated nut in a pear-shaped husk is scarcely edible. Red hickory (*Carya ovalis*), with a more palatable nut, is now generally considered a variant of this species. The range is through the deciduous-forest life zone, from extreme southern Ontario and adjacent Michigan and New York south from northern Florida to eastern Texas.

Carya ovata, SHAGBARK HICKORY, leaves a lasting impression with its great vertical strips of bark that peel back.

Hybrid Hickories

Hybridization is occasionally found among the hickories, but because the individual species are often so similar, the hybrids are generally not noticed. The common hybrids are between shagbark and shellbark (*C.* × *Dunbarii*), pecan and bitternut (*C.* × *Brownii*), and shagbark and bitternut (*C.* × *Laneyi*). Although hybrids between shagbark and pignut and between pignut and bitternut are also found, I know of no name assigned to them. All of this hybridization — uncommon though it may be — leads one to suspect that hybrid swarms might be found with a considerable range of intermediate forms. Nuts from good-sized, healthy trees of any of these hybrids would be ideal candidates for planting in the local area.

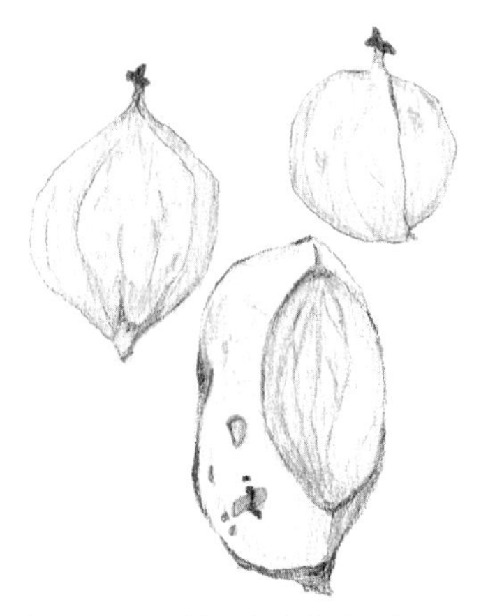
Carya ovata *(shagbark hickory)*

Carya laciniosa *(shellbark hickory)*

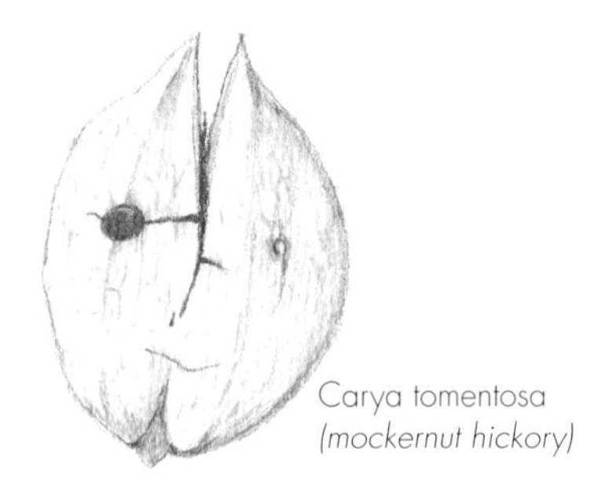
Carya tomentosa *(mockernut hickory)*

In the Carya species, whole fruit is the most reliable identification characteristic, whereas buds, bark and leaves only add to the confusion associated with the presence of natural hybrids.

The fruit is almost spherical, containing a flattened, four-ridged nut that is very sweet and highly prized. This is a very difficult species to establish. Shagbark hickory grows in a wide range of soils, from damp sand to moist woodland slopes, and from heavy clay to a distinct population on granitic rock outcrops in the area of Perth, Ontario. It is found in the Ottawa Valley, extending down the St. Lawrence, southern Maine through New England and the Midwest to southern Minnesota, and south from eastern Texas to northern Georgia. (*See* photo p. 238.)

Carya laciniosa, SHELLBARK HICKORY. The bark of this large species is shed in large, narrow plates similar to shagbark hickory. Lower branches characteristically retain the leaf petioles for a year or more — looking almost like a chimney brush on some branches. The fruit is the largest of all the hickories, with a thick husk and a thick-shelled, elongate and flattened nut that yields a very tasty kernel. Shellbark hickory is associated with floodplains, low woods and poorly drained soils. The ability of some populations to reproduce may be threatened by drainage. Its range is from extreme southern Ontario and adjacent Michigan and New York through the Mississippi valley south from Arkansas to northern Georgia. (*See* photo p. 238.)

Carya tomentosa, MOCKERNUT HICKORY, has leaflets with tufts of hairs on the underside and a nut with a very thick shell. It occurs in dry upland habitats from south of the Great Lakes to eastern Texas and northern Florida.

Seed Production and Propagation

Hickories produce flowers in the spring and they are wind pollinated. Although both sexes occur on the same tree, very isolated trees tend to produce empty nuts, indicating they require cross-pollination with neighboring trees to produce a viable crop. Like many species, hickories will shed aborted, premature fruits. Hickories that are along a fence line adjacent to hickory woodlands will produce good crops that are less likely to be stripped by squirrels.

The hard-shelled nut is enclosed in a tough husk that has four seams. Crops of nuts are produced about every two years. The fruits are filled only in the last few weeks before maturity. Occasionally entire crops are devoured by weevils — the fruit will still mature and the nut will be of normal size but the kernel will be all or partly eaten. Always crack open a few nuts to determine their soundness before gathering part of the crop. The kernels of most hickories are highly prized, and baskets of the edible species (in addition to pecans) occasionally turn up at farmer's markets.

Hickory nuts are mature in late September just as the fruit is shifting color from green toward yellowish tan prior to turning brown when fully ripe in October. Squirrels will often cut nuts from the tree at the mature stage and drop them to the ground. Do a cut test to confirm seed quality. Harvest freshly dropped hickory nuts or pluck them from low branches; keep them in an open bucket topped with wire mesh to protect the seeds from rodents. Within a week, you should notice a change in color of the husks and they should split open. If the husks peel off easily, this indicates a good-quality seed. If some husks are difficult to remove, there is a good chance those nuts are not filled. Soaking them for several hours helps separate the husk from the nuts. Hickory seeds are easily damaged from drying out, although nuts can remain in partially open husks up in the tree for a week or two before being dislodged by wind, squirrels and, occasionally, blue jays and crows. Hickory nuts can either be planted in the fall or cold-stratified in a refrigerator for a minimum of 120 days and planted in the spring.

Soak the seeds for a day before planting. Plant about ¾ to 1½ inch apart in a fully enclosed outdoor seedbed, or plant (as the squirrels do) in their permanent location. Even the largest seeds should be buried to two times their thickness, then thoroughly watered and mulched with several inches of loose leaves. They should be protected from seed predators with wire mesh, in seedbeds or in field plantings.

Seedlings grow 4 to 20 inches tall in their first two years. At the end of their second spring, take care digging them out, because their roots are amazingly long. I once lifted a three-year-old shellbark hickory seedling and managed to pull up a straight root that was 43 inches long. We used a posthole auger to dig a deep enough hole for it. The sapling grew poorly, though, and I suspect it would have performed better had I stimulated new root growth by severing the root at about 16 inches.

Hickories are excellent backyard and natural area trees. Newly planted hickory seedlings will benefit from some shade. In my experience, shagbark and bitternut are the most difficult to transplant, whereas shellbark and pignut are the easiest. Saplings that are not growing a vigorous terminal shoot in their third year should be cut off just above ground level in March to force a vigorous shoot from dormant basal buds.

Meadow voles occasionally take to the bark of young hickories. In areas where squirrel numbers are very high, the expanding terminal buds of shagbark hickory are eaten, forcing weaker lateral buds to produce the season's growth. Deer are also fond of the terminal growth. This abuse can potentially kill a young tree after a few years of severe defoliation.

Castanea dentata, AMERICAN CHESTNUT

Castanea mollissima *(Chinese chestnut)*

> **Exotic Alert**
>
> Chinese Chestnut, *Castanea mollissima*, and its hybrids with the native species have been planted on many farms; they may occasionally be found in the wild as escapees from cultivation. Chinese chestnut is a lower, spreading tree with pubescent twigs. Its small (3 to 6 inches long) leaves are glossy on the upper surface and pubescent (usually white) below, at least on the veins. Many of the commercial varieties are hybrids that do not always produce intermediate leaf characteristics, making hybrid identification difficult. Seedlings of the Chinese chestnut, because of their broader soil tolerance, can be used for archival orchards as the root stock for chip budding healthy American chestnut survivors.

American chestnut is a tall interior forest tree with long (5 to 9 inch) glabrous leaves. (*See* photo p. 241.) It thrives in humus-rich, acidic soils on sand plains and in moist but well-drained upland woods. Chestnut was once unquestionably one of the most important foods for animals that needed to accumulate winter fat. Almost every species of mammal — native peoples and settlers included — and many birds such as the wild turkey, wood duck, red-headed woodpecker and perhaps even the passenger pigeon came to the "table" set by this remarkable species. Natural germination takes place in the first spring, from individual seeds planted by squirrels. Chestnut was a signature species of the eastern deciduous forests from southern Maine through southern Ontario and southern Michigan and south from Louisiana to Florida.

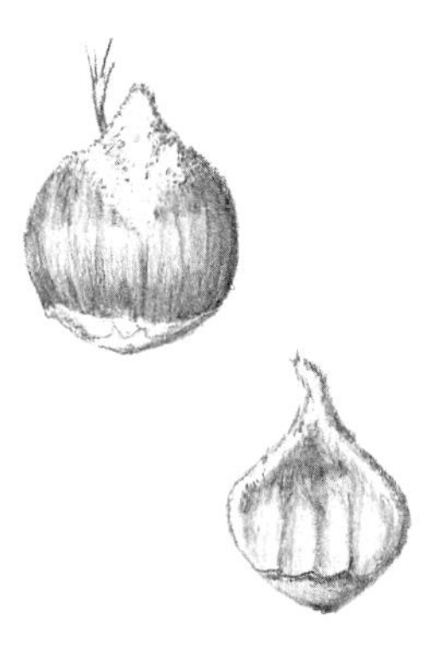

Castanea dentata *(American chestnut), aborted seed*

Creamy white flowers, visible from the ground as a white haze, are produced from late June to early July and are insect pollinated. Although the long flower chains have one to three female flowers at the base and male flowers along the rest of the stem, the female flowers require pollen from a neighboring tree to produce plump, viable seeds. Mature chestnut trees produce fruits every year but produce only empty seeds if they are isolated. The fruit is a spiny, sea-urchin-like structure protecting two to three pointed, partly flat-sided brown seeds (as differentiated from the seeds of horse chestnut, which are rounded).

Castanea dentata *(American chestnut), bur*

The green fruits, or burs, mature and split open to release the seeds from late September into early October, after the bur has turned yellowish green and finally brown. (*See* photo p. 241.) Burs that have dispersed their nuts and those with aborted seeds drop to the ground in October. Handpicked whole mature fruit, or those gathered as squirrels are cutting them from branches, are hard to open but ensure that you will at least get some seeds. A daily morning visit may be required to pick up a few seeds as they naturally dislodge from canopy trees. The seeds are easily damaged by a very short period of drying out, so keep them refrigerated until they can be planted in a rodent-proof frame, preferably in October.

Seeds should be planted about 1 inch apart in small blocks so that digging out the seedling clumps is easily managed. Seedlings grow to 8 to 12 inches tall in the first year and can be planted in a permanent location the next spring. Chestnut has a deep, fibrous root system and is relatively easy to transplant. If the seedling doesn't grow well in the second year, cut it off an inch or so above ground level in April to force a new shoot from a dormant basal bud. This seemingly heartless treatment mimics an important evolutionary adaptation and will likely be accomplished by rabbits during the winter anyway. Deer will browse on the succulent foliage, often killing basal sprouting stumps. Thus protection is important where browsing is a concern. If chlorotic leaves are evident, your soil is too alkaline to support this species.

The Chestnut Blight

The chestnut blight was one of North America's first observed plant disease introductions. The fungus disease arrived from Asia around 1904 with the thousands of Chinese chestnut seedlings that were imported for commercial nut orchards. (Orchard growers prized its slightly larger nut on shorter trees.) All accounts of the blight suggest that, by 1950, it had virtually wiped out the American chestnut, leaving only stump sprouts that grew to pole size before becoming blighted again.

Since about 1918, citizens, foresters and scientists had hoped to find a way to preserve the wild American chestnut. Reluctantly, scientists shifted emphasis away from the elusive search to find blight resistance in our native trees and began promoting hybridizing in an effort to produce an American chestnut that carried the Chinese chestnut's genetic resistance to the blight. However, the hybrids — no matter how many generations of "back-crossing" to native stock they do — are no longer American chestnuts. Even with 90 percent native genes, they are still hybrids. The breeders, then and now, are seeking hybrids with resistance to the blight as well as the upright form important for a canopy tree. Even with the best of intentions, there remains a concern that hybrid tree improvement may not be appropriate as a recovery strategy.

The blight played a major role in chestnut decline but the final blow may yet be the result of human error. Prevailing opinion was that the chestnuts were all going to die anyway, and so the tree wasn't cloned and farmers reluctantly, or with zeal, cut this valuable wood — diseased and surviving trees alike — for timber and lumber. The assumption, which persists in some sectors to this day, was that just because a disease exists, all hosts will get it and die from it. This is not how nature works. Two billion years of evolution did not create a forest full of wimps. Genetic diversity ensures the survival of the species — not the survival of all individuals.

Researchers have found natural blight survivors with healed cankers in a few locations. They are not blight-resistant but blight-tolerant trees. Some trees seem to have escaped infection or perhaps show no outward signs of it. Both types of survivors, along with innumerable stump sprouts, are all that remain in a highly fragmented forest landscape. The few trees that express tolerance or have escaped infection are mostly too far apart to exchange pollen now. I have stood beside a chestnut tree 1½ feet in diameter that was in perfect health but produced only empty seeds because there wasn't another chestnut nearby to pollinate it.

What can and is being done? First, we must resist the temptation to look for absolute resistance and instead contemplate the combined genetic resources of surviving American chestnuts. Disease tolerance is complex. It involves a relationship between the rare tree, some individuals of which may have a slightly stronger immune-system response to disease invasion, and the evolution of viruses that weaken the blight fungus (hypovirulence) and reduce the ability of the fungus to kill its host. Some of these trees are characterized by healing and healed blight cankers. In Europe, hypovirulence is now spreading naturally, protecting European chestnut trees from severe blight infections. This has not yet been observed in North America. Researchers here are investigating the possibilities of introducing the hypovirulent strain that disabled the chestnut blight fungus so successfully in Europe.

We must not throw out the original species because of what we do not know. Due to forest fragmentation and the great distance between the few relic chestnut trees, a pragmatic approach is required. We must clone the last of the surviving American chestnuts before they die of age-related stress. Fortunately, this has been initiated in many regions. Archival clone orchards will serve doubly, to preserve genetic diversity in the event that hypovirulence introduction is successful and as breeding orchards — a "dating service" for lonely chestnuts. The seeds produced from archival orchards will inevitably perpetuate a small percentage of blight-tolerant seedlings, meaning every seed from the orchard must be planted in an area free of the introduced and hybrid chestnuts but not necessarily free of the blight. The new trees must continuously be challenged by the disease anyway. This long-term work must be done with the knowledge that it could take a few generations of chestnut to select for the natural tolerance to blight by constantly eliminating trees with virulent cankers and keeping those with healing and healed cankers.

If you are going to plant chestnut, either American or hybrid, be sure to contact your local chestnut council and let them know the species/variety, seed source and location of your planting. This will ensure that your trees don't inadvertently have a negative impact on archival or breeding orchards.

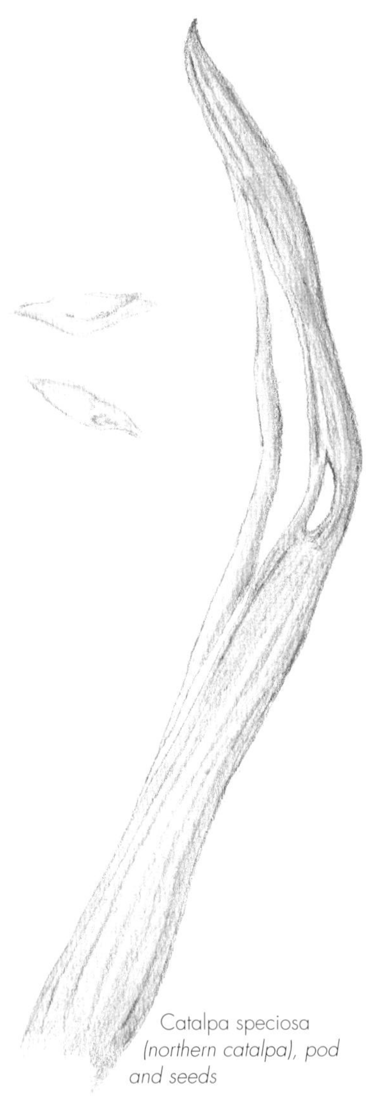

Catalpa speciosa (northern catalpa), pod and seeds

The native catalpa's long pod contains hundreds of flat seeds.

Catalpa speciosa, NORTHERN CATALPA

Catalpa speciosa is a moist-forest and embankment species that is natural only south of the Great Lakes region. Its natural range extends from central Illinois to southern Indiana, south from northern Arkansas to western Tennessee. It has naturalized sporadically in a number of areas as far north as southern Ontario. Northern catalpa is an insect-pollinated tree, once favored by beekeepers, with perfect flowers appearing in late May and early June. The long, cylindrical pod contains hundreds of flat seeds that require no pretreatment to germinate. It is a fine yard tree, but since it is not an integral part of the flora of the Great Lakes watershed, it is not appropriate for naturalization projects there. (*See* photo p. 240.) Less common in our range is the SOUTHERN CATALPA (*Catalpa bignonioides*), native from south Georgia to Mississippi but planted well to the north.

Ceanothus, the RED ROOTS

There are about 50 species of *Ceanothus*, mostly in western North America. It is a sun-loving genus of mostly low shrubs growing in dry, rocky and sandy locations. In the Great Lakes region, they are associated with sparsely vegetated landscapes and savannah ecosystems, occasionally in the partial shade of trees in areas where fire suppression allows infill.

Ceanothus americanus, NEW JERSEY TEA, is an inconspicuous, low, sprawling shrub that spreads partly by underground stems to form small colonies. It is usually abundant but widely scattered on dry, sandy sites in open woodlands, along the edges of woods and clearings, or in prairie fringes. New Jersey tea is a fire-responsive species, meaning that its seed have adapted to, and may to some extent require, the surface soil steaming associated with low-intensity ground fires. Its range extends from Maine through southern Quebec and southern Ontario to Minnesota and south from Florida to Texas.

Note the intricate shape of the structure that ejected the seed of C. americanus.

Ceanothus herbaceus, PRAIRIE RED ROOT, is a dense shrub that forms hemispherical mounds from a tightly multistemmed base. I have only observed it growing in the fissures of open and exposed limestone balds on Manitoulin Island in Lake Huron. It is reported to grow on sand and rocky soils from Maine and southern Quebec to southeastern Manitoba and south from Georgia to Texas.

Reproduction and Propagation

The pure white perfect flowers of the red roots are produced in July and appear as long-stalked clusters in the upper leaf axils. These flowers are visited by many species of insect, including most butterflies in flight at that time of year. The fruit is a three-chambered brown capsule containing three small seeds. Each seed is held by a fascinating structure that, as the capsule dries, suddenly snaps open to eject the seed a distance of about 10 feet. Seeds are thus scattered very locally from September to early October. The blackened base of the fruit remains through the winter. The northward migration of this species remains a mystery to me. It is likely that sparrows assist in dislodging any seeds that haven't been ejected from the fruit, but they head south in the fall, not north.

Whole fruits can be harvested when the capsules mature in late August. Dry the fruit in a cloth bag or under a sieve. Final seed extraction may require rubbing the capsules to dislodge any seeds not yet sprung free. The small, yellowish, hard seeds can then be stored cold and dry until early February. Seed pretreatment requires the hot water soak followed (uncharacteristically) by a 60-day cold, moist stratification. Seeds are planted at ¼-inch spacing in a warm location. I have consistently achieved 100 percent germination with this method.

Seedlings should be grown with partial shading and will attain 1 to 3 inches in the first year. I recommend plunging the entire pot full of closely spaced seedlings into the soil to grow for two seasons. Some seedlings will grow rapidly in their second year, outcompeting the weaker ones. Early in the third spring they should be dug up carefully to avoid breaking the thick, almost fleshy roots with their numerous nitrogen-fixing bacterial nodules. Some seedlings may have their roots so intertwined that they should be left together. *Ceanothus* transplants easily. Rabbits will eat the twigs in winter but older plants readily recover and flower because they bloom on current season's growth. Both species are well suited to xeriscape, rockery and prairie gardens.

Celastrus scandens *(climbing bittersweet)*

Note the large terminal cluster of fruit on the indigenous species (above). In contrast, Asian bittersweet fruit clusters are smaller and found at the leaf nodes (below).

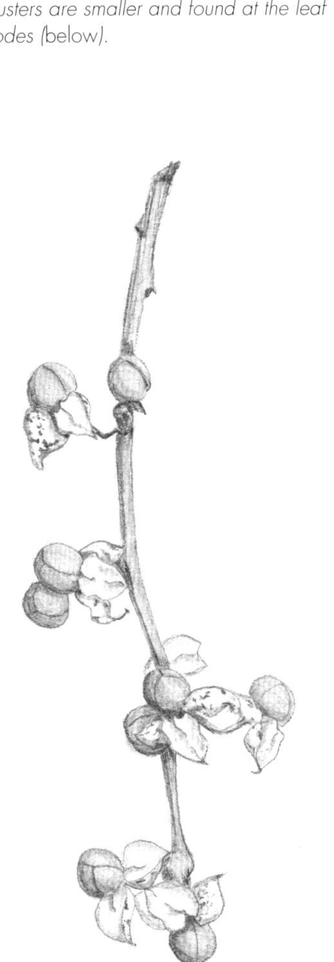

Celastrus orbiculatus *(Oriental bittersweet)*

Celastrus scandens, CLIMBING BITTERSWEET

Exotic Alert

Climbing bittersweet, *Celastrus scandens,* is often confused with its Asian counterpart. The nursery industry sells, almost exclusively, Oriental bittersweet, *Celastrus orbiculatus.* This is now the most likely species to be found naturalized in the near-urban landscape, and in some locations it is an increasing concern. *C. orbiculatus* is easily mistaken for the native bittersweet, and many seed collectors have not made the distinction. Naturalization projects will undoubtedly contain Oriental bittersweet from such nurseries, and accurate identification and exotic removal needs to be seriously considered as a part of the follow-up management of any planting project. Homeowners might inquire (and some will insist) about returning the exotic species to their nursery. The exotic will hybridize with our native bittersweet.

What a thrill these bright orange and red fruits can deliver even on the darkest overcast days of November. There are about 30 species of bittersweet, found mostly in the warm temperate and tropical regions, with only a few in temperate regions.

Climbing bittersweet is a high-climbing vine of dry to moist thickets, fence lines and open woodlands. It is found on sand, glacial till, clay, limestone outcrops and igneous rock. The natural distribution of *C. scandens* occurs from Quebec through southern and western Ontario to southern Saskatchewan and south, Georgia to Oklahoma.

Bittersweet sometimes forms extensive colonies with shoots rising from a wide-spreading root system and here and there scrambling into the shrub and tree community. Fruits are eaten in late autumn and throughout the winter by cedar waxwings, cardinals and mockingbirds. Fruits persist well through the winter and can be harvested at any time up to mid-March, when the spring migrant birds (primarily robins and bluebirds) finish off what is left of the energy-rich berries. Dispersion over a wide area is thus achieved, as the relatively soft seed is passed through digestion without being damaged. It is the oily flesh of the orange berry that is digested by the birds. Natural germination is delayed until after the seeds have been at the soil surface, ungerminated, for a full summer and winter.

Flowers are greenish yellow, unisexual, and in terminal clusters on separate male or female plants and are pollinated by flies in early June. The species will flower in the fourth year if growing vigorously. Each globose fruit is a capsule with an outer, dry husk that splits into three (rarely four) sections. As the fruit dries, the husk opens to expose three to four sections covered with a reddish orange, fleshy aril. Each aril contains one and occasionally two light brown seeds with a relatively soft seed-coat.

The fruits can be picked at maturity, when the capsules are still closed but have changed color to yellowish orange. Avoid fermenting the fruit, by spreading it to dry until the husks split open and the arils have dried enough (about two weeks) to have the consistency of a raisin. Long-term storage is best as dried arils.

When ready for pre-germination treatment or sowing, the dried arils must be soaked for 24 hours and then the seeds can be removed from the flesh (as well as from the husk if not already done) using the grit bag. **Caution**: the seeds have a relatively soft seed coat and can be torn open if you grind the fruits too hard; go gently with this species. Bittersweet fruit is quite oily and many seeds will float with the pulp. Add a little natural soap to the pulp to facilitate separation of the pulp and seeds. Some seed lots just won't separate, and in this case, the whole works can be spread to dry. After a few hours, when it is mostly dry, the pulp/seed mixture should be rubbed to prevent the skins from sticking to the seeds. Before the seeds dry out, they should be rubbed again so that most of the dry pulp can be fanned off.

Pre-germination treatment requires a 120-to-150-day cold, moist stratification. Seeds can be almost touching when planted and must be protected from mice. Germination is rapid and seedlings are very susceptible to damping-off fungus. Seedlings can be transplanted at the four-leaf stage to produce sturdy one-year plants about 8 inches tall.

Bittersweet has a dense, fibrous orange root system that is very easy to transplant. The bark and twigs are highly favored by rabbits during the winter months and established plants are occasionally reduced considerably. In nature, they resprout from ground level or from the roots, but young seedlings do not recover if chewed down to ground level when there is no snow cover. Climbing bittersweet is declining in the wild and that may be due to the loss of herbivore predators.

It is important to plan for the establishment of several seed-grown plants in the area, in order to establish both male and female plants. *Celastrus* will spiral around tree trunks, producing a beautiful visual effect on the tree, but it rarely kills the support. This is a really lovely species that has a place in almost any natural landscape or on a fence or old fruit tree. The horticultural industry sells cloned males such as 'Hercules' and 'Indian Brave' along with their partners 'Diana' and 'Indian Princess', but restoration planting should be done with wild seedlings.

Celtis, the HACKBERRIES

About 70 species of *Celtis* are found in the northern hemisphere, with two in our area. Hackberry is in the elm family, despite the completely different fruit. It has been suggested as a replacement for American elm after Dutch elm disease has emptied our city streets. We must be wary of any one-species solution, and never again plant a monoculture of trees.

The local dwarf and tree-form species produce virtually identical fruits but on stalks of different lengths. The fruit matures in October and persists through winter. Seeds are dispersed by birds; I have observed cedar waxwings eating the fruit and, occasionally, only the fleshy part, leaving the seed still attached to the dwarf hackberry. Local dispersal of fallen fruits is accomplished by rodents. The thin fruit covering of the large seed is nutritious, and dried fruits were carried by native peoples on their journeys — possibly accounting for the interrupted distribution of the two species.

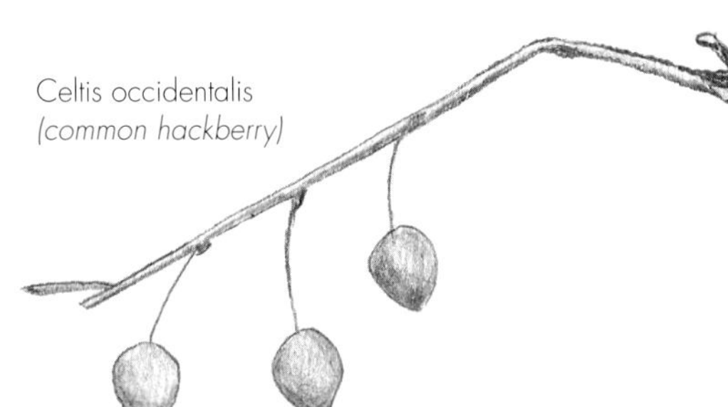

Celtis occidentalis, COMMON HACKBERRY, can become a large tree, thriving on moist but well-drained rich soils of floodplains, stream and river embankments, inter-dune beach ecosystems and, occasionally, drier sites on sand or limestone. It is notable as forming nearly pure stands in areas of Point Pelee National Park and its counterpart, Fish Point, on Pelee Island, in Ontario. Often locally abundant in an otherwise widely scattered distribution, it ranges from the southern Great Lakes south from eastern Oklahoma to northern Georgia.

Celtis tenuifolia, DWARF HACKBERRY, is a dense, irregular shrub or small tree to 16 feet. It occurs in much drier conditions than the common hackberry, such as sand dunes, rocky hills and shallow soils over limestone. One might think these severe conditions account for the stunted growth, but seedlings of dwarf hackberry remain stunted in cultivation alongside common hackberry. While officially threatened in Canada, it is locally abundant in very widely scattered populations in Ontario, at the Pinery Provincial Park on the shores of Lake Huron, in the eastern Erie islands and Point Pelee, and near Belleville, on Lake Ontario, with additional scattered

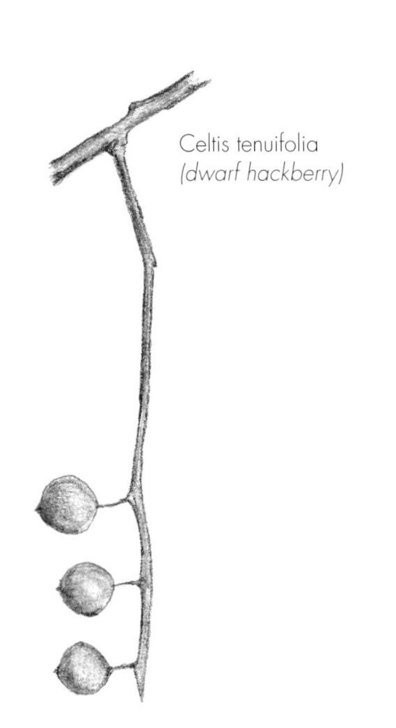

A very thin pulp layer is distinctive for the genus Celtis.

populations in southern Michigan (where the species is listed as of special concern). Its range extends south to eastern Texas to northern Georgia. The distinctly larger size of specimens in the Belleville area and fruit on intermediate length stalks suggests that natural hybrids are present.

Celtis laevigata, SUGARBERRY, is a more southern species occurring in bottomlands of the lower Mississippi valley to the east coast; from Iowa and southern Indiana to Virginia and south, Florida to Texas, with outlying populations in northern Mexico.

Reproduction and Propagation

Hackberries produce small, yellowish green, unisexual flowers on the new growth in late May. The male flowers are in clusters in the lower leaf axils, and solitary long-stalked female flowers are produced in the upper leaf axils. The flowers are wind pollinated. The fruits of common hackberry ripen to a dark purple-black while those of dwarf hackberry mature to a salmon-orange color.

The fruit can be picked from low branches at any time after it has fully ripened in late October. The seed is a very hard "stone" with distinct vein-like markings. The seeds are readily removed from the fruits by scrubbing them in the grit bag. Cleaned seeds or whole fruits can be dried and stored cold for a number of years.

Hackberries are best sown in an outdoor seed frame to accommodate the tough, long, often deep root. The seeds can be sown at ⅓-inch spacing in the fall or in the spring after stratification; I have always had excellent germination with spring sowing. The seeds require a 90-day cold, moist stratification for common hackberry and a longer period of 120 days for dwarf hackberry. Seeds must be thoroughly protected from mice and will germinate rapidly in warm conditions.

Hackberry transplants easily but digging seedlings requires a good sharp spade and some effort to get at least 12 inches of undamaged root on a two- to three-year seedling. Both the tree and the dwarf species are excellent urban trees, the latter well suited to prairie gardens or any droughty hot summer conditions. Winter protection from rabbit browsing is definitely required for young plants and deer will browse on new growth.

Cephalanthus occidentalis, BUTTONBUSH

Buttonbush is most often a shrub to 7 feet growing quite crowded in the rich, moist soils of buttonbush swamps or along creeks. In Lambton County, Ontario, I have found it on

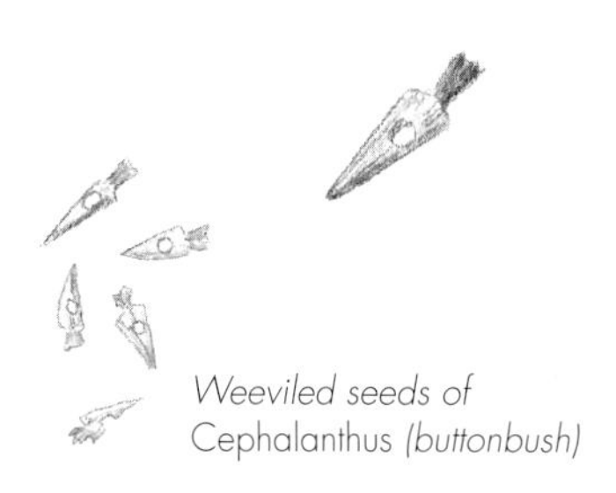

Weeviled seeds of Cephalanthus (buttonbush)

a dry embankment, with one specimen growing tree-like to about 13 feet. This shrub has a significant distribution in mixed and deciduous swamps from Nova Scotia through southern Ontario to Minnesota and south to California and Mexico across to Florida and the West Indies.

Migrating sparrows and winter finches feed on the seeds in late fall and winter, causing some local dispersion if seeds fall into water or drifting snow. Natural germination occurs the first spring from locally scattered seed. Since the seeds float and some are still dispersed in March, it is possible that they may be carried by the northward migration of marsh ducks.

Pure white flowers, produced in July, are densely crowded into round heads in open-branched clusters terminating the current season's shoots. The perfect flowers are an important midsummer source of pollen and nectar for insects, including butterflies. The fruits are persistent, ½- to ¾-inch round balls, with each ball containing up to 400 seeds. (*See* photo p. 240.) The seeds are an elongated nutlet that is easily shucked from the seed ball.

Fruits mature in early October, shifting to a reddish green before turning brown. Seed weevils may devour the contents of a very high percentage of the seeds in each fruit ball, leaving one seed to enter the adjacent seed throughout the fruit. Mature to ripe balls should be inspected for weevil entry before deciding how many to gather. Once the ball is shattered, the weevils are exposed and can be removed by hand before placing the seeds in dry, cold storage until ready for spring sowing. Seeds of buttonbush require no pretreatment. Sow the seeds at 1/16- to ¼-inch spacing in a pot and cover them lightly. They must be kept quite damp until they germinate.

Seedlings can be transplanted at the four-leaf stage and set about 1 inch apart to produce four to six inch seedlings in the first year. Growth is about 4 to 12 inches per year; with their very fibrous root system they are easy to transplant. Buttonbush can be planted in the second fall, when it is easier to access the moist ground that it thrives in. I have yet to notice any feeding on the bark and twigs of this species. They will tolerate a moist, well-drained location in cultivation. It is a perfect backyard, pond-edge species, and may be pruned to form distinctive small trees.

Canadian Redbud

In spite of the specific epithet *canadensis*, natural plants are found only on the American side of the Great Lakes. Plants in Ontario are naturalizing from planted stock, including escapees from the old tree nursery in Point Pelee National Park. There is one record of a wave-washed, half-dead tree observed on the south shore of Pelee Island back in 1892. The whole fruits could easily have blown across a frozen lake. Still, this can hardly be called a species that is native to Canada. The specific name *canadensis* (published by Linnaeus in 1753) likely refers to the former extent of Canada, not the present extent of redbud. After the American Revolution, a small part of Canada changed hands when England had to relinquish all the land south of present-day Canada, north of Florida and east of the Mississippi River, as agreed in the Treaty of Paris (1783).

Cercis canadensis, REDBUD

No other plant can match the stunning display of purple-pink flowers that cover the branches of this tree before the leaves have expanded — a welcome spring sight indeed. Worldwide there are several species, another two or three species occurring from Texas to Mexico and a few from southern Europe to Japan, so most temperate cultures know the plant. Redbud is a sprawling large shrub or small tree growing in well-drained, moist to dry soils at the forest edge, in clearings and on floodplain terraces and embankments, in full to partial sun. The seed pods hang on through the winter and are dispersed as whole fruits, blowing considerable distances over crusted snow or barrens. The seed has a very hard, impermeable seed coat. Natural germination may be possible with seeds that have been nicked by mice, but this is likely a fire-response species. Seeds can remain viable for a few years at or near the soil surface; the seed coat proteins decompose or are denatured by the steaming resulting from ground fires or intense solar radiation. Redbud is found from Connecticut to southern Michigan and Nebraska and south from Florida to New Mexico.

Flowers are perfect and open in mid to late May from exposed and overwintered lateral flower buds on recent growth, as well as from spurs on very old wood. Bees are the major pollinating insect. Isolated trees and even some individuals within a stand produce few fruits; cross-pollination seems to increase fruit set considerably. The fruits require a long growing season and seeds are destroyed by a heavy September frost if they have not yet matured.

Pods contain three to six seeds that can be removed by peeling them open or thrashing/scrunching them in a cloth bag. The small, round and flattened seeds can be separated from the chaff by winnowing. Those with allergies to legume foods should avoid inhaling the dust from the shattered pods. Seeds can be stored cold and dry for many years.

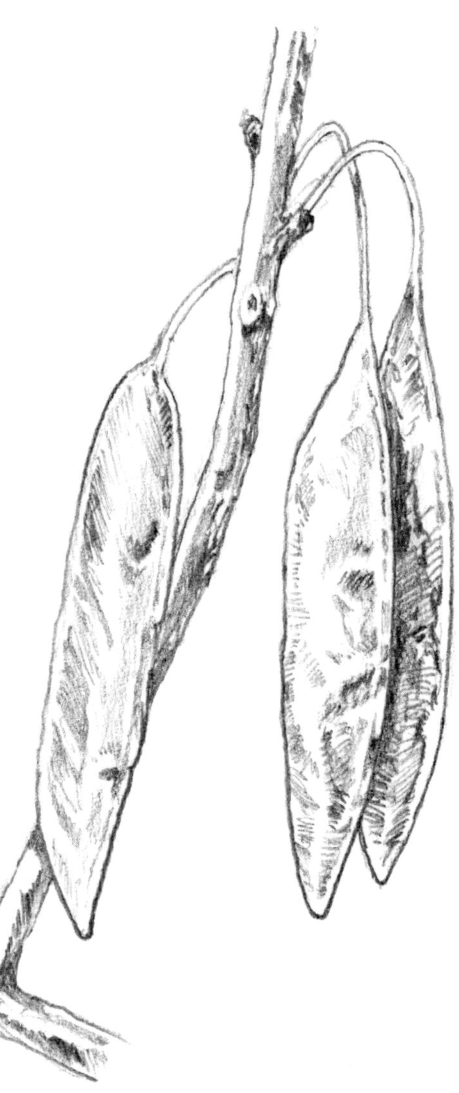

Check the pods of Cercis canadensis, *redbud, to make sure they are plump and filled with seeds.*

Germination is quite high when the seeds are treated with hot water. After several hours of immersion, the water should turn quite dark; it should be changed a few times. Seeds that swell to almost three times their dry size germinate in about ten days. Seeds that did not swell from the hot water tend to germinate in the second and even the third spring after sowing of treated seeds. Although I have not tried this method, a soak in 7 percent acetic acid (pickling vinegar) for 10 minutes is reported to produce favorable germination two weeks after planting the seeds.

Seeds are planted at ⅓- to ¾-inch spacing to produce 12-inch tall seedlings in the first year. The shoot tips of this rapid-growing species tend to freeze back, and those that die back the most should be discarded. Should you desire to grow the species well north of its natural distribution, the hardiest of these can be planted into the nursery for another year of testing. Redbud produces extensive tough roots that require a good sharp spade to cut. The otherwise fibrous root system is transplanted easily in the spring. Rabbits will trim the shoots back during the winter. With its low, sprawling branches, redbud is an excellent species for dry embankments.

This species and a southern cousin, *Cladrastis lutea*, the yellowwood, are prone to splitting at branch junctions due to a tendency to produce very narrow branch angles, which result in bark inclusions. Select for wide branch angles and prune out one of two branches exhibiting the beginnings of a bark inclusion. Note that yellowwood will literally pour sap out of an early spring cut; you can avoid this by pruning in early June.

Cladrastis lutea *(yellowwood)* pods

Clematis, CLEMATIS

The large-flowered horticultural hybrids shamefully upstage the delicate and subtle beauty of the nearly 300 wild species of clematis found throughout the world. About 15 species are found in eastern North America, with two in the Great Lakes region. Clematis are intolerant of deep shade but will grow in partly shadowed areas, especially if protected from midday heat.

Clematis virginiana, VIRGIN'S BOWER, is a dense, scrambling plant, clasping with twining leaf petioles; it will cover several yards of ground and grow 10 to 16 feet into the shrub layer. It is commonly found on moist embankments and fence lines as well as in clearings. *C. virginiana* spreads extensively by shoots arising from a relatively deep, spreading root system. This species is distinctive in the genus in having unisexual flowers, with entire plants being either male or female. In July/August, masses of small flowers are without petals but owe their showiness to the white sepals. The species is found from Nova Scotia to Manitoba and south from Florida to Texas.

Clematis occidentalis, PURPLE CLEMATIS, is a loose, open, scrambling plant, rarely a large tangle. It seems to be quite particular to rock outcrops, both limestone and granitic, ascending rarely 7 feet by clasping with leaf petioles. I have not noticed this species growing from root shoots, but know it only from two locations in Ontario.

In late May, with the unfolding leaves, the perfect flowers are without petals, but the four large, purplish sepals make quite a display. Its range is from eastern Quebec to Manitoba, south through New England to West Virginia and west to Iowa.

Clematis virginiana
(virgin's bower)

As can be seen from the illustrations here and on p. 111, the density of clematis fruit heads varies considerably, especially in more southern species.

Reproduction and Propagation

Clematis flowers are a rich source of nectar and are bee pollinated; the fruits mature in October. Natural seed dispersal takes place during the winter as the long-awned seeds break away from the fruit cluster or are scattered by finches as they are feeding. The embryo is not fully developed when the fruit is ripe and natural germination is delayed until the second spring.

The long, feather-like remnant style is attached to each seed in the globose fruiting head, a very striking aspect of the beauty of many clematis. The styles can be rubbed free to facilitate seed handling. Seeds are stored cold and dry until ready for pre-germination treatment, which requires a 24-hour soak and a 30-day warm, moist stratification followed by 60 days cold and moist.

Clematis columbiana *(rock clematis)*

Seeds can be planted quite close together and germination may take up to three or four weeks. It is important to be patient with clematis. Seedlings grow best if transplanted to wider spacing when they are at the three- to four-leaf stage. The young plants should be set into a garden bed for two growing seasons, where they can be trimmed back if needed. Two-year plants are ready to be set out into the landscape and transplant easily in the spring or early autumn. Rabbits will nip off the stems of *C. occidentalis*; this may partly explain its scarcity in the wild.

Comptonia peregrina, SWEET FERN

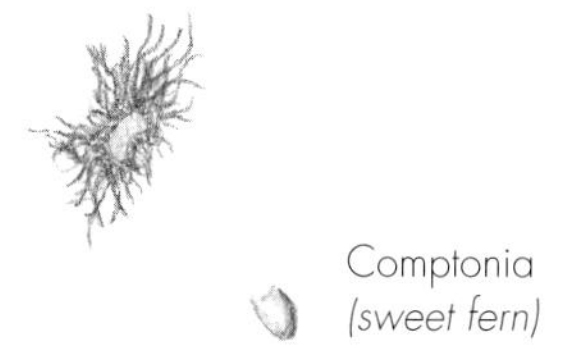

Comptonia *(sweet fern)*

Sweet fern is a monotypic genus — there are no other woody plants even similar to it.

Comptonia is a lovely aromatic shrub with gorgeous textured foliage. It is a dense, colony-forming shrub, spreading extensively by underground stems (rhizomes). It produces what amounts to a ground cover, usually 1½ feet and, rarely, up to 3 feet high, thriving in dry, sandy and rocky soils of barren landscapes or partly shaded, open woods. It is found from Nova Scotia to western Ontario and Minnesota and south from Illinois to Georgia.

This is a dioecious species, with the male and female flowers produced in catkins on separate colonies in May. The fruit is a bur-like cluster that matures in September and holds two to five hard nutlets. Never having had more than a few seeds to try to germinate, I have always grown this species by divisions from small pieces of the rhizome. It is related to *Myrica*, and I can only suggest that a 90-day cold stratification is likely all that is required. Fall-planted seeds in a cold frame should produce a spring germination.

Sweet fern is difficult to grow even by divisions of the rhizome. For best results, after division, cut the shoots to ground level and then plant the rhizome piece. Select a partly shaded location with acidic soil; the species does not tolerate alkaline conditions. Plants in cultivation at the University of Guelph Arboretum are growing very well in the partial shade of a red pine, on a slightly raised bed of peat. The plant remains untouched by herbivores.

Cornus, the DOGWOODS

Of the 50 species of dogwoods in the northern hemisphere, the 8 or 9 local species are of significant ecological importance and are distinct to several habitats. No landscape should be without at least a few species. Seed dispersal is by birds such as cardinals, cedar waxwings and thrushes (including the American robin). Although most species bloom in early summer, the fruit ripens in succession over a period from mid-July, with *C. alternifolia*, to November, for *C. drummondii*. The staged fruit maturation of the various dogwood species over four months undoubtedly coevolved with preferential feeding by birds. Natural germination periods for dogwoods are little understood, with the exception that *C. florida* germinates in the first spring after seed dispersal.

Identification of the various dogwoods is actually relatively easy for most species. *Cornus amomum*, though, is quite variable across its range and intermediates result from extensive hybridizing. It's also of little help that the botanical names of some (*C. amomum* for one) have been changed from time to time. Attention is best focused on the habitat, fruit color, fruit maturation period and growth habit.

The growth habit takes three basic forms: some species are colony forming by vertical stems rising from the roots and some by branches falling horizontal and rooting where they touch the ground; some species are single, independent shrubs or small trees. Divisions can be separated from the clone-forming species without affecting the parent plant. The species are described below in the order of fruit maturation.

Cornus alternifolia, ALTERNATE-LEAVED DOGWOOD or PAGODA DOGWOOD. Single- to multiple-trunk trees are found commonly in moist woodlands and forest edges, but occasionally on exposed fence lines in relatively dry and sunny conditions. The species should be grown from such stress-adapted seed sources if you intend to plant them in the open. It has an extensive range from Newfoundland across the north shore of Lake Superior to southern Manitoba and south, from Missouri to Georgia. In July and August, the deep blue-black fruits are often stripped by robins and starlings. The stone has a deep pit at the summit end. This tree should be used much more extensively. It will regenerate from seed once locally established. (*See* photo p. 242.)

Cornus obliqua *(silky dogwood)* *fruit and seeds*

Cornus amomum ssp. *obliqua*, SILKY DOGWOOD, is a variable species that some authorities want to divide into a northern and a southern race. It is an often-overlooked shrub growing in sunny conditions on moist to mesic soils along streams and in meadows and open woods, but rarely on dry sites. It is a non-suckering shrub that occasionally branch-layers and can be found from western New Brunswick to North Dakota and south from Oklahoma to Georgia. The fruit ripens between August and September to a striking porcelain-blue on the sunny side with a creamy white underside. (*See* photo p. 244.)

Cornus stolonifera, RED OSIER DOGWOOD, is the classic red-twigged shrub of wet meadows, open woods and shores. Impenetrable tangles are formed as the sprawling branches take root or layer to initiate a new plant. This species ranges across much of North America, extending north of the Great Lakes watershed from Newfoundland to Alaska, and south from California to West Virginia. A form with yellow twigs is commonly found in cultivation. The abundant fruit is pure white and matures in September. This species is sometimes considered a subspecies of the Eurasian *C. alba* (*see* Exotic Alert on p. 114), which would then make it part of a circumboreal species complex giving way to the virtually identical *C. alba* in Europe. (*See* photo p. 245.)

Above, clockwise from left, enlarged about 600 percent, seeds of Cornus stolonifera *(red osier dogwood),* Cornus drummondii *(rough-leaved dogwood) and* Cornus obliqua *(silky dogwood).*

Cornus florida, EASTERN FLOWERING DOGWOOD, is a gorgeous tree that does best in a cool location along a forest edge or understory in humus-rich, moist to mesic or dry soils ranging from sand to clay. It grows from southern Maine throughout the deciduous-forest zone of southern Ontario to Kansas and along the eastern states south to Florida and Mexico. Occasional branch layering (likely due to heavy snow load combined with rabbit browsing) in the woodland understory will produce small colonies of a single tree. This is a striking tree with stunning white floral bracts in May, fabulous fall color and crimson fruits between September and October. A number of horticultural selections with pink or red flowers are propagated by cuttings and grafting. (*See* photo p. 243.)

Cornus florida *(Eastern flowering dogwood) fruit and seeds.*

Cornus foemina ssp. *racemosa*, GRAY DOGWOOD, is so named for the older 5- to 7-feet tall, gray stems that rise prolifically from the spreading root systems to form very dense, mounding thickets in open fields or thin colonies in an open woodland understory. This species occurs on moist soils, but occasionally thrives on dry embankments. It is found from Maine to southeastern Manitoba and south from Maryland to Oklahoma. It produces a pure white fruit on bright red stalks in September to early October. (*See* photo p. 244.)

The beautiful flower of Cornus florida, *Eastern flowering dogwood.*

Cornus rugosa, ROUND-LEAVED DOGWOOD. The purplish brown marks along the twigs and stems are a distinctive feature of this colony-forming shrub. Its stems rise 5 to 10 feet high from a wide-spreading root system. Its habitat is primarily associated with well-drained soils and rocky sites in semi-open woodlands and along the forest edge in cool, partially shaded conditions. Occasionally it occurs in fully exposed locations. It ranges from Nova Scotia to southeastern Manitoba and south from Iowa to Virginia. The creamy white, porcelain-blue-tinged fruits are ripe in late September through October and are held on red stalks. (*See* photo p. 244.)

Cornus drummondii, ROUGH-LEAVED DOGWOOD, occurs along dry forest edges or produces dense thickets up to 13 feet tall; it seems to have a higher tolerance for drier sites than most other dogwoods and that may be associated with its downy foliage. It is primarily a prairie species in the western portion of the region, extending from southwestern Ontario to Nebraska and south to Mississippi and Texas and into Mexico. It has the last of the dogwood fruits to mature, with creamy white berries ripening in October and persisting into November. Gray squirrels will sit for days in these shrubs, cracking out the tiny kernel in each seed. (*See* photo p. 243.)

Cornus canadensis, BUNCHBERRY. While hardly a woody plant, this unique and lovely species forms a groundcover, with its annual herbaceous shoots rising from a wide-spreading woody rhizome. It is an acidic soil–loving species that is very rare in cultivation because it is such an ecological specialist. Bunchberry is found in moist and mossy conifer woods from southern Greenland through Canada to Alaska and eastern Asia, and south to New Jersey across to California.

Bunchberry resembles a herbaceous version of *C. florida*, with its very similar, pure white flower bracts displayed in mid to late May and the clusters (bunches) of orange-red single-seeded fruits in September. (*See* photo p. 243.) I have not germinated the seeds of this species but to do so I would clean them, plant them in a clay pot of acidic soil, and set them into the cold frame until they germinate.

Exotic Alert

Cornus alba, Tatarian dogwood, a Eurasian species, is so mixed in the nursery trade with our native red osier that provenance can only be assured by wild collection, and even this is becoming more problematic as Tatarian dogwood escapes cultivation or is mistakenly used in restoration plantings. The taxonomy of the two species is equally confused. Any plant with variegated leaves or brilliantly colored stems is suspect and worthy only of rejection.

Reproduction and Propagation

Dogwood flowers are perfect and insect pollinated. Flowering time is slightly variable for each species from mid-June to early July, the notable exception being the early display of *C. florida* in late May. The spherical fruits (elongated in flowering dogwood) are held in clusters. Each fruit contains a single, hard-coated seed with two embryo compartments, of which usually only one is filled. Fruits can be picked when they are fully ripe, colored as indicated with each species description.

A Destructive Fungus

Dogwood anthracnose is such a ravaging fungus disease that it was named *Discula destructiva* after it was discovered a few decades ago in the U.S. It has spread throughout much of the range of *C. florida*, including Ontario, and has killed many trees, especially those in deeper shade. The fungus tends to start on lower leaves as tan spots with purple edges, with eventual leaf death. The dry, shriveled leaves persist as the disease spreads down the petiole, infecting the twigs and branches. Ultimately trunk cankers form where branches die; the continued reinfection of sucker sprouts that grow from below the cankers finally leads to death.

True to genetic diversity, some individuals do not die from the anthracnose. Counteracting the cry that the flowering dogwood is doomed, healthy survivors are found. We are likely to see a habitat shift in this species from forest understory to edge and open space. Individuals in the open are the ones showing the greatest ability to survive despite infections, likely due to the greater energy reserves produced by leaves in full sun and the increased air circulation. We particularly should be on the watch for the survivors in shade for promising genes. In order to maintain genetic richness and disease tolerance traits, it may even be appropriate to raise cuttings or grafted stock of several of the best shade survivors to plant in close proximity for seed production.

Contrary to warnings about not planting flowering dogwood (because it might die) and choosing the oriental kousa dogwood as a substitute, I suggest that we do plant flowering dogwoods — because they might live. Select open but sheltered sites, east- or north-facing woodland edges, good soils without the debilitating effects of nitrogen fertilizer, and best available seed sources.

Seeds are readily extracted from the fruit with a firm grinding using the grit-bag method or a low-speed blender. The round-seeded dogwoods are an exception to the rule that good seeds will sink. Sometimes many viable seeds will float because one side of the seed is empty and the hollow cavity has enough air to float the seed, even though the other side is filled. You can check this by cutting a few of the floaters in half to look for evidence of the bright white embryo. Floaters can be separated from the pulp by drying the whole works, rubbing the seeds between gloved hands and blowing the pulverized pulp from the seeds.

All dogwoods can be stored as dry, cleaned seeds. Extraction will achieve some of the seed-coat scratching that I believe significantly increases the percentage that germinate. I have heard of whole fruits being planted in outdoor seedbeds and tried it, without success, for *C. obliqua* and have not studied this method further. With the exception again of *C. florida*, most of the dogwoods benefit from warm, moist stratification for 60 days prior to a 60- to 120-day cold treatment to produce high germination rates. Fall-planted seeds do not germinate reliably in the first spring.

Seeds of all dogwoods can be planted at about ⅓- to ¾-inch spacing. *C. florida* has very low germination rates when planted outdoors (north of its range) in the fall, yet I consistently have 100 percent germination with the controlled long cold stratification of 150 days. This is the only dogwood species that exhibits radicle emergence in cold conditions, and it seems to be an important indicator that the pre-germination treatment is completed. I remove and plant seeds of *C. florida* as the emerging radicle splits the seed coat and place the unsplit seeds back into cold stratification, to be checked a week later.

Germination is quite uniform for all species. *C. florida* is very sensitive to damping-off fungus. Dogwood seedlings can reach 8 to 12 inches in the first year and be planted out the following spring. Due to their very fibrous roots, dogwoods are easily transplanted. As long as you cut the shoot back to a single pair of buds above ground, all of the root-sprouting and layering species can be propagated by severing a rooted stem from a colony.

Since *C. florida* is a highly desired winter food for meadow voles and rabbits and summer browse for deer, I am inclined to plant only specimens that are 3 to 7 feet tall and put them into mesh enclosures with the base well protected. Without snow cover, unprotected seedlings are chewed to the ground, with no buds left for re-sprouting. All other species of dogwoods seem to be left untouched by herbivores once they are established. Farther north, moose and deer are known to browse on *C. stolonifera*.

Corylus, the HAZELS or FILBERTS

Of the ten species of hazels in the northern hemisphere, the two local species are quite distinct but still have many similarities. Both of the indigenous hazels form colonies by underground stems, which readily resprout after they have been heavily browsed or after a forest fire. Hazelnuts have a very high food value and are consumed by small mammals, squirrels in particular, as well as birds, such as blue jays. One should not pass on the thrill of eating either fresh or dried native hazelnut. Dispersal is achieved where a rodent cache of hazelnuts germinates in the first spring. Long-distance dispersal of these relatively heavy seeds is likely by birds such as blue jays, which will carry seeds away to cache for later use.

Corylus americana
(American hazel)

Corylus americana, AMERICAN HAZEL, is a shrub that sometimes forms dense thickets, spreading rather slowly from short underground stems. It thrives on dry, sandy or rocky ground and moist, well-drained embankments at the forest edge and in open areas. The range of American hazel covers a few life zones, extending from Maine and southwestern Quebec through southern and western Ontario to Saskatchewan, south to Oklahoma and across to Georgia. (*See* photo p. 244.)

Below, female catkins (top right) and male catkins (bottom left) of Corylus americana *(American hazel).*

Corylus cornuta, BEAKED HAZEL, forms very extensive open colonies in moist, open forests, clearings and edges. It thrives in well-drained, humus-rich soils throughout several life zones across North America from Newfoundland to southern British Columbia, south to central California and across to Georgia.

Reproduction and Propagation

The hazels flower in early April. As the male catkins enlarge, they become bright yellow and shed pollen into the wind to fertilize the two minute but exquisite red stigmas that protrude from lateral buds. Although both male and female flowers are produced on the same plant, isolated individuals do not produce fruit, indicating a mechanism that prevents self-fertilization. Since beaked hazel is a shade-tolerant species, a single plant colony may cover huge areas over hundreds of years.

Fruits are mature when the color of the floral bract (the leaf-like structure that surrounds the seeds) shifts a little toward yellowish green. Pick the fruits when you notice the first signs of jay or squirrel activity, or the crop will be consumed or stashed away in relatively few days. Dry the fruits indoors for a few days before extracting the hazelnuts.

American hazel fruits are sticky but the nuts do peel out rather easily unless the fruit was picked too soon. Beaked hazel fruits, however, are covered with very small, spine-like hairs that can stick into and irritate your hands (even through light gloves). These spines become more intimidating as the fruit dries and can make the extraction of seeds from the fruit a most difficult task. For small quantities of seed, which is the most that anyone really needs, I hold the fruit with a thick leather mitt and use pruning shears to cut away the bract in a manner similar to the way squirrels might tear it off.

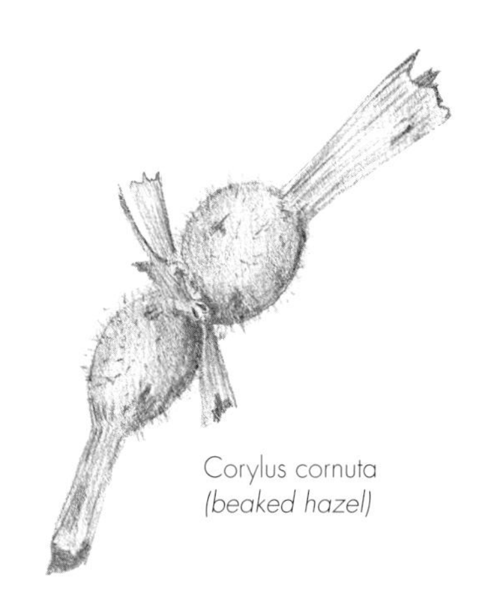
Corylus cornuta
(beaked hazel)

Hazelnuts must not be allowed to dry for more than a week or two. Planting the seed in an outdoor frame should be done soon after dehusking. Alternatively, a cold stratification of about 150 days should precede a spring sowing. Nuts are planted about ⅓ inch apart to impose a little competition on the seedlings. Seedlings will reach 12 inches in the first year and up to 28 inches in the second year, at which time they must be transplanted out of the seedbed. Hazels have a dense, fibrous root system and are easily transplanted bare root before bud break or after leaf fall. Rabbits heavily browse on the twigs and bark during the winter.

Crataegus, the HAWTHORNS

Few groups of plants are so ignored as the biologically important genus *Crataegus*. More than 100 species in the northern hemisphere are at their greatest diversity in eastern North America, where more than 80 species have been suggested. Hawthorns are usually dense, mounded or horizontally branched thorny trees to around 16 feet but occasionally higher, with trunks to 20 inches and rarely to 40 inches across. (*See* photo p. 245.) The fruit is usually dry and often insipid, but hawthorn jelly is not to be shunned. What the early settlers often called "thorns" are met to this day by a reluctance to grow or plant this genus. It provides part of the diet for many species, including grouse, quails, turkeys, deer, rabbits and foxes. The fruit of hawthorn once made a significant contribution to cedar waxwings' diet but now I see the waxwings feeding in the omnipresent European buckthorns (*Rhamnus cathartica* and *R. frangula*).

No other group of woody plants in this watershed is so extremely rich in species and hybrid diversity, which is so great that only a few have seriously tried to make sense of it. Even listing the numerous species in this region would be a challenge well beyond this manual, let alone describing their ecology and propagation individually, even if I knew them myself. I have observed distinct species in a range of habitat regimes from sunny rich alluvial floodplains to open

Hawthorns and Cedar-Apple Rust

Hawthorns are an alternate host with *Juniperus virginiana* (eastern red cedar) for cedar-apple rust, which defoliates its host by mid-September. The disease does not kill the host tree.

The rust makes hard, dark brown fungal growths of 1/3 to 1 inch on the twigs of Juniper. Spores are released from beautiful orange fruiting bodies in late May. The spores infect the leaves of *Malus* and *Crataegus* (apples, crabapples and hawthorns), producing a circular orange-yellow, ultimately dark blotch on the leaves with fruiting bodies on the underside. This disease is a natural part of the ecology of both hawthorn and juniper trees and need not be interrupted by pesticide use. Orchard growers now select rust-tolerant apple varieties. This same selection criterion can be applied to hawthorns, but for natural planting it is appropriate to raise the local race of *Crataegus*, regardless of, and perhaps because of, how thin the foliage is by August.

woodland and moist clay embankments to dry south slopes in sand dunes. I have chosen to describe only a few of the more common species and suggest that you at least grow the local geographic expression of *Crataegus* species and hybrids as they are found in your area.

> **Exotic Alert**
>
> The English hawthorn, *Crataegus monogyna*, was planted extensively for rural fences and has naturalized in some areas. Perhaps the greatest concern is that it will be collected as a native species and grown for restoration plantings. Become familiar with its distinctive deeply lobed leaf and the one-seeded fruit (as implied by the name *monogyna*).

Crataegus crus-galli, COCKSPUR HAWTHORN, is perhaps one of the easiest hawthorns to identify with its dark green, glossy leaves. This dense, non-suckering tree with spreading branches and thin, 1½-inch long thorns is found principally in drier conditions, including hard clay embankments. The nutlet (which encases each seed) is quite heavy, with perhaps the thickest wall of any hawthorn. It is found in mixed and deciduous forests from southern Quebec across southern Ontario and Michigan's lower peninsula into Minnesota, south to the Carolinas and across to Texas.

Crataegus mollis, DOWNY HAWTHORN, a species with slender thorns, occurs in open woodlands, often on rich, moist sites. The red fruit are fleshy and each seed is encased in a relatively thin nutlet. Its range extends from southern Ontario and Michigan into Minnesota and south from Alabama to Oklahoma.

Crataegus punctata, DOTTED HAWTHORN, is a striking savannah-like species with its open, horizontally spreading gray branches. It has sharp, stiff thorns and grows in open, gravelly ground, typically along fence rows and woodland edges. It has an extensive range from Newfoundland through New England, southern Ontario and much of Michigan to Minnesota and south from Kentucky to Iowa.

Crataegus succulenta, LONG-SPINED HAWTHORN, is a stiffly upright-branched shrub with very stout 1½ inch thorns. It spreads by shoots rising from the roots to form open colonies in fairly dry, gravelly ground. The nutlet has a deep concave hollow on one side and very thin walls. It has an extensive range from Nova Scotia to southern Manitoba, through the Great Lakes region, and south from New York to Iowa.

Hawthorns in northeastern North America thrive primarily in open space, with a few species tolerating forested conditions. As forest succession advances, even these species decline in a heavily shaded understory. Many species become

Crataegus monogyna *(English hawthorn)*

thin-foliaged by late summer due to rust, and as such are excellent early-succession or nurse cover under which a number of tree species can flourish without being browsed on by deer. A few species (for example, *C. succulenta*) spread by sprouting from a wide-spreading root system. Most hawthorns are nonsuckering trees and spread only by fruit, dispersed by animal associates. Natural germination is usually in the second spring, after the bony nutlet surrounding each seed has decomposed.

Full bloom is in mid to late May. Where the genus is abundant, a fantasia-like snow-in-summer effect is created in the countryside. Hawthorns are closely related to apples (*Malus*) and pears (*Pyrus*) and are attractive to many insect pollinators, including honeybees. The fruit resembles a miniature apple and contains three to five very hard, bony nutlets, each containing a seed. Fruits of all *Crataegus* mature in October when they have changed color — red, orange or yellow in most species, or rarely near black. Even though the fruit of many species is quite solid when ripe, like apples, they can be kept in a bag to partly ferment and soften for a week. The whole fruits (even if still firm) can then be crushed by walking on them or smashing them with a rubber mallet before placing the whole works in a blender set on Chop. The relatively heavy nutlets separate readily from the pulp, which is then easily poured off. The action of the blender will chip the hard nutlet walls and may help to promote first-spring germination for many species. Dried nutlets or dried ripe fruits can be stored for many years.

Seed germination is complicated by the hard nutlet enclosing each seed, but possibly also by an immature embryo; most seeds will germinate only in the second spring. Provision of two to three months of warm conditions prior to five months of cold stratification often yields partial first-spring germination, with the rest germinating in the following spring. For those who have the facilities, sulfuric acid pretreatments can increase first-spring germination but sometimes only marginally. I have found that the need for a sulfuric acid bath relates to the thickness of the nutlet, which I visually gauged by cutting the fruit in the middle using pruning shears. *C. crus-galli*, with the thickest nutlet wall, requires a two-hour acid bath, and *C. mollis*, with the thinnest nutlet wall, requires no acid bath. The nutlet thickness of other species was judged in relation to the two extremes and the acid bath duration was adjusted accordingly. All seed lots attempted yielded some germination in the first spring. Success might have improved with a longer warm treatment, since I allowed only one month of warm followed by five months of cold stratification. Once again, the

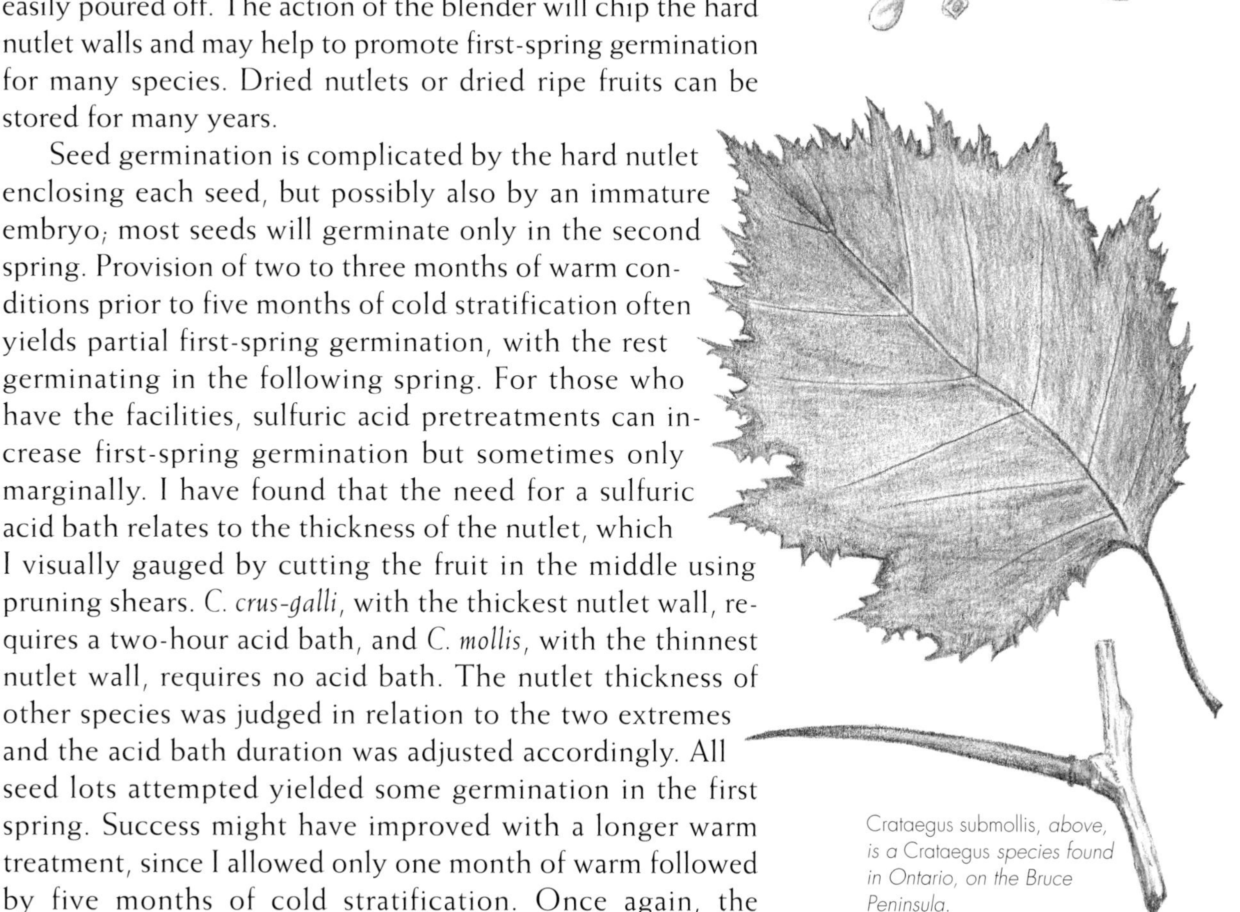

Crataegus submollis, *above, is a* Crataegus *species found in Ontario, on the Bruce Peninsula.*

ungerminated seeds ultimately came up the following spring — I managed to keep the mice out!

Seedlings grow to about 4 inches in the first year and can be kept in close proximity for another year before digging and separating them to plant in their permanent location. Seedlings are easily transplanted in spring, but once the plants are older they are best planted in autumn. Deer will browse on the foliage in summer and rabbits will strip bark and eat twigs in winter, from which the hawthorn is usually able to recover. You will want to protect the base of your trees from meadow voles, since even 4-inch diameter hawthorns are killed if trunks are completely girdled by them in peak years. Hawthorns are important succession plants, providing critical protection for maple where deer might otherwise prevent their establishment.

Diervilla lonicera, BUSH HONEYSUCKLE

Bush honeysuckle is a low shrub in the partial shade of open woodlands and forest edge communities. It spreads toward clearings by underground stems (rhizomes) to produce extensive colonies up to 3 feet high. It thrives in dry or rocky soils, on hillsides and in mesic woodlands south of the boreal forest from Newfoundland to Saskatchewan and south from Iowa to North Carolina.

The unique fruit capsule of Diervilla lonicera, *bush honeysuckle, remains closed unless ripped open by a sparrow or junco.*

Sexually perfect, yellow orange flowers last for a few days and new flowers are produced over an extended period in July and August, providing a continuous source of nectar and pollen for their chief pollinator, the bumblebee. Fruits are elongated capsules in terminal clusters that mature in October and persist through winter and remain closed. Each dry capsule holds up to 20 seeds that are scattered in the immediate vicinity of the plant through winter as they are torn open. Dispersal over much greater distances might be explained by the eventual northward migration of sparrows and juncos.

Seeds germinate readily, without pretreatment, but a couple of weeks of cool nights will yield a more uniform germination. Seeds are scattered on the surface of the growing medium, partly covered by a thin layer of grit, and kept quite moist. Germination is rapid, and seedlings can be transplanted to ¾- to 1¼-inch spacing to grow for the first year. This species is also easily grown by divisions, which, if taken from more than one source, will provide the pollen diversity for continued seed production. This plant is such a rapid colonizer that only a few installations are really needed. In the north, deer or moose may browse on it lightly.

Diospyros virginiana, PERSIMMON

This tree is native to the eastern U.S. from New York west to Kansas, through the lower Great Lakes basin and south from Florida to Texas. Its current distribution in the wild was no doubt influenced by native peoples and then early settlers, who cultivated it for its desirable fruit.

D. virginiana is a medium-sized tree growing to 66 feet in well-drained soil. The tree produces fragrant flowers in summer. The trees are dioecious, so both males and females are needed for seed production. The fruit of the American persimmon is spherical to ovoid, ¾ to 2⅓ inches in diameter.

Seeds should be collected as the fruit mature in the fall. Separate the flesh from the seeds and then either plant directly into an outdoor seedbed or stratify for three months and then plant into a seedbed the following spring.

Dirca palustris, LEATHERWOOD

I had read about this plant in my youth but nothing could have prepared me for my first sighting of this lovely shrub, which resembles a small tree. Perhaps the interest in this plant lies in the fact that it is so unusual. It occasionally grows to 10 or 13 feet but mostly to about 5 to 7 feet tall. It has curiously flexible stems with fibrous bark that tears off in long, "leathery" strips. The name palustris suggests "swamp" but this is far from leatherwood's reality. It thrives in upland sugar maple woodlands on rich, moist, well-drained loam, or on rocky limestone outcrop soils and sand-dune forests. Leatherwood is found throughout eastern North America from Nova Scotia across the north shore of Lake Huron to Minnesota, and into the southern states of Louisiana and northern Florida. (*See* photo on p. 245.)

> **Exotic Alert**
>
> In the same plant family as leatherwood is the purpleflowering *Daphne mezereum*, the paradise plant, which blooms in March. It has naturalized in local pockets throughout the region and there are many who would like this lovely 3-feet high, neat and tidy shrub to be a "native" plant. Rest assured that plants found scattered along fence lines and rail beds in dry to mesic locations originate from cultivated plants. No doubt this European plant will increase in the wild, it so readily self-seeds in gardens.

Dirca is often found in significantly localized, seemingly even-aged groupings or on paths close to old plants, indicating little requirement for the seed to pass through a digestive tract and a possible

Timing is crucial to obtaining the greenish yellow fruits of Dirca palustris *(Eastern leatherwood). Seeds are shown at twice life-size.*

requirement for a disturbance leading to removal of the leaf cover. Both trail use and ground fires will reduce the leaf cover enough to allow germination of these very slow-growing seedlings. Dispersion over any distance (considering its occurrence on Manitoulin Island in Lake Huron) must be associated with birds such as thrushes or grosbeaks. Undoubtedly mice and chipmunks will harvest and disperse some of the crop but I have yet to find a germinated leatherwood seed cache.

Leatherwood is in full bloom in the leafless spring woods, presenting tiny yellow points of light in a distant shrub. The early to mid-April flowers are bee pollinated. Although they are reported to be a perfect flower, I have determined that some plants do not bear fruit even though adjacent plants bear fruits every year. The fruit is a small, one-seeded berry that is bright green and changes to a yellow green before being easily dislodged from its branch in early to mid-June. The gray-brown seed is about half the size of the fruit. Most of the fruits just drop to the ground, turn black in a few days and become very difficult to locate after that.

Timing your seed gathering is very critical with this species, as it disappears within a week of being mature. I have often said half-jokingly that the fruit is ripe on June 3 around 2 p.m. To collect the seed, I place a small tarpaulin or inverted umbrella under the shrubs and dislodge the fruits that are ready to fall by lightly raking my hands along the underside of the branch. Fruits must not be enclosed even for a day, as they ferment rapidly. Spread the whole fruits three to five seeds deep in a dry location to ripen further. When they begin to turn black, sow them immediately to avoid excessive drying.

This is one species for which seed extraction from the fruit seems unnecessary or even detrimental. Blackened fruits can be planted at ⅛- to ¼-inch spacing in a pot or seedbed in a cold frame and covered with up to ¼ inch of rich compost. They should be watered and mulched with sugar maple leaves to keep them moist throughout the summer, and remulched in the fall to protect them from freeze/thaw fluctuations in winter. The leaf mulch must be removed in early May so that emerging seedlings are not smothered.

The seedlings are very prone to damping-off fungus and slugs. One-year seedlings are ¾ to 1¼ inches tall and reach 2¾ to 4 inches in the second year. Plant them out in small groups of twelve at 2- to 2¾-inch spacing to grow for a couple more years. The whole group should be dug up as a unit to avoid tearing the leathery roots. Seedlings are easily shaken apart and can be planted in woodlands and gardens in early spring. Watering is crucial through any drought in the first growing season. Plants that are several years old can be planted in full sun, where they develop very dense growth rather than the more interesting open, irregular growth in the shaded woodland garden. Individuals in a population are singled out and heavily browsed on by rabbits and deer in prolonged winters. The plants resprout from heavy browsing and, within a year, self-prune the truncated branch in a most unusual fashion (almost like leaf senescence).

Euonymus, the BURNING BUSHES or SPINDLE TREES

Some 200 species of *Euonymus* are found throughout the world, with the highest concentration in Asia. The evergreen-leaved mound-forming exotic *Euonymus fortunei*, in all of its variegated leaf color and size variations, is well known. The exquisite beauty of our native, deciduous species is virtually unrecognized. Natural seed dispersion is primarily by the thrushes (including American robin and bluebird), which eat the whole berry and pass the soft-coated seed, intact, to a new location, while cardinals tend to carefully strip the pulp from the berry and drop the seed. When I cleaned the seeds, I found the pulp noticeably oily; perhaps the richness makes *Euonymus* quite desirable to birds in the cold season. Natural germination of *Euonymus* appears to be delayed to the second spring.

Exotic Alert

Two "ornamental" deciduous shrubs that are very common in the horticultural landscape have become naturalized (extensively in many places) in close proximity to older settlement and parks. Beside these two, garden escapees of the evergreen-leaved *E. fortunei* are also beginning to finding their way into the wild.

E. europaea, European spindle tree. The identification of this close relative of *E. atropurpurea* can only be made with certainty at flowering time. The European species has greenish white flowers in late May as compared to the maroon flowers (in June) of the native wahoo. The European spindle tree holds its leaves much longer into the fall, and the twigs tend to be thicker and more ridged than wahoo. However, due to variability in both species, the two are occasionally so similar vegetatively (including the root sprouting) that without knowledge of the flowers, the nearly identical fruit should be left alone — or the flower color should be verified in late May.

E. alata, winged burning bush. A dense, outward-arching bush originating in Japan. It has small orange fruits on short stems that persist into winter. Twigs on the species are prominently four-winged, a characteristic that is more diminutive on a more compact-growing horticultural cultivar but still gives this species a distinctly square-stemmed appearance.

Euonymus atropurpurea, WAHOO or BURNING BUSH. Few plants can begin to rival the stunning display of this upright shrub in November when it is leafless, with hundreds of pinkish purple fruits dangling from the branches. The species is

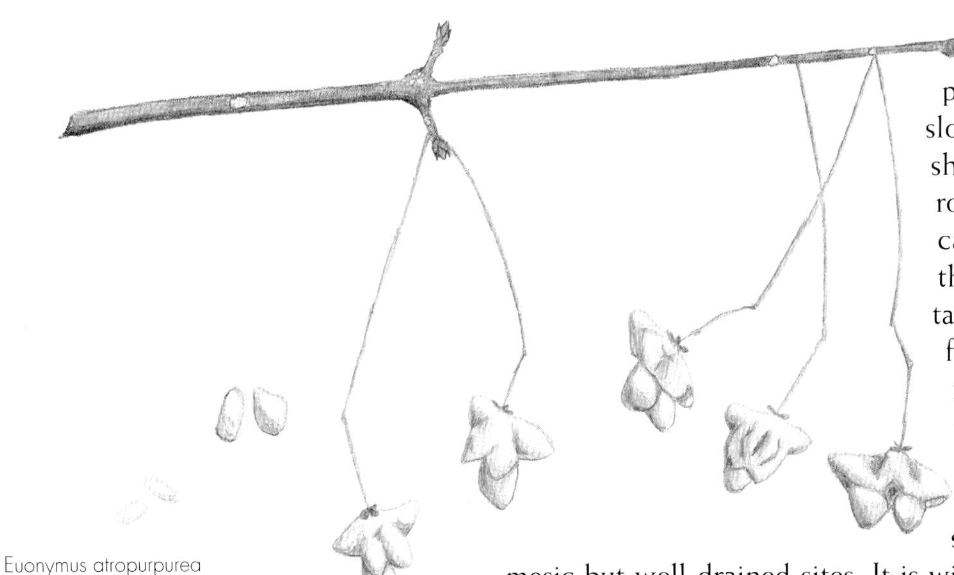

Euonymus atropurpurea
(wahoo)

found in full to partial sun and spreads slowly, by way of shoots rising from the roots. In this manner, it can produce dense thickets up to 13 feet tall in clearings, on the forest edge and in fence-line communities. Wahoo thrives in a range of clay, sand and rocky soil types in moist to mesic but well-drained sites. It is widely scattered in deciduous forest through the lower Great Lakes region from New York through southern Ontario and Michigan to Montana and south from Alabama to Oklahoma. (*See* photo p. 245.)

Euonymus obovata, RUNNING STRAWBERRY-BUSH, is a very common shade-loving woodland groundcover, yet I am always surprised and delighted by the fantastic display of warty pink fruits with orange berries. It is a prostrate shrub with green, thin twigs, growing on well-drained, moist, rich soil of closed-canopy woodlands. The branches root where they touch the ground to form wide-spreading colonies about 6 inches high. It is mostly restricted to deciduous forest in southern Ontario and from western New York to Illinois, and south from Tennessee to Missouri. Woodland gardeners should not rest until this species is established in their gardens.

Euonymus americana, STRAWBERRY-BUSH, is a densely upright relative of the running strawberry-bush. It is similar in flower and fruit, with equally green, thin twigs. Strawberry-bush just approaches the Great Lakes watershed, ranging from the southern states of Florida to Texas and northward from southern Illinois to southeastern New York.

Euonymus flowers are fascinating in their minute architecture; they are produced in May to early June and pollinated by flies. *E. atropurpurea* has a showy four-petaled reddish maroon flower and four segmented pinkish purple fruit. *E. obovata* has a five-petaled greenish yellow flower that is quite hidden under the new leaves, and a three-segmented pink fruit.

The entire genus is reported to have perfect flowers but the native species seem to be irregularly tied to that rule, as only certain individuals produce fruit. In the extreme, individuals of *E. atropurpurea* in the Guelph Arboretum gene bank are without fruit in the proximity of trees from a dozen other seed sources, some of which have only a few fruits and others

Euonymus americana
(strawberry-bush)

a full crop. An isolated *E. atropurpurea* three miles away must be self-fertile, as it has been producing a massive crop of viable fruit for years.

A single soft-coated white seed is encased in each orange fleshy aril, up to eight of which (usually two to four) are further encased in a fleshy pinkish purple capsule. When mature, the capsule splits and folds outward as it dries. The seed-bearing arils then drop out of their chamber but remain attached and hang by a thin, white, thread-like fiber. Dried whole fruits can be refrigerated for a few years.

Fruits should be watched when the capsules turn color and can be picked any time after they start to naturally split open. Like its relative the bittersweet, the capsules will open after a few days of drying and release the arils. Whole fruits or hand-separated arils can be gently ground in the grit bag to extract the relatively soft seeds. If the mashed pulp is too oily for the seeds to sink, first try adding a few drops of detergent to free the seeds from the floating pulp, or dry the whole works for a few hours (not long enough to damage the seed) before rubbing the pulp free of the seeds so it can be blown off.

Germination is best achieved with fresh, cleaned seeds, which should be rehydrated with a 24-hour soak and placed immediately into warm stratification. *E. obovata* is kept warm for three to six weeks until the seeds swell to near double in size before a shift to 120 days of cold, after which they will germinate shortly after planting. *E. atropurpurea* is easy to germinate if you wait. The safe method is to plant the seeds in a pot bedded into the cold frame and mulched with leaves, to germinate in the second spring. Less certain (due to the hazards of a long warm stratification) is to try a 90-day warm stratification prior to 150 days of cold in an attempt to obtain a first-spring germination. If they do not germinate and the seeds are firm, they will germinate the following spring. A completely screened enclosure is required to keep mice away from this seed. Germinating seeds of both species are planted at about ¼-inch spacing. They are sensitive to damping-off fungus.

One-year seedlings are 2 to 4 inches tall and 6 to 10 inches in the second year. *Euonymus*, with their dense, fibrous root systems, are very easy to transplant. They are easily lost to mice and rabbits when young and are best grown in a wire enclosure. Separate the two-year seedlings and grow them on for two more years before planting out. Wahoo is an absolute jewel of a plant but not for the inattentive gardener, for it is relentlessly consumed by herbivores and newly established plants can easily be killed. Deer heavily browse on it in summer, the twigs are consumed by rabbits in winter, and meadow voles will debark the base. Older plants can resprout from the roots again but only if well established; need we guess why it is a rare plant? Euonymus webworm, a.k.a uglynest caterpillar, is a classic insect pest that devours foliage in June while protected by a web of gray silk. The webworm can be controlled on young and newly planted specimens in peak years by handpicking the webs when they first appear. Once established, the shrubs can easily resprout after the feeding frenzy and, where suitable habitat diversity exists, natural predators seem to keep the pest numbers low.

The two native species are usually propagated asexually from softwood cuttings or (rarely) from the spreading, rooted shoots; however, single clone plantings, aside from a few exceptional plants, will tend not to fruit well.

Fagus grandifolia, AMERICAN BEECH

> **Exotic Alert**
>
> *Fagus sylvatica*, European beech. This species is the parent source of every horticultural beech: cut-leaved, pendulous, twisted, columnar, dwarfed and of course the never-ending parade of beech trees with colored leaves. The species (both green-leaved and purple variants) is occasionally found in the near-urban wilds, where it may produce a viable nut crop. These are particularly good to eat.

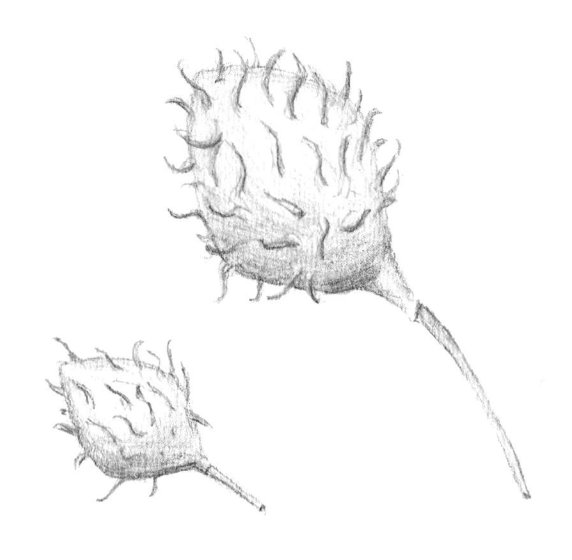

Fruit of Fagus sylvatica *(European beech) and* Fagus grandifolia *(American beech).*

Stately and memorable in many ways is the beech tree, especially since the bark records the history of bears and local lovers. Nine species are described from England to Japan, with only slight variation across the huge continents. The single North American species is similarly variable across its entire range. Though not taxonomically significant, subtle differences are attributed to var. *caroliniana*, white beech, on moist to wet soils to the south and east of the Great Lakes watershed, "gray" beech in rocky woods at the northern edge of the species range; and var. *grandifolia*, red beech, on rich upland woods almost throughout the range. Beech is found from Nova Scotia to north of Lake Huron into Wisconsin and south from Florida to Texas.

Beech trees spread by shoots rising from the shallow, often exposed, roots to form immense colonies. It is possible that every beech in many woodlots has spread from the same tree, with the original now long decayed. Perhaps then, because the genus is shade tolerant, it is theoretically possible that some beech trees in our region are 7,000 years old, since the first seeds were carried here by passenger pigeons after the glacier retreat.

Always take the opportunity to eat a few beechnuts.

Female flowers are produced in pairs in the more terminal leaf axils of the current season's growth, and separate male flowers in clusters from the basal leaf axils as the shoots are elongating in the spring. Beech is reproductively mature at about 30 years; the flowers are wind pollinated. Beechnuts rank very high in energy and food value and are consumed with remarkable intensity. Local dispersal of fallen fruits is by grouse, turkeys, chipmunks and mice. Black bears make "nests" — aerial platforms — in beech tree crowns by pulling branches back to stand on them, often breaking them to get at the nuts.

Squirrels, porcupines, blue jays and crows also feed in the crowns. I have observed single seedlings in the woods far from any beech trees. Long-distance dispersal must be accomplished by seeds being dropped by birds as they attempt to break them open. I have also noticed whole germinated caches containing dozens of seedlings not far from a beech grove. Natural germination takes place in cool soils the first spring.

Fruits are usually single, in the leaf axils. All fruits will mature and ripen even if the entire seed crop is empty. In late September, the soft-spined husk splits into four to expose two triangular seeds. In good crop (mast) years, both seeds are plump, with convex sides. In average crop years, a good seed is often paired with an empty seed that has slightly concave sides. You will soon become familiar with the shape of good seeds after cutting a few open. This visual evaluation of seed quality will allow you to pick a few good seeds from the ground with a little searching. The kernels are bright white and firm and delicious, especially after being left to dry for a few days. Seed crop years are very sporadic but generally at two- to three-year intervals.

The fruits are mature when still closed but turning yellowish. Gather seeds by picking them from a low branch or from the ground; check first to see that some are plump, filled seeds. Keep the fruits in a cloth or paper bag until they can be spread out to dry. The husks will spread wide open within a few days and make seed removal very easy. Beechnut viability drops dramatically if they are left to dry any longer. Dump the freshly extracted seeds into a bucket of water for a few hours and discard the floating empty seeds.

Sow the filled seeds in a very well protected outdoor frame. Or, alternatively, store freshly dehusked, surface-dried beechnuts cold until late November to stratify them for sowing in workable soil in late March. Soak the stored beechnuts for 24 hours and then place them in cold stratification for 120 days. Seeds will initiate radicle emergence at refrigeration temperatures. Plant seeds about ¾ inch apart. Close spacing helps to keep the young seedlings cool. Add some soil obtained from under an old beech tree to your

Beech Bark Disease

Beech bark disease is a *Nectria* fungus that can only invade and kill the bark after it has been altered by the feeding of beech scale insects. Both arrived in Nova Scotia from Europe around 1890. The disease spread southwest, becoming well established in eastern Quebec, eastern Pennsylvania and West Virginia by 1980 and entering Ohio and Ontario in the early 1990s.

Although some forest colonies seem to be resistant to the bark disease, it is fatal to many beech trees. The self-cloning nature of beech makes every tree within a clone equally susceptible or tolerant. Reports indicate that on several sites in Nova Scotia, bark epiphytes (lichens) provide significant levels of protection to deter the scale insect, while trunk-foraging predators, such as brown creeper and the nuthatches, are effective in reducing scale populations. Seeds from healthy clones where the bark disease has passed through might be used to establish beech in woodlands that have lost all of the beech (either through the outdated practice of removal to favor sugar maple or from the disease).

Open husks and seeds (beechnuts) of Fagus sylvatica *(European beech), above, and* Fagus grandifolia *(American beech), below. Shown at half-size.*

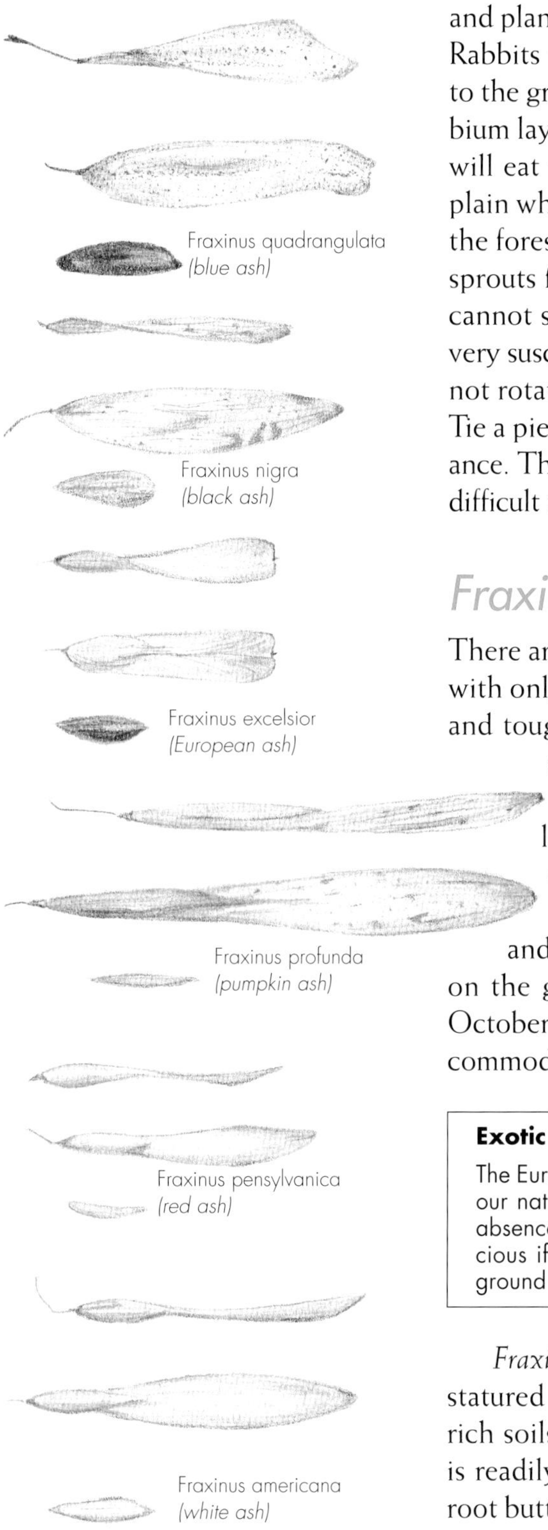

seed frame to inoculate the soil with an appropriate mycorrhizal fungus. Enclose your seed frame with ¼-inch hardware cloth cages, denying entry to mice and chipmunks; they desire this seed the most. Be sure to provide about 50 percent shade.

Two- to three-year seedlings have either not developed if your soil is unsatisfactory or they are vigorous stems 12 to 16 inches tall. These can be lifted in early spring, separated and planted at 4-inch spacing to grow for a couple more years. Rabbits will browse on the winter twigs and chew seedlings to the ground. Beeches can generate new shoots from the cambium layer and overcome occasional browsing. Meadow voles will eat the basal bark if groundcover is dense; this may explain why new recruitments of beech are rarely found beyond the forest edge. Do not even attempt to transplant beech root sprouts from the wild. They grow from thick root pieces that cannot survive the shock of being severed. Beech tree bark is very susceptible to sun scalding, and it is essential that trees are not rotated from a north to a south side during transplanting. Tie a piece of cloth to a branch that is oriented south for guidance. The American beech is suitable for large yards but it is so difficult to grow that it is rarely available for planting.

Fraxinus quadrangulata
(blue ash)

Fraxinus nigra
(black ash)

Fraxinus excelsior
(European ash)

Fraxinus profunda
(pumpkin ash)

Fraxinus pensylvanica
(red ash)

Fraxinus americana
(white ash)

Fraxinus, the ASHES

There are about 60 species of ash in the northern hemisphere, with only five species natural to our area. Ashes are important and tough early-succession trees, and few of the millions of saplings that fill disturbed sites will reach great age. Ashes occupy a range of conditions from swamps to lone trees on outer Georgian Bay islands to old field succession and mesic woodland canopy. Finches, grosbeaks and squirrels consume seeds in the canopy and wood ducks, quails, grouse and mice consume seed on the ground. Seeds usually fall close to the parent from October through to midwinter, when ice and crusted snow accommodate long-distance dispersal of seeds across open areas.

Exotic Alert

The European ash, *Fraxinus excelsior*, is very similar in appearance to our native black ash and occasionally substituted for it. Look for the absence of leaflet petiole hairs on European ash and become suspicious if the seeds are notched at the end, or if the tree is on dryer ground or obviously planted.

Fraxinus nigra, BLACK ASH, is a rather striking medium-statured tree, due to its thick and sparse twigs. It occurs on rich soils in swamps, wet woods and stream edges. Black ash is readily identified by the prominent white lenticels on the root buttress. It adapts well to cultivation on moist sites. Black ash is scattered but locally abundant in open areas and widely distributed through the deciduous and mixed-forest zones, completely encompassing the Great Lakes watershed with the exception of the area south of Lake Michigan; it ranges from Newfoundland to Manitoba and south from Delaware to Iowa.

Fraxinus profunda, PUMPKIN ASH, is a swamp species, growing in the rich soils of swales, wet woods and stream edges, occasionally attaining considerable size and producing a large basal thickening — the "pumpkin." Because ash tends to be ignored (as in "oh, that's an ash"), this species was not known in the Great Lakes watershed before 1992. Since its "discovery" in Windsor by citizen-botanist Gerry Waldron, the range of this species was revealed to extend through northern Ohio and southern Michigan and across southern Ontario. It was previously known along the eastern coastal plain from New Jersey to northern Florida and across to Louisiana, and up the Mississippi Valley to southern Indiana, Illinois and Ohio.

Fraxinus americana, WHITE ASH, is usually noticed for its gorgeous purple fall color. This stately tree is the typical ash of interior woodlands on moist, well-drained loam soils. Occasionally I have found huge white ash growing in very exposed, extremely dry glacial deposits and sand dunes. Such extreme seed sources should be investigated for restoration efforts. This tree reaches perhaps the largest size of any ash and grows throughout this region. It ranges from Nova Scotia, along the St. Lawrence and Ottawa rivers to Michigan's upper peninsula and west to Iowa then south from Texas to northern Florida, where a pubescent form called *F. biltmoreana* is described.

Fraxinus quadrangulata, BLUE ASH, has the curiously "quad" or four-angled twigs formed by corky bark ridges that are easily seen on close inspection. They may serve to reduce twig desiccation by wind. Although most trees with a color name are likely named for the wood coming out of the sawmill, blue ash was used for a blue dye made from the inner bark. This medium-sized tree is often densely "cluttered" with numerous internal shoots growing straight up through the crown from the base of limbs. It is principally a tree of heavier soils on banks, floodplains and moist but well-drained woods, as well as the thin soils over limestone sites (alvars) of the Erie islands. It is restricted to extreme southwestern Ontario, through western Ohio to Indiana and south from Tennessee to eastern Oklahoma. A number of four-angled-twig relatives are found in the American southwest.

Fraxinus pensylvanica, RED ASH. Of all the ashes, this is perhaps the most significant early-succession species, variably growing in imperfect drainage and moisture extremes. It is often found inundated by water in the spring and extremely dry by late summer, or growing in dry gravelly or rocky embankments. Twigs and leaves are densely hairy. A form with glabrous stems and

Blue Ash

Blue ash is relatively uncommon and recruitment seems to be rare. The intensity at which the meadow vole consumes the basal bark of this species may account for this. One controlled-burn site in Lambton County, Ontario, had significant regeneration of blue ash, perhaps due to the destruction of meadow vole cover. I know of no other species that is so relished by the meadow vole (field mouse). In several locations at the University of Guelph Arboretum, 15-year blue ash trees with 5- to 6-inch-diameter trunks have had all the bark chewed off the trunk for a couple of inches above and even below grade! This makes blue ash a very difficult species to establish in open, natural areas with good meadow vole cover.

Such feeding starts in early October. The application of Skoot rodent repellent to the base of the tree worked for years but recent feeding indicates that meadow voles may develop a tolerance or even a taste for Skoot. Wrapping the trunk with a spiral tree guard is equally ineffective since the vole eats bark below grade. The University of Guelph Arboretum has resorted to using half-inch chickenwire mesh, loosely wrapped around the trunk and flared out over the ground for a few inches; consider protecting the trunks until they are about 10 inches in diameter and canopy shade suppresses the herbaceous cover that attracts and protects the meadow voles. The wire will likely rust enough to fall apart as the tree trunk thickens.

leaves, called green ash, is found on heavy clay soils as well as sand ridges that are both very dry in summer; it is more common as one goes west, becoming the dominant form. The two forms are not taxonomically distinct, and intermediates in twig hairiness are very common and not to be overlooked. This generally abundant species ranges from Nova Scotia west to eastern Alberta and Montana and south from northern Florida to Texas.

Pest Alert

The recent establishment and outward migration of the emerald ash borer, an insect from Asia, has many people worried (*see* p. 67 and photo p. 246). The response that commands a halt to planting ash trees is outdated and does not recognize genetic diversity and ecosystem health.

Urban trees and natural trees are compromised by numerous constraints on ecosystem health. Among these stresses are nitrogen loading from acid rain that causes many species to grow at a rate beyond the capabilities of their root systems; leaf removal in windswept areas and leaf pickup that robs the soil of valuable organic acids from the decomposition process; storm sewers and drainage ditches that exacerbate the impact of drought cycles. But above all is the virtual absence — until recently — of deadwood habitat for predatory wasps, small mammals, bacterial diseases and woodpeckers, the hairy and downy woodpeckers in particular.

Green ash has recently colonized the landscape in great numbers, filling the gaps created by Dutch elm disease and sugar maple decline. Many urban brownfields and abandoned farms are covered with ash regrowth. A flush of growth of this nature without competition is highly prone to being ravaged by some disease or insect, because an early-succession species is programmed to decline anyway. A combination of natural predation in healthy woodlands and genetic diversity in ash should yield a number of trees that tolerate and survive the invasion of the EAB. Some ash trees could have sufficient chemical compounds in the bark or wood to make them unpalatable. The risk of losing the best trees — those with rare alleles in the genetic makeup — to indiscriminate cutting is a concern with any species that is being challenged by new pests and diseases. In ecological forestry practices, the healthiest trees of any species are left standing to reseed the forest with the best gene pools. We must be cautious about allowing high-grade logging or clear-cutting every host tree in advance of an invading pest. I even have reservations about removing infected trees, the most likely place for predators to establish and increase in numbers.

Reproduction and Propagation

Ash flowers are mostly unisexual on separate trees (dioecious) but occasionally trees with perfect flowers are found. After noticing a heavy seed crop on a 25-year-old 'Marshall's Seedless' ash, I wonder if some trees are similar to the red/silver maples in going through a sex change, or perhaps the graft died and the rootstock formed the tree. Ash seed crops are so infrequent, and records of individual trees so few, they may easily flower inconspicuously male at first and morph toward female as the tree ages — although large trees are found with no fruit in "seed years" — room for further study. Flowers mature in late spring with the unfolding leaves and are wind pollinated, producing fruit (keys) in dense pendulous clusters.

The fruit is a dry and elongated key, with a seed encased at the end of or within an elongated wing, the whole of which is sown intact. Ripe fruits are light brown and readily stripped from low branches or shaken from trees. Seed collection, especially in this genus, should consider the oldest natural trees, and also old planted trees for their demonstrated longevity. Collecting from these trees is even more urgent with the arrival of the emerald ash borer in the Windsor-Detroit area. The seeds should be kept dry if they are fully ripe. They can either be sown immediately, close enough together to be partially overlapping, or kept dry, in refrigeration, and sown in spring. Either way, they will mostly germinate in the second spring. The exception is pumpkin ash, which appears to have no inhibition to germination; fruit will germinate in the fall with adequate moisture and warmth, which then requires a cool greenhouse to keep them until spring. Better, keep the keys dry over winter, or delay outdoor seeding until cool weather has arrived. Red ash produces a seed crop almost every year, while all other species may wait for three to seven years between seed crops, an event that is not to be missed.

Obtaining germination in the first spring is possible with all species of *Fraxinus* by collecting fruits when they are at the mature yellow stage. Keep the fruit in a plastic bag and sow immediately in a cold frame with a good mulch of straw or leaves so that the seed receives about one month of warm soil conditions prior to frosts. Alternatively, the fruits can be placed in warm, followed by cold, stratification with durations outlined in Appendix III. A few seeds may germinate during the warm stage but those remaining still must go through a cold period, with the exception of pumpkin ash. Plant the seeds by scattering them densely enough that the wings are partly overlapping. Most seed lots will produce good germination in the first spring, but if it seems thin, leave seed frames intact so that any dormant seeds can germinate the following spring.

Seedlings are 10 to 20 inches tall in two years depending on the species (blue and black ash are slower) and can be planted to a permanent location or into a nursery. Dense fibrous roots make it generally easy to transplant in spring or fall. Rabbits and deer seem to avoid the ash enough to be of little concern, but meadow voles are unpredictable; aside from the meadow voles' ravenous appetite for blue ash, unprotected red, green and white ash up to 6 inches are also often girdled by meadow voles during the winter.

Ash species occupy diverse habitats and may present site-matching challenges. Young planted trees seem to be quite adaptive and are widely utilized, often from obscure seed sources. A general mismatch of seed source to planting site plus their early-succession nature may be the reasons that ashes occasionally suffer serious crown dieback and decline. Overall, ash ecology needs to be taken more seriously.

Due to very infrequent seed crops, black, blue and white ash are often grafted on green ash rootstock by the horticultural industry, thus making them a tree suited to the drier green ash ecology and unacceptable in naturalization work. I have seen nursery-grafted black ash on a green ash rootstock planted in a floodplain area! It is extremely crucial that the ash are grown from seed and not presented as clonally grafted trees with no genetic diversity.

Gleditsia triacanthos, HONEY LOCUST

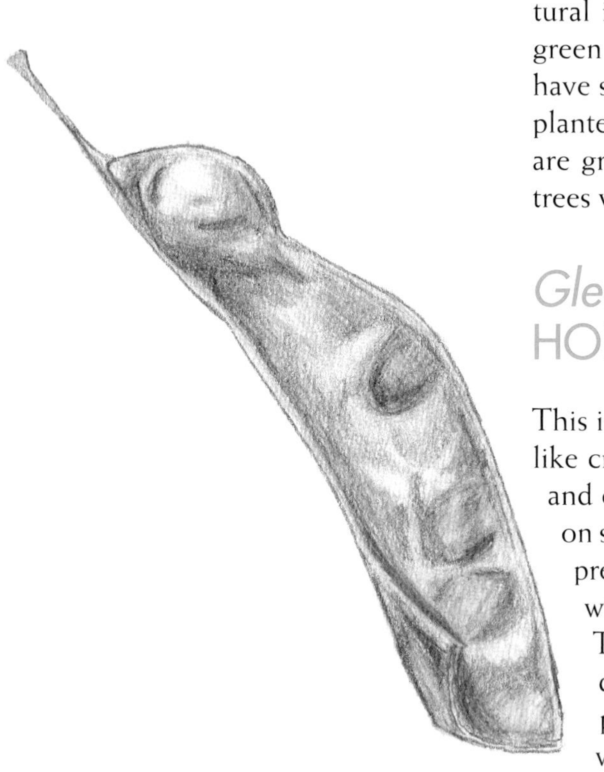

The large, thin, spiraled pods of the honey locust (Gleditsia triacanthos) are 8 to 12 inches long and very distinctive. About one-third of a pod is shown here.

This is a classic legume tree, ultimately with a broad, acacia-like crown. There are fourteen *Gleditsia* species in the world and our single species is quite variable, with thorn colonies on some trees that are singularly stunning. Interestingly, the present-day proliferation of thornless honey locusts in the wild may be associated with the extinction of mastodons. They likely ate the bark of this species and could have diminished the thornless expression in the early breeding pools. The ability of trees to hold on to the fruit through winter until the seeds are extremely hard was a simultaneous coevolutionary adaptation to get past the molars, as mastodons would have devoured the nutritious pods.

Honey locust is a large tree of rich, moist, open woodlands and river embankments. It is found naturally from Pennsylvania, a few locations in extreme southwestern Ontario, and southern Michigan to Nebraska and south from northern Florida to Texas. Within and beyond its native range it was planted historically for hog food (pods), living fences and, more recently, shade. It occasionally spreads from old plantings. The species can spread by sprouting from its roots, especially if a live tree is cut down. Honey locust is thought to be tolerant of very poor soils because of its acacia-like appearance, but trees in very dry conditions are compromised and prone to pest invasions and a shorter life. The "seedless" horticultural cultivars are male selections of the thornless variant, *G. triacanthos* var. *inermis* ("without thorns").

The small, greenish yellow flowers are unisexual in clusters on the same tree but occasionally on separate trees. It is possible that *Gleditsia* starts blooming male and some trees gradually switch to female flowering with hormone changes associated with age or stress. Flowers open with the unfolding leaves and are pollinated by bees and other insects. The magnificent fruit pods drop out of the tree during late winter; natural germination may have been commonly induced when seeds were steamed in the soil by a low-temperature ground fire. Honey locust always responds to the hot water treatment.

About 20 to 30 seeds are enclosed in flattened, twisted and bent pods up to 12 inches long; they contain a sweet edible pulp, thus the name "honey" locust. Pods mature in October — they turn from green to reddish in color and, finally, brown, when fully ripe and dry. They can be picked during the winter or gathered from the ground in spring. Dry pods can be thrashed in a heavy cotton bag or peeled open to release the very hard seeds. Dry seeds can be cold-stored for many years. Germination is readily attained within 10 days by subjecting the dry seeds to the hot water treatment, after which the expanded seeds must be planted immediately, at ⅓- to ¾-inch spacing.

Two-year seedlings are 16 to 24 inches tall and sometimes already thorny. They have fibrous roots that are easy to transplant and can be installed in their final location or grown further for another year or two. Groundhogs and rabbits have been observed eating the foliage and twigs. Relegate the thorny races to the wilder areas of farms and yards — the thorns are fascinating but you don't want to be driving over them.

Gymnocladus dioica, KENTUCKY COFFEE TREE

The leaflets of the Kentucky coffee tree turn clear yellow before falling, leaving the bright red petiole stalks for another week or so. Clusters of short, thick pods hang in female trees until April, creating the impression of something distinctly tropical and out of place in the winter landscape. Two other species occur in southeastern Asia. Mastodons, before they became extinct, undoubtedly consumed the nutritious fruit just as they consumed the fruit of the honey locust.

This is a species of floodplains and rich, moist woodlands but seemingly tolerant of open dry sites or partial shade in cultivation. A clonal species, it forms an open colony by sprouting from the vast, spreading root system. It is found from central New York across several counties in extreme southern Ontario to southern Michigan and Minnesota and south from Tennessee to Oklahoma. It is most abundant in Ohio to Missouri.

Fragrant, greenish white flowers are insect pollinated and produced in June, with male and female flowers on separate tree colonies. In natural conditions, the entire colony may be only one genetic individual, in which case fruit is rarely produced if the colony is isolated.

The fruit is a thick pod some 4 to 8 inches long with three to seven rounded, somewhat flattened seeds about ¾ inch across. The seeds are swollen and soft into November and only become rock-hard when they dry out by midwinter. Pods can be

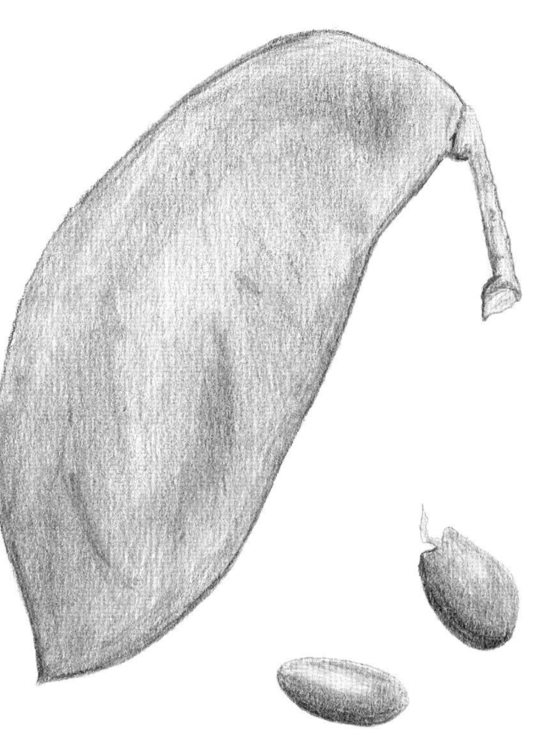

The thick, leathery pod of the Kentucky coffee tree, Gymnocladus dioica*. In order to have consistent germination, the hard seed must have a patch of the outer seed coat filed off.*

harvested from low branches in December when they turn dark brown, or gathered when they drop to the ground in late winter. Dry seeds can be stored for years in refrigeration. (*See* photo p. 247.)

Seeds will not germinate if only soaked in water, and my experience using the standard hot water treatment for legumes has produced very poor germination. They germinate readily after filing a patch off the seed coat. I carefully hold the seed up to a grinding wheel and remove a small patch from the outer layer of the seed coat. A 24-hour soak in cold water will cause the seed to swell to two to three times its dry size. Plant the swollen seeds ¾ to 1 inch apart in a cold frame, avoid over-watering them, and they should germinate within two weeks.

Two-year seedlings are 8 to 12 inch tall and have a very thick fleshy root that breaks relatively easily. Care should be taken in handling this species, although it transplants easily in the spring, perhaps because it is one of the latest to leaf out. It should be planted into a nursery for a couple of more years until it is at least 3 feet tall before setting it out. I have observed groundhogs feeding on young plants and I imagine deer would munch on the expanding shoots in the spring. Open-grown specimen trees should be pruned to prevent weak narrow-angled branch crotches. It can be maintained as a specimen tree only with continuous mowing to stop the root sprouts. *Gymnocladus* deserves planting in more remote "back forty" sites, to grow wild.

Hamamelis virginiana, WITCH-HAZEL

Take time in October to savor the heavy, sweet scent of the blossom of this large shrub or small tree. Witch-hazel is not a true hazel (*Corylus*); it just has leaves that sort of look like those of hazelnut. The "witch" part comes from the English word "wych," used to describe pliant branches; the forked branches of witch-hazel were used for water-witching (dowsing). It is perhaps best known for the astringent extract from the bark and leaves.

Six species of *Hamamelis* are known, from North America to Asia. Hybrid witch-hazels in the cultivated landscape usually have pubescent twigs, a contribution from the spring flowering Chinese witch-hazel, *H. mollis*. (*See* photo p. 247.) The hybrids are grafted onto the native *H. virginiana*, which will usually sprout from below the graft union and provide a combination of the two; this occurrence is readily confirmed at flowering time. Although the hybrids bloom in early spring, their often larger fruits mature in October, close to the flowering time of the native species. Autumn colors on witch-hazels are sometimes more spectacular than the flowers.

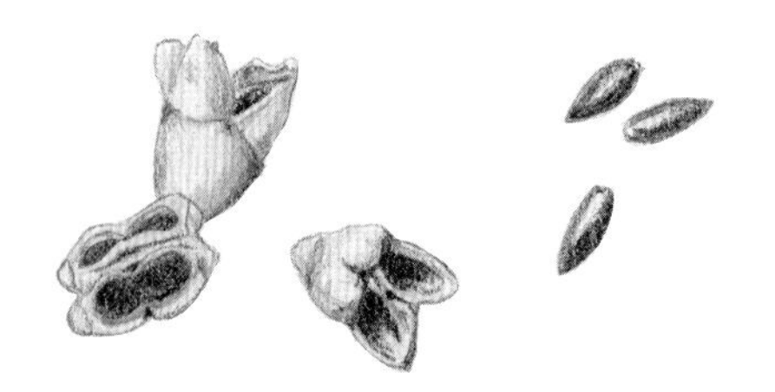

The woody capsules of the witch-hazel, Hamamelis virginiana, *eject the seeds up to 30 feet away.*

Hamamelis virginiana is a woodland and forest-edge species of dry to mesic well-drained soils. Natural dissemination takes place at flowering time in late September, the previous year's capsules having taken an entire year to mature. Natural germination likely takes place in the second spring after seed dispersal. It is commonly found in deciduous forest and the southern edge of mixed forest from Nova Scotia across the north shore of Lake Ontario to Goderich, Ontario, through southern Michigan and Minnesota and south from Georgia to Missouri.

The perfect flowers — with ribbon-like yellow petals — of witch-hazel are insect pollinated, but due to the late bloom time, fertilization is not completed until the following spring. The fruit enlarges during the summer and matures to a greenish yellow color before opening and turning brown. As the very hard, woody capsules open, they develop an intense pressure at the base that ultimately ejects the one or (usually) two smooth, elongated black seeds as far as 33 feet from the parent plant. The ejection force comes from a pinching at the base of the capsule as it is opening.

Good seed crops occur at two- to three-year intervals; sometimes there is no fruit in the interim years, despite the annual bloom. The fruit capsules must be picked at the mature (yellow) stage around late September, just as the shrub comes into bloom each fall. Place the capsules in a paper or cloth bag in a warm, dry location. These fruits provide a little entertainment over the next two weeks, as you will occasionally hear the loud snap of an ejected seed hitting the paper bag. A few partly opened fruits placed in your pocket will have ejected their seeds by the end of the day. Dry seeds can be stored cold for several years.

Germination can be attained in the first spring. First soak the seeds for 24 hours (and continue for that time even if they sink within the first few hours), followed by a 60-day warm, then a 120-day cold stratification. Check the seeds weekly near the end of the cold stratification period to look for evidence of the tough seed coats splitting in half lengthwise as radicle emergence takes place. Plant germinating seeds at ¼-inch spacing in a clay pot or shaded seedbed plot and cover them with soil rather than sand. Young seedlings are quite susceptible to damping-off, so water with care. If germinated in a greenhouse, the young seedlings should be set into a cold-frame nursery in late May to grow as a clump for the first two seasons.

Two-year seedlings are 6 to 12 inches tall and are easily transplanted. They produce a very dense fibrous root system and, as larger specimens, can be moved in the fall with a soil/root ball intact. This understory large shrub develops an open and irregular form, but can adapt to full sun conditions, where it remains dense and compact. This species is an essential component of deciduous woodland gardens but rabbits browse heavily on even older plants, and mice will girdle the stem. Unprotected plants do not recover if they have been fed on heavily for a few years.

Hydrangea arborescens, WILD HYDRANGEA

A delightful plant to come upon in the moist woodlands south of the Great Lakes, wild hydrangea is also an excellent accent in the shade garden. This shrub grows to 7 feet tall and flowers in midsummer. The flat-topped inflorescence of white flowers is accentuated with outer sterile flowers with enlarged, showy bracts. This species occurs from Massachusetts, across the lower Great Lakes basin to Kansas and south from Florida to Louisiana.

The small, dry seeds can be collected during the fall, stored dry and sown in a greenhouse in late winter. No seed treatment is required.

Hypericum, the St. JOHN'S-WORTS

Hypericum kalmianum
(Kalm's St. John's-wort)

Hypericum prolificum
(shrubby St. John's-wort)

Identification of the St. John's-worts is easiest when you look at the fruit capsules, which persist for over a year.

There are two woody shrub species of St. John's-wort found in the region. Seed dissemination is from fully opened, upright capsules maturing in October. Local scattering of seeds takes place over an extended period, through the entire winter and into spring. Long-distance dispersal northward is likely accomplished by migrating sparrows picking up the tiny seeds in March. Seeds germinate readily on open ground.

H. kalmianum, KALM'S ST. JOHN'S-WORT, is a charming low, bushy shrub, sometimes resembling a bonsai tree, 8 to 12 inches tall on moist to mesic open rocky shores. It is readily distinguished by the mostly five- (occasionally four-) chambered fruit capsule. This is primarily a Lake Huron and Georgian Bay endemic shrub with a remote population along the Ottawa River and scattered populations around Lake Erie into Ohio and through Michigan to Illinois.

H. prolificum, SHRUBBY ST. JOHN'S-WORT, is a small, dense shrub 3 to 5 feet tall. My only experience of this species is a single plant in a moist prairie remnant near Windsor, Ontario, but it is reported to grow in a range of habitats including swamp edges and cliffs. This species is identified by the mostly three (occasionally four) chambered, elongated fruit capsule. It is a more southern species, ranging from New York to Michigan (with only a few local sites in Ontario) and south from Georgia to Louisiana.

Both of the woody *Hypericum* species are in bloom over a considerable period through midsummer. Bees constantly visit the bright yellow flowers. The fruit is a thin, woody capsule about ¼ to ⅓ inch tall that contains dozens of elongated seeds about 1/16 inch long. Dry seeds are easily stored for many years and no pre-germination treatment is required. Seeds can be sown in early spring indoors and require light for germination. Scatter the seeds over the soil surface and cover the container with a Plexiglas sheet or plastic bag. Small seedlings are readily transplanted at the four-leaf stage, spaced about ¾ inch apart in trays and then planted into a nursery frame in May as 1½- to 2-inch plants.

The fibrous-rooted, woody species are easy to transplant and should be grown for two to three years before planting out. They will both occasionally naturalize once a few plants are established. Neither species is bothered by our common herbivores. They are lovely shrubs and deserve more use in sunny yards, including at the edge of pools and stream gardens.

Ilex verticillata, WINTERBERRY

The brilliant red fruit of winterberry are breathtaking, both close up and covering whole shrubs in the distance. Of the more than 300 species of holly throughout the tropical and temperate northern hemisphere, our lone, highly variable species is even that much more a gem.

Winterberry is a shade-intolerant, large shrub characteristically reaching around 7 feet tall and rarely 13 to 16 feet high and half as wide. It thrives in slightly acidic sedge and peat soils at the edges of cattail marshes and in bogs and sphagnum clearings in wet woods. It is a tight, many-stemmed shrub of deciduous and mixed forests from Newfoundland across southern Quebec and into Ontario, considerably north of Lake Huron, to Minnesota and south from northern Florida to Louisiana.

I have not been in holly swamps often enough to notice seed dispersal. The fruit is reportedly highly favored and dispersed by songbirds such as the bluebird, hermit thrush, brown thrasher and catbird. Based on the difficulty in obtaining first-spring germination in the nursery, natural seed germination is likely to take place in the second spring.

Male and female flowers are produced on separate plants in late May to early June. (*See* photo p. 247.) The small white flowers are usually solitary in the leaf axils of the current season's growth and are pollinated by a great variety of small bees, flies and other insects. The fruits are a bit variable in color, from bright orange to scarlet to dark red. They can persist beyond December, when deep freezing will ultimately cause the fruit to shrivel and turn brown. The round berries each contain

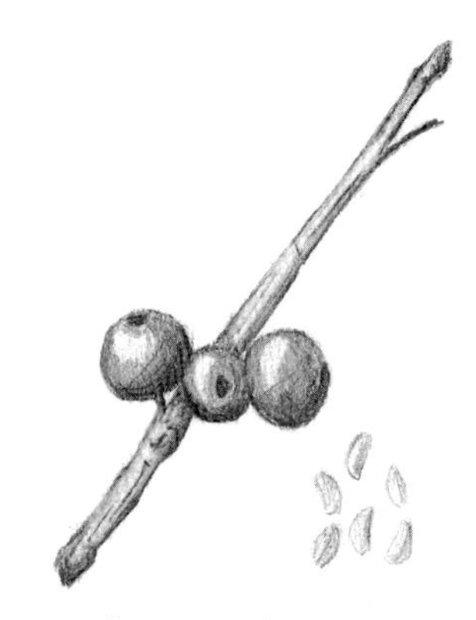

Ilex verticillata *(winterberry)*

six to seven elongated seeds that are easily extracted from the fruit with the grit bag method. They have a very tough seed coat and can take a firm grinding in the grit, which will also scratch the seed coat and contribute to their ability to germinate. Dried seeds can be cold-stored for many years.

Fresh seed (and stored seed that has been soaked for 24 hours) will germinate in the first spring with a 60-day warm and 60-day cold stratification. Spread the seed and stratifying mixture so that seeds are spaced approximately ⅛ to ¼ inch apart. Germination is usually even. Damping-off fungus is a risk because the soil needs to be kept moist. Always allow the soil surface to dry between each heavy, thorough watering.

One-month-old seedlings, at the four-leaf stage, can be transplanted into a sedge peat or peat moss soil mix or be thinned out to ¼-inch spacing and left in the seedling container. Either way, they can reach 1 to 2¾ inches in the first year. It is best to set the whole root mass into a peat-enhanced soil in a damp location, but be mindful of frost heaving in early winter, which is typically more extreme on organic soils. Transplant the young plants at two- to three-year intervals. Four-year-old plants can be 12 to 16 inches tall and ready for establishment in marshy locations that have a good humus soil. This species is best set into the landscape in the fall when moist sites are more easily accessed. The dense fibrous root system makes them very easy to transplant. Herbivores leave *Ilex* unscathed.

The horticultural industry has produced many selections, among which are the requisite males such as 'Holden', but more often with macho names ('Jim Dandy', 'Southern Gentleman', etc.). These "guys" are obviously required for the dozen or so "gals," which usually have color in their name, such as 'Scarlet O'Hara' and 'Red Sprite'. Naturalization planting should obtain plants from the local seed, and even gardeners can try to obtain seedling plants. Several plants, of course, are required to ensure that there are at least a couple of male and female plants, since one cannot know until they flower.

A less well-known native holly is *Nemopanthus mucronata*, mountain holly. It occurs in bogs and tamarack swamps from Newfoundland to Minnesota and south to Illinois and Virginia. (*See* photo p. 250.)

Juglans, the WALNUTS

Two natural walnut species may be confused with an Asian species, and a butternut-heartnut hybrid complex makes it even more difficult. The illustration shows the dry seed coat characteristics after removal of the fleshy green husk.

Walnut Lore

Walnuts are known for their ability to reduce the options in vegetable and ornamental gardens. This is due to the effect of shade and the release of a chemical compound called juglone, released from the roots, hulls and, to a lesser extent, the leaves. It works as a growth inhibitor and will prevent many vegetable seeds and transplants from growing near a walnut. Juglone persists for a few years after walnuts are killed. The keen observer, however, will realize there are other options than removing a garden walnut. They will notice that many indigenous plants grow in association with walnut, including red and black raspberries, wild roses, alternate-leaved dogwood, gray dogwood, nannyberry bittersweet, Virginia creeper and red cedar as well as a host of woodland herbs. Thus, creating a native woodland landscape under a black walnut is a reasonable gardening objective.

Highly valued but now rare are the logs that come out of natural old-growth woodlands. Eager to leave a profitable legacy for their grandchildren, farmers would plant rows of walnut. No one knew back then that plantation trees tend to grow too fast on good soils to obtain a high-quality veneer log, or so badly on poor soils that they produce nothing better than firewood. Perhaps worse yet, the millions of compromised walnuts on poor soils produce trees with a tendency to become heavily cankered. Such a large presence of this otherwise weak canker could perhaps create the conditions for it to mutate into a race strong enough to affect healthy trees.

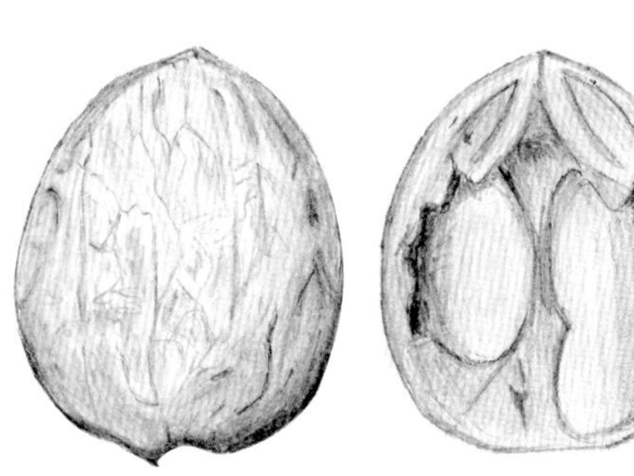

Juglans regia *(English walnut)*

Exotic Alert

An Asian species called heartnut, *J. ailanthifolia* var. *cordiformis,* has naturalized in a number of locations and somewhat resembles butternut with the sticky glandular stems and fruits. The fruits are in longer chains and the nuts are quite smooth and almost heart shaped, as the common name suggests. Heartnut also hybridizes readily with butternut, and a cross known as buartnut is found throughout the southern part of the butternut range, presenting a potential threat to the genetic integrity of butternut. In the lower Great Lakes area, one might also come across English walnut, *J. regia,* a tree with smooth stems and leaves and the smooth, thin-shelled nut that is so familiar in markets.

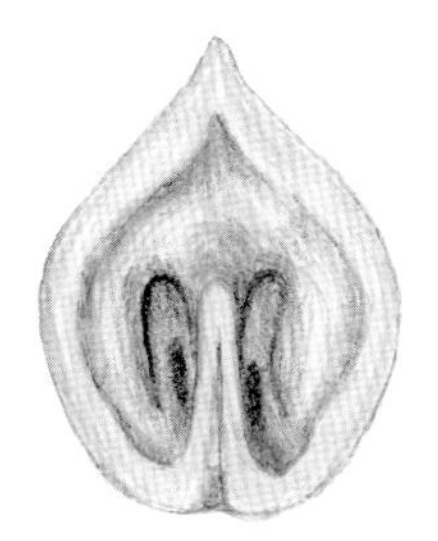
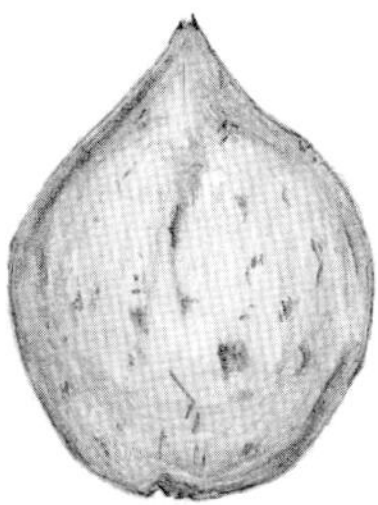

Juglans ailanthifolia *var.* cordiformis *(heartnut)*

Of the 20 species found across the northern hemisphere, we are well endowed with two magnificent species. The fruits of all walnuts are well known for their ability to produce a brownish black stain that can be used as a dye. Nuts are produced every year but usually a heavy crop alternates with a light crop. Natural dissemination is by squirrels, which usually remove the husks and bury the nuts or cache them in hollow trunks. Buried nuts germinate the following spring. After germination, the seed kernel is still rich in food and supports the young seedling for several weeks. During this time it is strongly aromatic and easily found by squirrels, which will destroy the seedling to get at the kernel. Black walnut is harvested for commercial use and produces a fine salad oil.

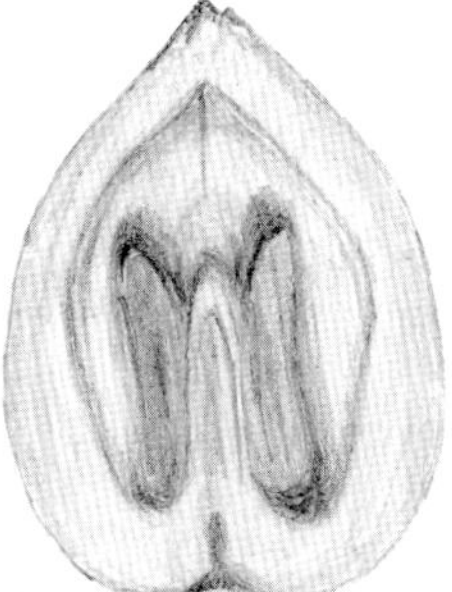

Buartnut, a butternut x heartnut hybrid

Juglans nigra, BLACK WALNUT, is a majestic, long-lived tree. The presence of black walnut was used as a guide by settlers to locate rich, deep, well-drained, moist soils for homesteading. It is native only on moist soils in the deciduous forest zone, where it can be expected to reach its greatest potential. Black walnut has naturalized from historical planting well beyond its natural range and habitat. Its natural range is from Vermont through New York to southern Ontario and Michigan, west to South Dakota and south from northern Florida to eastern Texas.

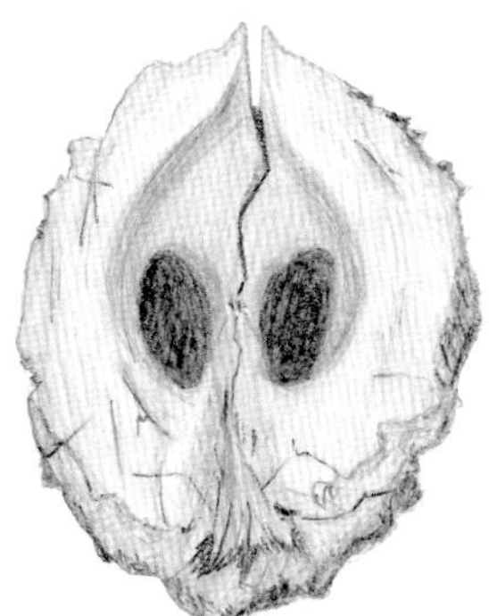

Juglans nigra *(black walnut)*

Juglans cinerea, BUTTERNUT, can become a large tree on the rich, very well-drained, mesic soils of gently sloping woodlands or on relatively shallow soils over limestone. Because butternut is quite fussy about moisture, it is found in relative abundance in widely scattered groves. It has a range that extends well into the mixed forest from New Brunswick through southern parts of Quebec, Ontario and Michigan to Minnesota and south from South Carolina to Arkansas. The kernel of this species is delicious and worth the effort to crack out. In 2003, the Committee on the Status of Endangered Wildlife in Canada (COSEWIC) gave butternut endangered status in Canada, due to the relative scarcity of canker-tolerant survivors.

Walnuts flower in spring, with male flowers in green catkins originating adjacent to the lateral buds and visible in

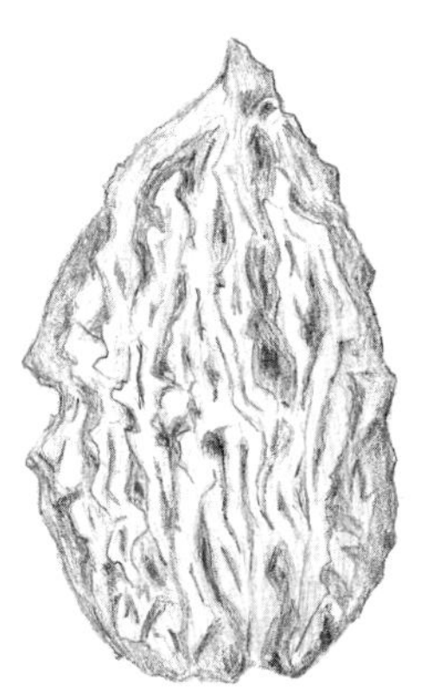
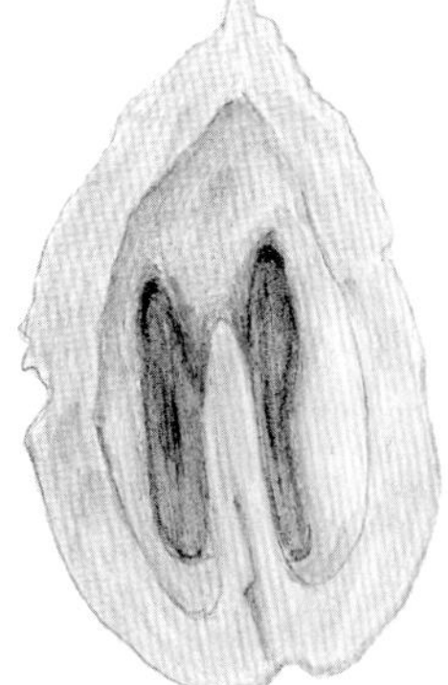

Juglans cinerea *(butternut)*

Butternut Canker

Butternut has been significantly reduced throughout its range by butternut canker, which arrived in North America about 50 years ago. The fungus disease likely originated in Asia and moved here from numerous imports of heartnut (*J. ailanthifolia*), which seems to be relatively tolerant of the disease. The disease is transported by fungus-feeding beetles and perhaps on the feet of migratory birds — both trunk and canopy feeders — as all parts of the butternut can support the infection. Active cankers are characterized by a black, oily ooze or stain emanating from vertical openings in the bark. Individual healthy older trees occasionally are seen in a stand of dying trees. They seem to have the genetic ability to tolerate the disease, and these trees need to be preserved at least long enough to obtain seeds and possibly even scion wood for grafting.

winter as a miniature cone-like structure, quite distinct from the foliage bud. The female flowers are within the unfolding lateral shoots and have a miniature nut at the base of two reddish-colored pistils. Pollination is by wind when the green catkins elongate, open up and shed the yellow pollen in late May.

Walnut trees defoliate early, often starting in late August or September. The nuts hang on bare branches for a short period, longer in heavy crop years. The fruit of black walnut is smooth and quite globose, while butternut is sticky, hairy and tapered at either end. They can be shaken from the branches using sticks or a rope or they can be gathered when fully ripe and naturally fall from the tree. Squirrels can remove an entire crop of walnuts in short order before the leaves are down, especially in lean crop years.

I have always cleaned the nuts before sowing them because squirrels do that. The husk contains a germination inhibitor that may influence the ability of the nut to germinate. Cleaning the seed involves leaving the whole fruits in a bucket until the husks turn black and then setting them on a gravel area and washing the husks away as I walk over them. Cleaned seeds are best kept moist and refrigerated; dried seeds will lose viability over time. Seeds can be sown with the nuts side by side in bands that are about 8 inches apart and covered with ¾ to 2 inches of soil. Mulch the seedbed with woodchips and enclose it in a wire cage, or lose all the seeds to squirrels and mice.

Cleaned seeds can be kept in refrigeration until ready to plant in the fall or be stratified for 120 days, and then planted in the spring. *Juglans* seedlings are 16 to 28 inches tall in two years and have a very thick deep root that must be dug out with some care and planted soon after. It is best to plant the seeds right where you want your trees to grow. If these individual planting sites are located in the range of rodents, protect them with a chicken wire enclosure, too.

Given the antler-like twigs on young trees, buck deer will occasionally pick an individual for a sparring partner; rodents leave walnut trees alone. Fall webworm will periodically establish in walnut but is of no consequence to the health of the tree so late in the season. If you need to prune walnuts, wait until mid-June to avoid sap loss from spring cuts

Juniperus, the JUNIPERS

The junipers constitute some 50 shrubs and trees in the northern hemisphere, generally well adapted to very dry sites in full sun. They sometimes reach magnificent form and great age in places where few other woody plants thrive. Red cedar, *J. virginiana*, has a darker red and more fragrant wood than white cedar (*see Thuja*), but neither is actually a true cedar of the genus *Cedrus*. Historically, "cedar" was a common term used for "evergreen" or pole-sized timber. The cedar waxwing is named for the red cedar (*J. virginiana*), and large wintering flocks of this colorful bird and its western relative, the Bohemian waxwing, are commonly seen consuming the berry-like cones of this tree.

Natural dispersal is carried out by many birds, including the cedar waxwing, evening grosbeak, cardinal, robin, bluebird and mockingbird. They consume the fruit during winter and pass the hard seed intact. Junipers are occasionally long-lived early-succession plants that provide a nucleus around which diverse communities of plants establish. The foliage catches the blowing seeds of plants like aster and milkweed; it provides cover for squirrels to plant oaks, and for birds to drop the seeds of other species such as dogwood or wild grape.

Juniperus chinensis *(Chinese juniper)*

Exotic Alert

Chinese juniper, *J. chinensis*, is occasionally found naturalized due to the prolific horticultural use of the species. It is a tall conical tree with fruit twice the size of *J. virginiana*. The two species hybridize to produce a broad, spreading shrub with slightly upward-angled branches seen occasionally along roadside embankments and in old fields. The hybrid is part of the large group of horticultural selections called Pfitzer junipers, the fruit of which is equally desired by birds.

Juniperus communis *(common juniper)*

Juniperus communis, COMMON JUNIPER. One has to admire the tenacity of a species that grows in such diverse and unpromising places as this juniper. It is a flat-topped, wide-spreading shrub to a maximum of 5 feet high. Common juniper is adapted to partly shaded or sunny, hot, dry environments in open woodlands, old pastures and rocky fields; on exposed gravelly slopes, shores, quartzite and granite balds and limestone pavement; and even protruding from cliff faces. It can be found often in pure stands, widely scattered right across North America through the boreal forest from Labrador to Alaska and south through the Great Lakes, and isolated populations down the Appalachian and Rocky mountains south into the Carolinas, Arizona and New Mexico. The species is circumpolar, continuing into northern Eurasia, where it is more typically columnar in habit.

Juniperus horizontalis, CREEPING JUNIPER, is a ground-hugging plant. Extensive mats are found on thin soils over bedrock, rocky soils in open woods, shorelines and dry sandy clearings. Rooting is commonly observed on the long branches radiating out from often twisted and gnarled branches. This is a transcontinental species ranging across the boreal zone from Newfoundland to northern Yukon, along the shores of the northern Great Lakes and North Atlantic, and scattered inland from Vermont to Montana.

Juniperus virginiana *(Eastern red cedar)*

The junipers can generally be distinguished by fruit size and habit.

Juniperus horizontalis *(creeping juniper)*

Juniperus virginiana, EASTERN RED CEDAR. The trunks of older trees are so interesting that it is curious that the species is not planted more often as a tree in the open rather than crammed up against a building. Red cedar reaches 30 to 50 feet (and occasionally higher) with a relatively broad crown and wide, spreading branches with dark green foliage. At the northern limits of distribution, the narrow upright form with bluish green foliage (var. *crebra*) is more common. The species is found in mixed and nearly pure stands in predominantly sandy or gravelly sites, on limestone pavement and even on clay soils. It takes over where white cedar fails on soils that are very dry in the summer, even if fairly saturated in the early spring. Red cedar occurs from southern Maine through the southern parts of Ontario and Michigan to South Dakota, and south from northern Florida to eastern Texas.

Cones of the female junipers are receptive in May and are wind pollinated. Male and female cones are produced on separate plants and this indicates the need for numerous individuals to be established on a site. The female cones are more likely to be called a berry when they are ripe, as the scales are fleshy and enclose one to five seeds, depending on the species.

The berry-like cones mature in October with a shift in color from green to whitish green to blue or dark blue-purple in *J. communis*, and are persistent well through the winter. The seeds are very hard, small and light brown in color. Ripe "berries" can be picked from the branches in late fall or early winter and dried for long-term storage or soaked in preparation for cleaning. Seeds are readily extracted by the grit-bag technique. A little soap can be used after the pulp is well ground up to remove the oily substance that can cause many good seeds to float. Cleaned seeds are best placed into warm followed by cold stratification as indicated for each species in Appendix III.

Seeds can be planted at a very close spacing of about ⅛ inch apart, either in a clay pot or in a seedbed patch. One-year seedlings are about ¾ to 1 inch tall and decked with juvenile leaves that are quite different from mature leaves for the first two years. Seedlings can grow close together for the second growing season, when they reach 3 to 4 inches tall. Lift them in groups to tease the fibrous roots apart (the fragrance associated with juniper is heavenly) and plant them into a nursery bed at an inch or more spacing to grow for two more years, to produce 8- to 12-inch plants that are large enough to establish in the landscape. Deer and rabbits will browse heavily on red cedar. Meadow vole can take up residence under horizontal and common juniper and may girdle some or all of the branches.

I have never grown *J. horizontalis* from seeds, as I have only needed one or two specimens, which are easily obtained as cuttings from the layered branches. Several dense, carpeting variants of creeping juniper, from bright blue to emerald green, are cloned for horticultural applications. Various compact, narrow selections of *J. virginiana* are propagated in the nursery trade as well; the most famous is the absurdly narrow cultivar 'Skyrocket'.

Kalmia, the LAURELS and HEATH RELATIVES

The heath family is a diverse family of about 70 genera, in which some 1,500 species are found in arctic and temperate regions of the northern hemisphere and cold mountains in the tropics, invariably on acidic soils. Their flowers are an important nectar source for insects. The mature fruit capsules begin to open in August and disperse seed well into the fall. The very small, dust-like seeds are spread by wind over large distances, and the seeds also cling to the wet fur of mammals. The dry-fruited genera are described together because their seeds are handled exactly the same way. The fleshy-fruited relatives, blueberry, bearberry and huckleberry, are covered under *Vaccinium*.

Kalmia angustifolia, SHEEP LAUREL, is a dense, colony-forming shrub that rarely attains 3 feet in height. Sheep laurel grows in soils that are moist and acidic. It is found in full sun or partial shade in the boreal and mixed forest from Newfoundland to northern Ontario and Michigan, south to Virginia, at the edges of coniferous woodlands, bogs and small lakes, including the mossy "gardens" on partly submerged logs. The eye-catching pink flowers are insect pollinated; the fruit capsules mature in late August and contain hundreds of dust-like seeds. The capsules persist on the plant for many years.

Kalmia latifolia, MOUNTAIN LAUREL, grows as a mounded evergreen shrub usually under 10 feet in height. It is surely one of our most beautiful native shrubs. The very attractive white to pink bell-shaped blooms appear in terminal clusters during June. (*See* photo p. 248.) These are followed by dry capsules of minute seeds in September. The species requires a strongly to moderately acid moist soil, whether coarse, fine or entirely organic. It is shade tolerant. Mountain kalmia's range enters the Great Lakes region only in New York State, western Pennsylvania and eastern Ohio, although it has an extensive range in the eastern states from Maine to Alabama.

Kalmia polifolia, BOG LAUREL, is a delicate small jewel with bluish leaves and pale pink flowers. It is found in mixed forests and barren lands from Newfoundland to Yukon and

Members of the heath family have very tiny seeds.

New Jersey to California. It grows in relict bogs in the eastern deciduous forest. This is a plant for the bog garden.

All three laurels are relatively easy to grow in moist, acidic soils with partial shade at midday.

Chamaedaphne calyculata, LEATHERLEAF. This intricately textured, evergreen-leaved shrub spreads by layering branches to produce 3-feet high thickets at the edge of acidic lakes and bogs. Small white flowers are bee pollinated from May to early June and the old fruit capsules persist for many years after maturing in October. It is a circumboreal species common in the mixed forest, sparse in boreal regions and only in relict bogs in the north portion of the eastern deciduous forest. It is strictly a bog garden plant in cultivation.

Epigaea repens, TRAILING ARBUTUS or MAYFLOWER, is a wiry-stemmed, evergreen, ground-hugging shrub of moist to dry acidic soils in mixed or deciduous woodlands. It is found from Newfoundland to Saskatchewan and into the southern U.S. The plant is deeply rooted in rock crevices or sandy soil at its origin and the spreading branches root shallowly into the humus mat. Insects pollinate the fragrant flowers in May, and the fruit, with hundreds of very tiny seeds, mature in early September. They are relatively difficult to raise from seed, and attempts at cultivation even from rooted branches rarely succeed because the plant is such an ecosystem specialist, best left in the wild.

Gaultheria procumbens, CHECKERBERRY WINTERGREEN, forms low, herb-like mats of glossy aromatic leaves. It seldom rises more than 6 inches from ground level. The scarlet berries ripen in late July but persist over winter. Birds and mammals, and even bears, eat them. Plant this species in partially shaded, acid sites ranging from highly organic, even boggy, soils to light sands. It is native from the maritime provinces to the Great Lakes and south along the Appalachian Mountains to Tennessee.

Ledum groenlandicum, LABRADOR TEA, is a sun-loving, evergreen-leaved shrub to about 3 feet in height. It is common in bogs and on shorelines of small lakes and streams across the entire boreal and mixed forest and in bogs at the northern edge of the deciduous forest. The showy white flowers are bee and moth pollinated in June. The fruit ripens in October, opening in a rather curious way, with the base of the capsule lifting away and being suspended by the pistil. This is a lovely bog and peat-garden plant. Seedlings are relatively easy to grow and are left unscathed by herbivores.

Rhododendron canadense, RHODORA. Unlike most of the 850 other rhododendron species, this is a deciduous shrub forming small, dense, 3-feet high colonies from underground stems. The stunning mauve-pink flowers are bee pollinated in

May as the leaves are unfolding. Fruit capsules are held upright and disperse seeds into the drifting snow through early winter. It is an eastern species, growing in moist to dry acidic soils of bogs, pine barrens and sandy barrens in the mixed forest, just reaching the Great Lakes region at one location in eastern Ontario and in central New York. This lovely shrub is worth cultivating in an acidic garden but be aware that rabbits will devour it.

Rhododendron maximum, ROSEBAY RHODODENDRON, has the largest leaves of the native rhododendrons, commonly 9 inches long and evergreen. The flowers are white or blush and spotted within. (*See* photo p. 248.) They grow in 4-inch wide trusses composed of up to 30 flowers. The fruit is an elongated, sticky capsule that splits in September to release tiny flattened seeds. In the north of its range in New York, Pennsylvania and southern Ohio, it grows sparingly in swamps, seldom reaching more than about 10 feet tall, but in the south of its range it can be decidedly tree-like.

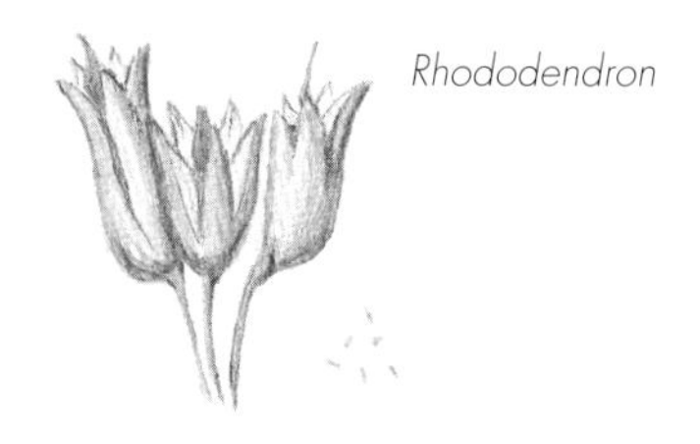
Rhododendron

Rhododendron nudiflorum, PINXTERBLOOM AZALEA, and *R. roseum*, ROSESHELL AZALEA. Both these species are globular shrubs, typically to about 10 feet in height. The foliage is deciduous. In pinxterbloom azalea, the pink, funnel-shaped flowers appear with the new leaves in late April to early May. The fruit is a dry capsule a little under ¾ inch long; when ripe in August, it splits and peels back like a banana. Roseshell azalea is similar but with larger flowers and a later blooming period. Pinxterbloom azalea grows in shaded, acid swamps as a rule but is sometimes found in dry woodlands. It ranges from Vermont to the Florida Panhandle but is not uncommon south of Lake Ontario to eastern Lake Erie, south of Buffalo. Roseshell azalea prefers a drier, less acid site. It has a more broken range centered in New York State and Pennsylvania. There are a few stations for it south of Lake Ontario, but it ranges farther west than pinxterbloom azalea, to south of Lake Erie in Ohio. The two species hybridize in their joint range. A third species with similar attributes, although later-blooming still, is *R. viscosum*, swamp azalea. It can be found in New York State south of the eastern end of Lake Ontario.

Fruit capsules of all the heath species can be picked when fully ripe and starting to split (generally October). They should be dried for a week and then kept refrigerated in a small envelope until ready to sow. Capsules can be crushed with your fingers and the seeds separated with a sieve. In spring, the seeds are sprinkled onto the surface of premoistened pure peat moss (it may be mixed with acidic sand) and watered with a mister to avoid washing out the seeds. Maintain high humidity under Plexiglas or with a daily misting in a sweatbox. They require no pregermination treatment and sprout in about two to three weeks.

Heath seedlings can be pricked out and planted a finger-width apart in a pot of peat mix that has about 50 percent acidic sand and 5 percent pasteurized manure added. Once they are established, the pot can be plunged into a peat section of your cold frame. When they need to be transplanted, their extremely fine roots are teased or even cut apart. Plant them at wider spacing in a moist, partly shaded acid sand and peat nursery bed for another year or two. These species are habitat specialists and can be established in an acidic bog edge or moist, acidic humus–sand habitat. Plants in alkaline soils will become chlorotic and fail. Transplants require daily misting and shade to recover and should be watered only with rainwater, especially during droughts in the establishment years. Chipmunks and squirrels may use your light, peaty soil to plant nuts in and completely disrupt small seedlings, so cage these plants until the root mat is well established.

Larix laricina, TAMARACK

The emerging, bright green needle leaves of tamarack in late May, and the late autumn glow as the leaves turn golden orange before falling, are truly stunning. Tamarack is a swamp, ditch and bog dweller, thriving on hummocks in wet sites — the base of the trunk always above the high-water line — and is usually abundant where found. It is relatively short-lived if planted on dry sites, where even established trees may die during droughts. Wind disperses the seed and germination takes place in the following spring. The species is found in the moist and relatively cool conditions of boreal and mixed forests into the northern parts of the deciduous forest from Labrador to Alaska, (where it overlaps with the range of western larch, *L. occidentalis*) south to New England and the lower Great Lakes to central Alberta. About eight other species grow in China, Japan and Russia and one in Europe — many of these are planted in the region.

Exotic Alert

European larch, *Larix decidua*, with significantly larger cones, has been established in plantations for decades. Although not as commonly planted, the Japanese larch, *L. kaempferi*, with an equally large cone, is also found naturalizing from plantations. It is likely that both of these species have been inadvertently planted into some natural areas due to substitution and the fact that these alien species are not easily distinguished from the native species. Fall color onset is perhaps the best identification clue for immature plants, since the exotic species will remain green while the tamarack is changing to yellow.

Larix laricina *(tamarack)*

Bright crimson red female cones stand upright among the bright green new growth of late May, while the male cones are yellowish brown and hang downward. Unlike many conifers, the female cones are distributed throughout the tree. Pollen is dispersed by wind.

Seed crops occur two to six years apart, on trees as young as eight years old. Cones mature in an upright position in late summer and open in September. Two seeds are held at the base of each scale. Good-quality seeds are brown rather than tan in color and are the first to fall from opening cones. Seeds are disseminated by wind and by finches as they tear cones apart in late summer and into September. In August of 1995, I watched a flock of white-winged crossbills in central Ontario, already tearing cones apart at a fantastic rate. The cone crop provides sustenance to crossbills and pine siskins for months.

Tamarack has a very small cone, compared to all other species.

Cones remain attached for many years and gradually weather to a silver-gray color. The current season's cones are on the previous season's growth and are distinguished by their bright tan color. Viable seeds are not found in the old cones. Cones are easily picked from low branches and, when dry enough that the cone scales open, they can be shaken in a cookie tin or screen box to dislodge the seeds. There are 16 to 20 winged seeds in each cone; they should be placed in an envelope and can be kept in refrigeration for several years.

Stored seeds must be soaked for 24 hours. The filled seeds should sink and can be separated from the empty floating seeds. Sixty days' cold stratification is required. The seeds can be scattered so as to be barely touching and covered with a thin layer of sand or soil. Maintain a moist soil (perhaps using a cover) until the germinating seedlings begin to push the soil open, after which they should be watered only when the surface soil dries — they are very susceptible to damping-off until they are four to five weeks old. Seedlings usually remain an inch or so tall but can reach 4 inches in the first year, depending on seedling spacing. The seedlings are best planted as a group, undisturbed, into a cold frame in late May. You can expect 8- to 16-inch plants after two years of growth and, at this size, they should be separated in the spring. Tamarack has fibrous roots and is relatively easy to transplant. Specimen plants can be moved in the fall, when moist sites are easier to access. Sturdy seedlings should be spring planted into their permanent location and smaller ones root-pruned and planted into a nursery bed for another year or two. I have observed rabbits browsing on young shoots. Tamarack grows best when planted in groves on moist sites and marsh edges.

Lindera benzoin, SPICEBUSH

Spicebush is so named for the spicy leaves and fruit. This shrub is in the same family as sassafras, and both (along with the tulip tree) support the spicebush swallowtail larvae. There

The brilliant and fragrant red fruit is unmistakable.

are about 100 species named in Asia (botanists are known to be serious splitters there) and only the one species in North America. Like so many other unusual plants, it is always a joy to chance upon this multistemmed, mid-sized shrub that often escapes notice when not in flower or fruit.

Lindera grows 3 to 10 feet, and occasionally 16 feet high, in both open and closed canopies of rich, wet woodlands and swamps. I have found several individuals growing on slopes well above swamps, and seedlings from these plants could be evaluated for tolerance to drier conditions. Spicebush spreads only by seeds to produce the thickets of numerous individual shrubs encountered commonly in the deciduous forest and rarely, the mixed-forest region. This fascinating shrub is found from Maine through southern Ontario and Michigan and into the southern states. *Lindera* has smooth twigs and leaves in the north and is generally pubescent in the south (Florida to Texas), with a range of intermediates between.

The bright color of the energy-rich fruit is an easy target for migrating birds. Thrushes and catbirds are among several bird species reported to feed on *Lindera*, but this I have not observed. Because the fruit dislodges so easily once fully ripe, many seeds fall directly under the shrub and are further dispersed by mammals. I once found over 100 seedlings in an abandoned rodent cache about 3 inches in diameter. The northward distribution in Ontario, along the eastern boundary of Lake Huron, is likely a result of the climate-moderating influence of the lake; however, the agent of northward dispersal remains a mystery to me.

Numerous round flower buds are preformed in late summer. During full bloom in early May, the flowers produce a yellow glow in spicebush swamps, just as the bright green leaves begin to expand. The tiny flowers are perfect and pollination is carried out by several insect species. A single dark brown, oval seed with a relatively soft seed coat is enclosed by the fleshy red fruit of the same shape, often in thick clusters along the two-year twigs. Abundant seed crops seem to come in alternate years. Natural seed germination is likely delayed until the second spring.

The fruits mature in late September while the leaves are still green; they remain attached to the stems until after the leaves have turned intense yellow and fallen in mid October — this is the time to gather them. Overripe fruits are easily cleaned using the grit bag, but avoid excessive pressure; keep in mind that *Lindera* has a relatively soft seed coat. I have not attempted to store this seed.

Immediately place cleaned seed into warm stratification for 40 days (a few seeds may germinate at this stage), followed by 150 days of cold stratification. Radicle emergence takes place in the cold and should be watched for to signal the end of the cold treatment. Germinated and ungerminated seeds can be planted at ⅓- to ¾-inch spacing and germination can proceed in warm conditions. Seedlings grown at this spacing for two years produce 8- to 12-inch plants. They are easily transplanted in the spring and have dense fibrous roots that are delightfully fragrant. Partial shade and moist soil are important for this species. Rabbits will browse spicebush seedlings to the ground and the seeds definitely need protection from mice.

Liquidambar styraciflua, SWEET GUM

Sweetgum is a deciduous forest tree of moist slopes and well-drained woods. It is found from Connecticut to southern Ohio and Oklahoma and into the southern states, with scattered populations in the mountains of Mexico and Central America. There are five other species in Asia. Seeds from the unusual fruit are consumed by canopy-feeding finches. Dislodged seeds are dispersed by wind and consumed on the forest floor by a number of ground-feeding birds — reportedly, the towhee and white-throated sparrow, both of which scratch like chickens and likely plant many seeds in the process.

Flowers are unisexual, with female flowers in solitary round heads in the leaf axils, and male flowers in dense terminal clusters on the same tree. This is a wind-pollinated species with fruits maturing in October. Greenish yellow fruits can be picked and will dry in 7 to 10 days to release hundreds of seeds. Open fruits found on the ground usually still hold a few seeds. The seeds require no pretreatment and should be stored dry until sown in the spring.

Seeds can be sown densely in pots and flats but they are susceptible to damping-off, and should be transplanted at the three-leaf stage. Seedling growth is rapid — they need the protection afforded by a cold frame during the first winter. The fibrous-rooted seedlings should be dug and pulled apart only in the spring to set the one-year-old plants at about 4- to 6 inch spacing in a nursery bed. Herbivores don't seem to care for this species but I have only limited experience with sweet gum. Only seed from northern populations will produce seedlings hardy enough to plant north of the natural range.

Liquidambar *(sweet gum)*

Sweetgum is a handsome species native to the southern edge of our region. Its stunning fall color and tolerance of urban pollution have created significant horticultural interest in this plant in the northern states and southern Ontario.

Evolution and the Tulip Tree

Evolutionary botanists consider the magnolia family, which includes the tulip tree, to be representative of the earliest true flowering plants — those making the transition from a naked ovary in a conifer cone to a perfect flower (male and female parts in one flower) with an enclosed ovary. The "cone" of the tulip tree, with the enclosed embryo at the base of each winged seed, is very similar in design to the cone scale that protects the external seed of a spruce or fir.

Liriodendron tulipifera *(tulip tree)*

The smaller seeds in the upper end of the cone are always empty.

Liriodendron tulipifera, TULIP TREE

Our tulip tree has a single relative in central China (*L. chinensis*) that has a more deeply lobed leaf and green-only flowers. Tulip tree is our tallest tree, producing majestic broad crowns, typically held aloft on straight trunks. The crowns are easily shattered in ice storms but recover rapidly, and old trees show the periodicity of ice storms. It is usually abundant where found and thrives in rich, moist to mesic but well-drained deciduous forests from Vermont through southern Ontario and Michigan to Missouri and the southern states.

Seed dispersal from annual seed crops takes place during high winds from late October through December as the upright, cone-shaped fruits open and shatter. Purple finches and cardinals will pick seeds from the shattering fruits. The wing-like seed can be carried considerable distances but much of it falls nearby. White-footed mice and squirrels are significant seed consumers. Natural germination takes place in the second spring in open, disturbed sites such as ditches, trail edges and clearings caused by tree mortality.

The tulip-like, orange-green flowers open in June and are pollinated primarily by ants and wasps. About 150 ovaries are present in each flower but most remain unfertilized, resulting in mature cones holding relatively few viable seeds in some years. There is some speculation that fruits produced at the top of the tree have significantly higher percentages of filled seeds, but I have found high numbers of filled seeds on low branches where trees are in close proximity. Seed collectors can evaluate the number of filled seeds by slicing through one "cone" (with pruning shears) at ¼-inch intervals and counting the bright white embryos that are exposed in each slice.

A single embryo, if developed, is inside the angular base of the ribbed, scale-like wing and no attempt should be made to separate the embryo from the seed. Whole cones should be collected at the mature stage, if possible. This can be achieved by keeping a close watch on squirrel activity. Squirrels will drop many whole, greenish yellow fruits to the ground. Pick up a few cones and keep them in a paper bag. As the fruits dry, the wings will begin to peel back. When the whole fruit will easily shatter when twisted, the seed is ready to be planted in a protected seed frame and watered. Immediate attention to the freshly separated seeds will increase the possibility of germination in the first spring. Sow the seeds thickly in the fall (seeds touching and overlapping, as relatively few will germinate) with ¼ inch of soil covering them (two times the seed thickness) and a 2- to 4-inch layer of leaves. The leaf mulch should be removed in early May.

Alternatively, try stratifying some seed. Soak the winged seeds for 24 hours. The seeds float, so I place them in a container with a sieve holding them under water. Place the seeds into warm stratification for 14 days followed by 150 days' cold stratification. With either direct fall sowing or stratified seeds, germination can be hit-or-miss. Still, some of the seeds usually germinate in the first spring, producing 1- to 3-inch seedlings, which grow rapidly in the second year to about 16 inches. In the second spring, you may also see a flush of germination from seeds that stayed dormant, so it is important to leave the seedbed undisturbed and covered by wire screen until after the second spring.

Dig all of the plants in the spring when vigorous two-year saplings can be separated out and planted into their permanent location. The one-year seedlings should be planted at 2- to 4-inch spacing to grow on for another year or two. They have thick, fibrous and pleasantly fragrant roots and are easily transplanted, but only in spring. They prefer moist conditions and will fail outright in dry sites. Seedlings can be browsed where deer numbers are high and I have seen many beautiful, 5-inch diameter trees completely girdled by meadow voles in peak years — even with the use of rodent repellent! Tulip tree is best planted in groves to take advantage of natural selection for trees that are most tolerant of wind and ice storms.

Lonicera, the HONEYSUCKLES

The honeysuckles comprise a huge genus of some 180 species, mostly Asian. A few of the exotic species have been widely used in gardens and wildlife planting and, shamefully, are better known than and still planted to the exclusion of our native species. Cedar waxwings and the thrushes, catbirds and thrashers quickly consume the fruits of native and exotic honeysuckles. Each berry contains two or three very hard-coated seeds, which pass intact but scarified through the digestive tract. Natural germination occurs on disturbed sites, likely in the first spring.

Exotic Alert

Of the seven exotic species that can be found in the Great Lakes region, three were historically planted for wildlife food and cover. Because the birds so relish the fruit, they have become significant invasive plants in many regions. The ornamental climbing honeysuckles with large terminal flower clusters occasionally show up as escapees from cultivation.

Japanese honeysuckle, *L. japonica*, has black fruits on a scrambling, almost evergreen-leaved groundcover that is capable of covering extensive areas. (*See* photo p. 248.)

Amur honeysuckle, *L. maackii*, and Tartarian honeysuckle, *L. tatarica*, produce white to pink or reddish flowers and stemless clusters of paired red fruits. (*See* photo p. 248.) These sprawling shrubs reach 7 to 13 feet in height and this, as well as their sheer numbers, makes them a challenge to remove from even small areas of the landscape.

Lonicera canadensis, FLY HONEYSUCKLE, and *L. oblongifolia*, SWAMP FLY HONEYSUCKLE. These shrubby species grow 3 feet high and form colonies by branch layering. They are easily distinguished by their long-stemmed red berries — elongated and diverging on *L. canadensis*, round and partly fused on *L. oblongifolia*. Both are intolerant of dry, sunny sites. They are generally abundant where found and are both common in the mixed forest. *L. canadensis* grows in mesic forest understory communities remaining in the deciduous forest, and *L. oblongifolia* in swampier ground, extending into the boreal region. The two can generally be found from New Brunswick to Manitoba and south to Pennsylvania and Indiana.

Lonicera dioica, GLAUCUS HONEYSUCKLE. The name *dioica*, inappropriately for a honeysuckle, suggests separately sexed plants, yet all honeysuckles have perfect flowers. This open, twining vine climbs 7 to 10 feet into thickets, forest edges and understory shrubs of clearings, and in open woodlands. The flowers are small and greenish yellow or reddish, producing orange fruits in dense terminal clusters. It grows on generally mesic to dry sites, preferring calcareous soils. It is very common in the deciduous and mixed forest from Vermont through Ontario to Wisconsin and rarely found in the boreal region. The distinctive large-leaved variety *glaucescens* is found from Michigan and northern Ontario to British Columbia.

Lonicera hirsuta, HAIRY HONEYSUCKLE, is generally a more vigorous twining vine than *L. dioica*, with hairy stems and leaves and very showy orangish yellow flowers. This species is found on mesic to dry gravelly and rocky sites, in clearings and open woods in the deciduous and mixed forests, and along rivers into the boreal and tundra region. Its range covers much of New England through Quebec and Ontario and west to Manitoba and Minnesota.

Lonicera involucrata, NORTHERN HONEYSUCKLE. This dense, upright, 3- to 10-feet high shrub is a cold-temperate species of damp, rocky slopes and the edges of bogs and streams. The fruit is a striking pair of glossy black berries held in a crimson bract. It is found from Quebec to northern Ontario, just touching Lake Superior and northern Wisconsin, to Alaska and south to the higher elevations of northern Mexico. It will not tolerate a dry or sunny location but can be cultivated south of its range in partly shaded cool, moist sites.

Honeysuckle flowers are renowned for a high nectar content — ample reward for pollinators like bees and hummingbirds. The paired tubular yellow flowers of the shrubby species bloom for a short period in early May and provide one of the first nectar supplies. The flower clusters of the climbing species bloom for a month or more in June and July. (*See* photo p. 249.)

Collect the fruits when fully ripe in midsummer before songbirds beat you to it. They are soft-skinned and very juicy, so transport them in a well-sealed container. A few days of fermentation are beneficial, as is a firm grinding to scratch the seed coat in the grit bag.

Sowing the seeds in midsummer, as soon as they are cleaned, will ensure high germination rates the following spring. Pre-germination treatment of stored seed is required for uniform germination, and all species respond well to a cold stratification. Honeysuckle seeds can be placed at about ¼-inch spacing to grow for one year. After this, the seedlings can be teased apart and planted into a shaded nursery to grow for another year before planting into the landscape. With its very compact thin-fibrous root, *Lonicera* is an easy species to transplant. I have never seen herbivore damage these plants.

Magnolia acuminata, CUCUMBER TREE

Some 80 species of magnolia grow in tropical and warm temperate regions, with several in eastern North America and the rest in Asia. This is one of many genera that suggest a massive extinction of North American species during glaciation when compared to the species diversity of un-glaciated Asia.

The fully ripe, bright red fruits take on a fantastic range of shapes in relation to the placement of seeds; however, cucumber tree takes its name from the pickle-like characteristics of the small, immature green fruit.

This is a rapid-growing tree and among the largest of the magnolias in the world. It is listed as an endangered species in Canada, due in part to a naturally small distribution in only two centers in Ontario. As well, human encroachment, combined with little regeneration success at most sites, places it at risk. It is intolerant of open, dry sites, preferring moist, well-drained, rich soils in the deciduous forest region. It occasionally forms nearly pure stands in widely scattered locations from New York and southern Ontario through Pennsylvania, Ohio and Missouri and into the southern states.

Blue jays and squirrels feed in the canopy of the cucumber tree and contribute to a natural drop of whole fruits. White-footed mice, squirrels and perhaps chipmunks, too, will clean up virtually every seed that falls to the ground. Natural germination occurs in the first spring in clearings and disturbed ground, usually in close proximity to the parent tree.

The green and yellow flowers in June are easily missed except by insect pollinators. It is normal for only one to several of the roughly 30 ovaries to develop a seed. The fruit is a

Above, the most unusual fruit of Magnolia acuminata (cucumber tree) when poorly pollinated.

fleshy structure, producing many contorted shapes in response to swelling that takes place to accommodate the randomness of fertilized seeds in poorly pollinated fruits. (*See* photo p. 249.) As the fruit matures in early October, it turns red. Then each swollen compartment begins to split open to release a bright orange "berry" (which is actually an aril) in which resides a single flattened black seed with a moderately hard seed coat. The orange aril remains attached to the fruit by a thin filament that allows it to hang suspended for up to a week. As the whole fruit ripens it dries, shrivels and blackens as the last arils fall away from it by late October.

Collection of seeds can be done by handpicking the unopened, mature fruit on low branches, or picking open fruits or berries from the ground in early October. Seed consumption by mice and squirrels is incredible on this species. Aside from peak dispersal time, I have yet to find an uneaten seed in the dense groundcover under a tree — they are all cut open with the kernel extracted. It has only been in clearings — footpath, skid trail and on a lawn — that I find intact seeds. Mice, by nature, do not foray readily into open spaces and this could partly explain the observation that *Magnolia acuminata* requires site disturbances to regenerate.

The seeds are easily separated from the orange coat using the grit bag. A little soap will help break up the greasy oils and any empty seeds will float with the pulp. Seeds should soak overnight and be placed in cold stratification right away. The seemingly long treatment of seven months has produced consistent results of 100 percent germination. Seeds planted outdoors in the fall yielded very low germination.

Seeds are planted at ¾- to 1-inch spacing and covered by twice the smallest thickness of the seed. Seedlings should be left alone for two years before separating and planting out. Magnolia (at least in the north) responds better to transplanting after the soil has warmed up but still before bud swelling in mid-May. The roots are well branched, somewhat thickened and strongly aromatic. After they are a few years old, it is best to move them with an intact soil ball. Traditional nurseries will burlap this species, now they are grown in containers, with associated possibilities of poorly structured roots. Once germinated, this species seems to be left alone by herbivores. Magnolia, even an old tree, tends to produce a second trunk from a basal sprout. If this happens on your trees, it should be removed at an early stage, to reduce the possibility of structurally-related tree-failure — splitting, in plain language — later on.

Malus coronaria, WILD CRABAPPLE

The crabapples of the northern hemisphere are renowned for their floral display and fruit diversity. The cultivated apple, often referred to as *M. domestica*, evolved through human-influenced selection from *M. sieversii*, the wild crabapple of Kazakhstan, but most of the 50 species of crabapple in the northern hemisphere are too small or tart to be eaten.

Exotic Alert

The Siberian crabapple, *M. baccata*, Japanese crabapple, *M. floribunda*, and Chinese crabapple, *M. prunifolia*, were all planted for wildlife food and cover, and we can be grateful that few survived (*see* photos p. 249). Those that remain produce clusters of small, round, reddish fruits on long, thin stems, thus distinguishing the crabapples from the closely related hawthorns (which have short, thick fruit stalks).

Wild crabapple is a not uncommon mid-succession, colony-forming species with suckers rising from a wide-spreading root system. Dense thickets are produced in open, sunny locations and trees decline if a forest is spreading over them. It is found in mesic and dry soils on embankments and in forest-edge communities. The fragrant pink flowers are 1½ inches across in clusters of three to five. They are pollinated primarily by bees and visited on occasion by hummingbirds. The apples are 1 to 2 inches wide, slightly flattened, with a smooth, very "greasy" skin. When mature, they shift slightly to a yellow-green from which there is no further color change in full ripeness. The fruit is edible but not what one might call palatable.

Ripe fruit drops from the tree in October, where it may lay untouched until the fruit has partly decomposed. Many mammals consume it; chief among them are the red fox, coyote, white-tailed deer and opossum. These mammals will pass many seeds undamaged through their digestive tracts. The scats of coyote are occasionally completely filled with crabapple skins and seeds — which mice inevitably find. Natural seed germination takes place the following spring.

Although a maximum of ten seeds per fruit is possible, three to five seeds are commonly produced in each fruit. The seeds turn from white to brown as the fruit matures. The fruits are quite firm at the mature stage, in late September, and can be left to ripen further for a week or two in an open container at room temperature. Once the fruits have softened a bit, they can be partly crushed (while in a bag) by stepping on them or bopping them with a rubber mallet. With fruits now opened, handpick some dark seeds from the chunks, or place the crushed fruits in water and watch the seeds sink. The seeds do not have a hard seed coat, so the crushing stage should not involve grinding or pulverizing.

Germination is very easily accomplished by fall sowing in a cold frame or by cold-stratifying the seeds for 120 to 130 days; radicle emergence then should be evident. Seeds are planted ⅓ to ¾ inches apart and can grow for two years, reaching 16 to 20 inches, before they are planted out. The seedlings with good fibrous roots are easily transplanted in early spring, before the buds swell; older plants (as with the closely related *Crataegus* and *Amelanchier*) are better planted in the fall. Try to keep defoliation by tent caterpillars to a minimum for the first few years, until the plant has an established root system and can easily recover from partial defoliation. Like the apple of commerce, the bark and twigs are a highly favored food for deer and rodents. Girdled tree trunks should be severed close to ground level so that new shoots will sprout from the base. This is a tree for remote parts of larger yards and parks. The thorn-like shoots make the branches difficult to manage.

Wild crabapple bark is smooth and gray for the first decade or so before splitting and peeling to reveal the rough, brownish mature bark. This change is not a disease or threat to the tree. *M. coronaria*, along with many species of *Crataegus* and some *Amelanchier*, is characteristically defoliated by early September by a fungus disease called cedar-apple rust. The rust fungi all have alternative hosts and this one resides on red cedar through the winter months. The rust is a natural part of the ecology of these species. If any trees are dying, the cause will be site-related constraints and not the rust.

The closely related and very similar prairie crabapple, *M. ioensis*, is virtually identical except that it is permanently tomentose on the twigs, leaf underside and flower stalk. It is not generally thought to occur east of Indiana, but native peoples may have introduced it to a few places further afield, such as Cross Village in Emmet County, Michigan.

Morus rubra, RED MULBERRY

Exotic Alert

The (Asian) white mulberry, *M. alba*, is what most of us regularly see in the landscape. It occasionally has white fruit but mostly reddish purple fruit — in fact all nine other Eurasian species have reddish purple fruit as well. White mulberry is a very common urban, rural and woodland weed tree that thrives in most soils and environments, ultimately reaching up to 40 feet tall with a trunk up to 30 inches in diameter. Landscapes everywhere are plagued by the ubiquitous weeping mulberry, a selection of the white mulberry, the fruits of which are no less capable of advancing the invasion.

Leaves and fruit of Morus rubra, *red mulberry*

Red mulberry is a medium-sized shade-tolerant tree in moist woodland openings. It is the only mulberry species native to eastern North America. As individual fruits mature over the course of a few weeks, birds such as robins, orioles, woodpeckers, cardinals and the "mimics" (brown thrashers, catbirds and mockingbirds) rapidly consume the crop and disperse the seeds naturally. The hard little seeds are passed intact and potentially germinate quite readily within two weeks, if conditions are good, or they may lie dormant until the following spring. The species is uncommon and widely scattered in the northern edge of its range from New England through New York, southern Ontario and southern Michigan to southern Minnesota, from where it is still relatively uncommon southward from Florida to Texas.

Kevin Burgess, a mulberry researcher at the University of Guelph's botany department, has documented the evidence that red mulberry is a sex-change species. Trees start out as male-flowering while young and then go through a transition or hermaphrodite stage in which male and female flowers are found on the same tree and even within the same flower catkin, and finally aging as a female tree. This supports the awareness that conservation of a species demands more than preservation of older individuals; it requires conservation of its ability to recruit new young plants.

Flowers mature just before the leaves unfold and are pollinated by wind. The mature fruit turns red in July and is fully ripe when it is dark reddish purple. Each fruit is an aggregate of some 100 small berries, each of which contains one seed. Grinding the whole fruits in the grit bag easily separates the hard-coated seeds. The filled seeds readily sink and the usually numerous empty seeds float with the pulp and can be poured off.

Cleaned seeds can be planted in July right after cleaning, or dry stored seeds can be planted in the spring. Germination is rapid in moist conditions, without any pretreatment for summer or spring sowing. Seeds are planted ⅓ to ¾ inches apart and will produce a 2- to 4-inch seedling before summer's end. One-year seedlings are extremely frost sensitive in the north and should be overwintered in the protection of a cold frame.

The seedlings should be planted in spring into a well-shaded and protected nursery bed. You can expect 1 to 2 feet of growth in the first year, and double again the next. I have never observed mouse predation, but it is clear that groundhogs are able to kill mulberries by continually devouring the leaves and young shoots. I suspect that deer and rabbits will also browse on the foliage.

Is the Red Mulberry an Endangered Tree Species?

Throughout eastern North America *M. rubra* is threatened by the introduction and extensive invasion of *M. alba*. Hybrids are commonly found and efforts are underway, in a number of reserves, to geographically isolate red mulberry by eradicating the white mulberry and hybrids in the area surrounding natural stands of *M. rubra*. Even with this effort the red mulberry seldom produces true-to-type seeds. Millions of years of geographic isolation and species evolution since the continental breakup of Pangaea have been reversed by two centuries of mixing with the horticultural escapees. Perhaps these closely related species are on the brink of becoming a single species again. Perhaps, too, this outcome of global trade is the inevitable destiny for other species. Since hybrid vigor is a commonly understood outcome of crossing, and certainly observed in the red x white mulberry, one is forced to ask if this represents the loss of a species or a revitalization of the adaptive ability of a genus. One thing is certain: we have rarely been able to eradicate a species that has established on a new continent. Natural selection always plays the last hand and will play a major role in what mulberry will look like in the future.

Myrica, BAYBERRY and SWEET GALE

Some 50 species of *Myrica* are found throughout the temperate and tropical world. The evergreen-leaved wax myrtle that grows along the coast from New Jersey to Texas is *Myrica cerifera*. Myricas have catkin-like wind-pollinated flowers, with male and female flowers on separate plants. Seed dispersal of bayberry takes place through winter by songbirds such as the bluebird and catbird and, curiously, the yellow-rumped (myrtle) warbler. Natural germination is in spring but seeds dispersed in late winter likely remain dormant for a year. Seed dispersal of sweet gale is accomplished primarily by ducks during both fall and spring migration.

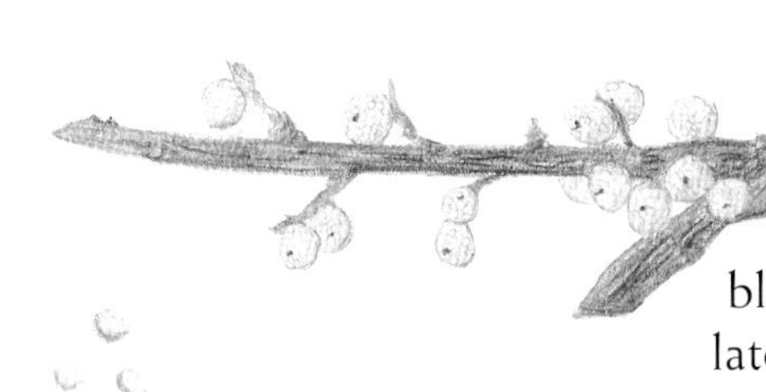

Myrica pensylvanica *(bayberry)*

Myrica pensylvanica, BAYBERRY. This near-evergreen shrub spreads by shallow underground stems to form extensive thickets to 7 feet high. It is a plant of mesic to dry sand in full sun, or partial shade of open woodland clearings, shores, embankments and dunes. It is chiefly a coastal species from Newfoundland to the Carolinas and widely scattered inland to Ohio and, at the extreme western edge of its range, southern Ontario, where it is found only at Turkey Point and on Toronto Island. Bayberry flowers in the leaf axils of the previous year's shoot. The fruits of bayberry are single round nutlets that are encrusted with a bluish white wax and are quite tightly pressed together in clusters that are persistent into the following summer.

Myrica gale

Both species of Myrica are aromatic in every part when bruised or lightly crushed.

Myrica gale, SWEET GALE, spreads slowly by underground stems to form a compact thicket reaching 5 feet tall in full sun. It is a common plant of wet soils along shores, bogs in granitic soils, quartz sands and organic deposits. The circumboreal distribution of sweet gale means that it is found across Eurasia and Canada southward to New Jersey, Pennsylvania, Michigan and Minnesota. *M. gale* flowers at the end of the previous year's shoots in early May; the small, bright crimson female flowers are exquisite. The fruit of sweet gale is a resinous yellowish brown nutlet, packed in dense clusters that persist through winter.

Fruits are gathered from mid-September to late winter and can be stored as whole dry fruits until ready to stratify them. The dry wax of bayberry seed can be removed in a dry grit bag, after which the powdered wax can be filtered off through a sieve. The seeds of bayberry require a 90-day cold stratification; sweet gale requires only a 60-day cold stratification.

Plant seeds at ¼- to ⅓-inch spacing and cover with soil. They should be left to grow for two years, after which the very fibrous-rooted, fragrant seedlings can be teased apart and transplanted. Two more years of growth will produce bushes that are about 12 to 20 inches tall. Since both species are fussy, requiring slightly acidic soil conditions, they are well suited to the edge of bog gardens. Deer may browse on the foliage and meadow voles will debark the stems of bayberry under the thick leaf litter of dense colonies. The massive underground root system provides ample resources for recovery from herbivory. Both species are easily propagated by divisions.

Nyssa sylvatica, BLACK GUM, TUPELO or SOUR GUM

There are four species of *Nyssa* in eastern North America and six in China, including the widespread *N. sinensis* and *N. javanica*. The presence of black gum in isolated mountains of Mexico and Central America is very interesting. Are they relicts from the retreat of the species during glacial times or are they a result of seed dispersal by migrating thrushes?

The stunning scarlet fall color is likely to be the reason you discover the location of these handsome trees. Black gum is a large tree that occasionally gives rise to root sprouts to form small colonies. It usually grows in the open in moist, but well-drained, marsh peat and acidic sandy soils at the edge of swamps and along roadside ditches in the north. Increasingly it is growing as a canopy tree on wet forested slopes further south. Black gum is widely scattered in the north portion of its range from Maine through extreme southern Ontario and central Michigan to Wisconsin — it becomes more abundant southward. From Florida to Texas it is found on drier ground than its cousin the water tupelo, *N. aquatica*, which thrives in swamps from Virginia to Illinois and south to Florida and Texas.

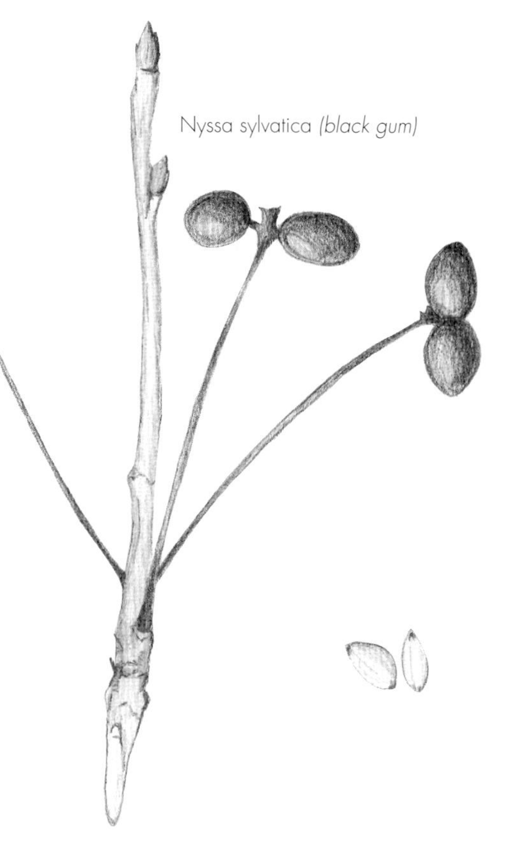

Birds will quickly pick a black gum tree clean of its fruit.

Once the leaves have expanded in mid-May to early June, inconspicuous small, greenish flowers on long stalks are produced on separate male and female trees. Flowers are pollinated by bees (the water tupelo in the southern U.S. produces the highly prized "tupelo honey" that was memorialized by singer-songwriter Van Morrison). The fruits mature with a shift to reddish purple and ripen to a lovely blue-black with a reddish orange stalk in early to mid-October as the leaves are turning color. The fruit is highly sought by birds, especially robins, other thrushes, woodpeckers, turkeys, brown thrashers, mockingbirds, and mammals. The fruits are ingested and the hard seed is passed undamaged to germinate in the first spring.

The seed is a single, slightly ribbed, oval stone within each thin-skinned fruit. Ripe fruits can be handpicked from low branches and fall-sown after being cleaned or dried and stored cold until early January, in preparation for a 90-day pretreatment. Soak the dried fruits for 24 hours and clean in the grit bag. The seed coat is very hard and all seeds produced are sound. They germinate evenly when cold-stratified for 90 days and spring-sown at ⅓- to ¾-inch spacing in a cold frame. The seedlings should be protected in a cold frame or similar structure through the first two winters in the northern parts of the range.

Two-year plants are 8 to 16 inches tall with dark, fibrous roots and can be separated and planted out in appropriate habitat in the spring. They tend to root deeply but transplant with relative ease. I have had success in establishing black gum in calcareous soil by planting it in the sedge peat of a marshy stream edge. Rabbits will chew the stems of young plants completely to the ground and consume the twigs and bark of saplings.

Ostrya virginiana, IRONWOOD or HOP HORNBEAM

Ostrya virginiana *(ironwood)*

A single seed of ironwood is enclosed in a papery sac with distinctly spiny hairs that can be quite irritating to the skin and lungs.

Thin strips of bark, curling top and bottom, make this tree easily identifiable, but it is the name ironwood that makes it so memorable. *Ostrya virginiana* is shade tolerant. It thrives in mesic to moist, well-drained mixed and deciduous woodlands on shallow soils over bedrock and in deep gravel and rich loam soils. Ironwood is common from Nova Scotia through southern Quebec, north of Lake Huron to southern Manitoba, then south to Florida and Texas, where it meets the western *Ostrya knowltonii*. Five other species are found in Asia and a single species in Europe.

Whole fruit sacs are dislodged during windstorms throughout early winter, and natural germination is normally in the second spring after seed-fall. Natural seedlings are found scattered in the woods or clearings adjacent to parent trees. The presence of holes chewed in the papery fruit sacs with the seeds removed indicates that small mammals are devouring much of the crop. Grouse are reported to consume the seeds as well and may be the principal agents for longer distance seed dispersal.

Ostrya, like other members of the birch family, has separate male and female flowers on the same tree. The male flowers are borne in catkins, which are preformed in late summer and exposed at the shoot tip through winter. The female flower is a modified catkin, with each ovary contained in a sac instead of under a scale (as in birches and alders). The female flowers are pollinated by wind as they protrude from the end of the unfolding leaves of lateral buds.

The clusters of small, bag-like fruits mature in October and persist on the branches into winter. Each sac contains a single elongated, pointed nutlet that has a fairly hard seed coat. Seeds can be extracted from the dried sacs by placing the whole fruits inside a thick plastic bag and forcefully rubbing them. Seeds can be shaken to the bottom and separated and can be stored for a number of years.

Be careful — the papery enclosures are covered with hundreds of minute, stiff, sharp hairs that readily embed into the hands and can be quite irritating. Wear thick, leathery gloves when picking and handling this fruit and wear a dust mask if you are thrashing the fruits to extract the seeds.

If fruits are harvested just as they are turning brown, they can be sown in the cold frame in early September and kept quite moist. Some germination may take place in the first spring, with the rest of the seeds germinating the following spring, or higher percentages of germinated seed can be obtained in the first spring if the cleaned seeds are warm-stratified for 60 days, followed by cold stratification for 150 days.

Stratified seeds can be sown at ⅓-inch spacing and left for two years of growth to produce 12- to 16-inch seedlings. *Ostrya* has a fine, fibrous root system and transplants easily but should be kept in light shade for the first few years. It can be planted in full sun to partly shaded conditions but does not tolerate extremely dry or wet conditions. Ironwood is one of the very few woody plants avoided by herbivores.

Parthenocissus, the VIRGINIA CREEPERS

The Virginia creepers are very important wildlife plants, providing dense cover and significant energy reserves for mammals and both resident and migrating birds. Most of the berry eaters, the thrushes and mimics (catbirds, thrashers and mockingbirds) are interested in this fruit; woodpeckers and even tree swallows are reported to eat the berries. Seeds are passed intact and germinate the following spring. Individual plants can cover extensive areas as a groundcover in relatively dry to moist sites or drape like curtains from old trees in moist to wet woodlands. As a groundcover plant in the understory, rooting at the leaf nodes assists the spread of both species. A dozen other species are found in North America and Asia.

Parthenocissus inserta, VIRGINIA CREEPER, is a very common groundcover or scrambling vine found on moist to dry soils in clearings, open woodlands, thickets and embankments.

Parthenocissus *(Virginia creeper)*

Its branched tendrils are without adhesive discs and cling by club-tipped twining tendrils. This species is commonly found further north than the following species, from New England across the north shore of Lake Huron to Manitoba and Utah and southward more sparsely from Florida to Texas.

Parthenocissus quinquefolia, VIRGINIA CREEPER, is a generally higher-climbing vine in moist to dry deciduous woodlands. The widely branching tendrils are terminated by adhesive discs. This is a more southern species, ranging from Maine to Iowa and south, though it has escaped from cultivation further north.

Fruit of both species ripen from mid-September through October and, despite its membership in the grape family, the fruit are not recommended for human consumption — they can cause minor symptoms of toxicity such as stomach upset.

Flowers, produced in great abundance every year, are pollinated by bees and flies. As in most perfect flowers, the stamens develop before the stigma is receptive and this favors cross-pollination. The green fruit matures with a shift in color to purple as fruit stalks turn bright red. The soft, round berry contains usually three seeds that resemble grape pits. The fruit clusters of the two species differ significantly, with *P. inserta* producing branching, open clusters and *P. quinquefolia* producing elongated clusters with more tightly packed berries on a central stem. Hybrids are not reported.

Fruit is easily picked from lower extensions of the vines and should be kept in a plastic bag to contain the juices, which will stain almost anything. In later stages of ripeness the fruits (with stalks) will drop onto a tarpaulin with little effort. The fruits can be allowed to further ripen a few days if necessary; otherwise they should be cleaned and dried soon after collection for cold, dry storage until the middle of January.

Presoak the seeds for 24 hours prior to a 90-day cold stratification, after which uneven germination is still likely. Seedlings produce a very stout crown and fibrous root system and a long trailing top in the first year. The long shoot should be cut off before being planted out in the following spring. They are easily transplanted and rodents seem to leave them alone. These are important species in the landscape but neither needs to be planted in great numbers, since they are so adept at colonizing. Cuttings are easily made from rooted stem pieces, but in the interest of biodiversity avoid getting cuttings from just one location.

Physocarpus opulifolius, NINEBARK

Roughly 10 species of shrubs relatively similar to ninebark are found across North America and Asia. Ninebark grows 7 feet tall and wide on moist ground along streams and shores as well as in dry meadows and alvars. Occasionally I have found ninebark on dry embankments or rocky outcrops. This species is truly a Great Lakes plant, crowding the shoreline virtually throughout. It is found from Quebec to Minnesota and south to the Carolinas.

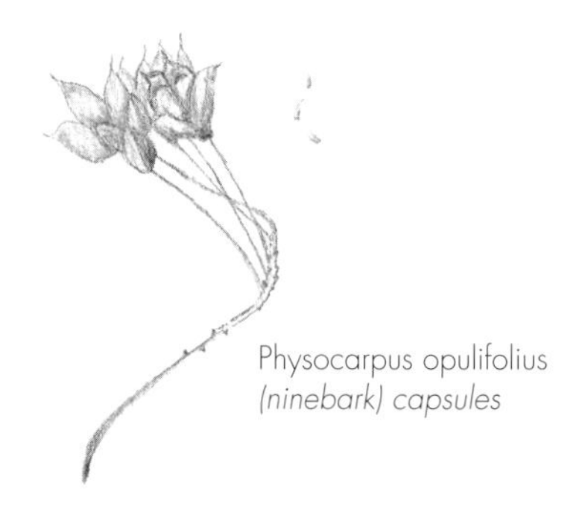
Physocarpus opulifolius (ninebark) capsules

Seed dispersal takes place from September to April and is enhanced by chickadees breaking the capsules open to obtain seeds. Part of the seed crop is picked from the ground during the fall migration of sparrows and juncos. Natural germination of scattered seed is in the first spring. The species is often localized but abundant, indicating a frequent local dispersal. The distribution so far north may someday be traced to sparrows feeding on ninebark during spring migration.

Abundant flower displays occur annually in June to July. (*See* photo p. 250.) This is an important nectar and pollen plant, as few other native flowers are found at this time of year. I have been in awe, on numerous occasions, watching how many species of insects take nectar from this summer-flowering shrub. Each flower produces a three- to four-chambered capsule containing several small, round, shiny, tan-colored seeds about 1⁄16 inch across. The fruits are brilliant red in midsummer and then turn yellowish brown as they mature in late August before ripening to a tan brown and opening in September.

Fruits can be harvested after they begin to split open, even though the fruits hang downward in order to shed the seed crop. Many capsules remain attached to the plant through winter and most still contain a few seeds. Fruits should be stored in a paper bag and allowed to dry until the capsules are rubbed to release the seeds, which can then be separated from the chaff by fanning or winnowing. These dry seeds should be cold-stored until February.

Germination is obtained by soaking the stored seeds for 24 hours followed by stratifying for 60 days. Seeds are scattered at about 1⁄4-inch spacing and barely covered with sand. Germination takes about 10 to 14 days and seedlings should be carefully monitored for damping-off. They can be transplanted to a larger spacing at the four-leaf stage or thinned out and left to grow for the first year. One-year seedlings are about 6 to 10 inches tall and can be set out into a nursery bed for another year of growth before planting to their final location. Ninebark is very easy to transplant due to a dense, fibrous root system. They grow well in sunny to partly shaded, moist but well-drained locations. I have observed no browsing on this species when other more desired foods such as raspberries are in the vicinity.

Due to the trading of dormant stock, the golden-leaved cultivars of this species occasionally show up in naturalization plantings and should be returned to the supplier. *Physocarpus* is easily grown from softwood cuttings. This primarily Midwestern genus is found across North America with loosely differentiated species. *P. intermedia* differs primarily in having stellate hairs on the capsules of a smaller-statured shrub. It grows west of Michigan and Ohio and likely hybridizes where ranges overlap. The nursery industry continues to provide the golden-leaved ninebark to fill the insatiable demand for naturalization. This hideously colored selection does not fit in the natural world and should be weeded out.

Picea glauca *(white spruce)*

Picea mariana *(black spruce)*

Picea rubens *(red spruce)*

Cone size is the easiest way to identify spruces. White, red and black spruce are natural in the Great Lakes; Colorado and Norway spruce are introduced. There are about 40 species found in cold and temperate regions of the northern hemisphere.

Picea, the SPRUCES

The female cones of spruces stand upright in the upper branches and male cones on lower branches as an evolutionary adaptation to reduce self-fertilization. The pollen must be carried in updrafts to pollinate flowers farther downwind. Once pollinated, the cones rotate to a downward hanging position and mature with a color shift from green to yellowish brown. During dry, sunny conditions in late August and September and continuing through winter, the cone scales open and release the winged seeds to spin earthward in great quantities on warm, dry afternoons. A stiff wind will carry the winged seed for considerable distances. Storms carry seeds high aloft and winter winds blow seeds with drifting snow.

Crossbills and finches open the cones to obtain seeds but red squirrels harvest great numbers of spruce cones, which they cache. Birds including grouse, sparrows and juncos eat seeds lying on the ground. Natural germination takes place in the first spring. When the seedlings first emerge, they carry the seed coat like a crown at the end of the seed leaves. At this stage they are particularly vulnerable to sparrows, which, while foraging, nip off the tops of the seedlings.

Exotic Alert

Two introduced spruces are frequently planted throughout the Great Lakes. The Norway spruce, *Picea abies*, with wide-spreading, arching branches and 8-inch long cones is often found naturalized. The Colorado spruce, *P. pungens*, with its very horizontal branches and a natural foliage color range from green to quite bluish (in *Picea pungens* var. *glauca*) is only occasionally found escaped. For some reason green just isn't good enough for some folks.

Picea glauca, WHITE SPRUCE. So complete is the shade under this handsome species that very few plants are able to grow in the blanket of cast needles beneath them. This species prefers a cool, moist but well-drained sandy or rocky soil. It will colonize open ground. The extensiveness of its range is a tribute to the adaptation of the species. It can be found in the boreal and mixed forests from Newfoundland to Alaska and from the tree line south through the upper Great Lakes and as far south as New York and southern Michigan. (*See* photo p. 251.)

Picea mariana, BLACK SPRUCE. The ability to form clone colonies by branch layering is unusual among spruces and shared only by the white cedar among tree conifers in our area. Unlike other spruces, the stubby cones remain firmly attached to the old branches for many years. Black spruce prefers acidic, moist, organic soils and is only found in sphagnum bogs in the southern Great Lakes part of its extensive range. This species, with its distinctive narrow crown, can be found from Newfoundland to Alaska and south to New Jersey, Pennsylvania, southern Ontario, Michigan, Wisconsin and the high mountains of New Hampshire.

Picea rubens, RED SPRUCE. This species extends eastward from the extreme eastern edge of the Great Lakes basin. It is rarely encountered in cultivation. It readily hybridizes with black spruce, providing a species complex that separates out by moisture gradient. Red spruce is an upland species, preferring the drier locations in rocky woods and mountains of Quebec and eastern Ontario, into Pennsylvania and south through the Appalachian Mountains to Tennessee.

Cone production in spruces is periodic and intervals of two years are common between cone crops. Small male cones hang downward on lower branches and drop off soon after the pollen is released. The young female cones of spruces are an exquisite violet-red and stand ⅓ to ¾ inches tall, depending on the species, pointing upward with the scales open to receive the wind-borne pollen. After pollination the scales close and turn green as the cones slowly bend to hang down. Cones mature within the first growing season when they have shifted from green to yellow, after which they soon ripen to light brown as the scales open to release the seeds. Two seeds are held under each scale. Filled seeds are dark black while empty seed coats tend to be tan colored. During periods of wet weather, the scales close again.

Cones can be picked from young trees as they shift from yellow to brown. Seed gathering from tall trees can lead to bold climbing adventures — not something for the inexperienced to attempt. Cones are best kept in a cloth or paper bag and spread out as soon as possible so the cone scales will open to release the seeds. Once they are dry, shake the cones in a bucket to dislodge the seeds. Place the dry seeds in a labeled envelope and store cold, until ready to plant.

Spruces have no germination inhibitors and will readily sprout within two weeks of sowing. Wait for the longer days of mid-March to sow the seeds. Scatter them at ¼-inch spacing and barely cover them with sand. If they are grown in a container, they are very susceptible to damping-off once they have sprouted. Mature seedlings can be plunged, without disturbing their roots, into the cold frame or a nursery bed in

Spruce Budworm

Foresters consider spruce budworm a serious threat to spruce and fir forests. Natural predators of the adult moth — bats, warblers, sparrows and vireos — all became inadvertent targets when, in the 1940s, spraying with heavy pesticides, including DDT, started. It was never considered that spruce forest management in the northeast was producing a monoculture of the budworm's two host plants, spruce and fir, nor was the option for continuous rotation of spruce monoculture in combination with "conifer release" herbicide spraying. To this day, the aggressive planting of newsprint forests and intentional suppression of other tree and shrub species have created a huge biodiversity gap, creating the perfect conditions for high pest numbers. Most of the highly toxic, carcinogenic aerial spraying has been phased out and a natural bacterial predator of moth larvae, Bt, *Bacillus thuringiensis*, is now used instead. This bacterium is said to be host specific; however, has anyone seen a luna moth lately?

Today, biotechnologists are exploring creating a spruce containing a Bt toxin gene to kill the budworms while they feed on plantation trees. This would be an absolutely irresponsible application of science, since budworm resistance against both pesticides and Bt toxin already has been observed. Many budworms are missed when Bt is applied from the air, and this maintains a non-resistant gene pool in the budworm breeding population. If every tree in a plantation contains Bt toxin, then a Bt-resistant budworm population will evolve rapidly. Further, since spruce is a wind-pollinated species and flowers as a young tree, the Bt pollen will spread into the wild populations of spruce, with unknown impact, all the time increasing budworm resistance to Bt — making Bt virtually useless. Will this again open the door to chemical pesticides?

With regard to plantation forests, William Bryant Logan said it best in his delightful book *Dirt: The Ecstatic Skin of the Earth*. While passing miles of conifer plantations in Georgia to show a colleague another composting station, he remarked, "and this is the forest on which America wipes its ass." The budworm "problem" is the only possible outcome of plantation spruce. There are other fibers (switch grass and hemp) from which paper can be made. The radical solution is to dramatically reduce demand for newsprint — what percentage of newspapers and advertisements are actually read every day? Recycling 100 percent of used paper for print, boxboard and other paper products is not out of our technological reach.

Genetic diversity of spruce must be combined with species and habitat diversity to support the natural predation of budworm. Reliance on pest control in any form seems always to ensure predator decline and pest success.

Pines Against Desertification

Land clearing in the mid-1800s was so extensive that by the late 1800s three desert-like areas were forming in southern Ontario. We owe much gratitude to the brilliant work of Dr. E.J. Zavitz. As forester at the University of Guelph, he led a team of associates to initiate pine planting to stabilize what were called the blow sands. This certainly represents one of the first times that shifting sands were stabilized by tree planting. His first white pine trial plantation of 1907 stands at the University of Guelph Arboretum.

May. Two-year seedlings are about 4 to 6 inches tall and can be lifted, root-pruned to 4 inches root length, and set into a nursery bed to grow for two more years, after which they are ready for planting out. Rabbits will browse on the one-year shoots of spruce during winter.

Transplant success is generally very high with the fibrous rooted spruces. This, along with their festive form, accounts for some of their popularity. Of the three native spruces, white spruce is the only species that is commonly planted, and it will grow well, despite heat and drought, in moist ground of the deciduous-forest life zone even where it is not native — just don't expect it to spread naturally in those conditions. Black spruce seems to adapt to moist alkaline soils but may be relatively short-lived in those conditions — I have observed it growing surprisingly well for at least two decades on glacial till.

Pinus sylvestris
(Scots pine)

Pinus, the PINES

In the same way that the Old English word *cedar* means evergreen, *pine* is derived from an ancient Roman name for tree and applied to almost anything with needles. This diverse genus of close to 80 species is classified by the number of needles in a bundle: the five-needle white or soft pines; the three-needle yellow pines; and the two-needle black or hard pines. They are found across the northern hemisphere, from near the tree line to the peaks of tropical mountains. The pine is a symbol of both natural beauty and economic exploitation. In the Great Lakes region and beyond, the controversy over protection or harvest of pines ebbs and flows to this day. The focus on what **is** old growth usually diminishes attention to what **will be** old growth.

Exotic Alert

Two of the European "two-needle" pines are now prominent in the landscape. Assuming the trees could handle the stress, both species were planted extensively in drier sites but they are natural to moist, well-drained, soils in Europe. The wide, spreading forms of Eastern red cedar are a lovely substitute for the two introduced pines.

The compromising conditions here have led to stresses that undoubtedly played a major role in the advance of *Diplodia* tip blight, a fungus disease that causes branch ends to die out. Under severe stress conditions (drought or poor genetics) the entire tree or plantation will die. A program of felling, removal and controlled burns has been successful in eliminating both mature and regenerating Scots pine from oak woodlands.

Scots (or Scotch) pine, *Pinus sylvestris,* was planted extensively for reforestation, but many seed sources proved to be ill-adapted. From both healthy and declining stands, Scots pine has none the less seeded out extensively.

Austrian pine, *Pinus nigra,* favored for urban, park and roadside planting and in sandy soil (like an oak savannah), has also become invasive.

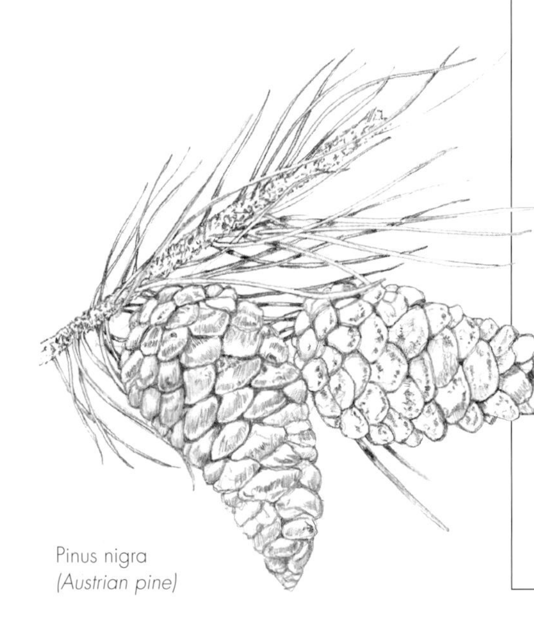

Pinus nigra
(Austrian pine)

Pinus banksiana, JACK PINE, is fussy as to soil type, with a preference for moist, well-drained, sandy, acidic soils, the humus layer over granitic bedrock, or the cracks of limestone alvars. Trees planted in dry, alkaline glacial till sites tend to form relatively short-lived, stunted trees — still of considerable aesthetic value. This pine is found north of the deciduous-forest life zone to the tree line from New Brunswick to Alberta and into the northern parts of Michigan, Wisconsin and Minnesota.

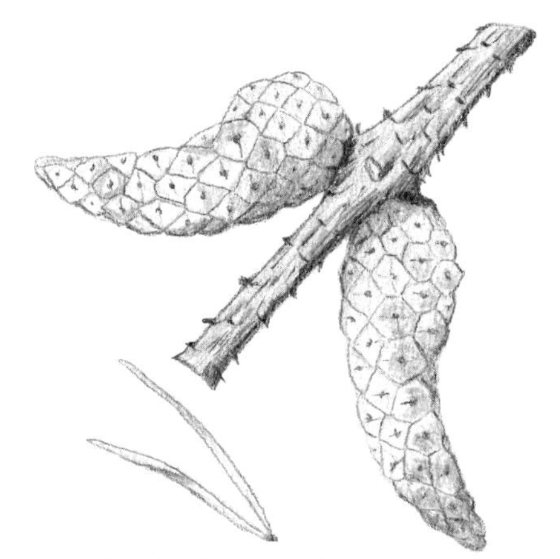

Pinus banksiana *(jack pine)*

Pinus resinosa, RED PINE or NORWAY PINE. This species, more than any other, reminds me of camping on the shores of small northern lakes. The tufts of needles, looking somewhat like chimney brushes, are so striking. Red pine is quite fussy as to soil type, with a preference for sandy, nutrient-poor, acidic soils. Plantations outside of this soil type fail within two to five decades. Red pine can be found north of the deciduous-forest life zone from Nova Scotia across southern Quebec and north of Lake Superior to the Manitoba border, and south through Minnesota, Wisconsin and northern Michigan and the New England states. There is a large segregated population in the mountains of New York, Pennsylvania and West Virginia. The species is known to be genetically uniform throughout its range, indicating that its numbers were greatly reduced by glaciers advancing south.

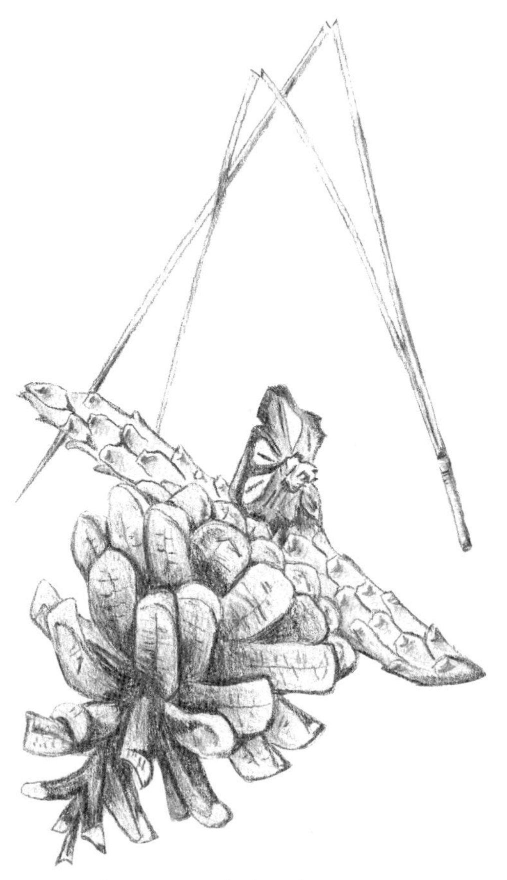

Pinus resinosa *(red pine)*

Pinus rigida, PITCH PINE. This uncommon and primarily eastern species requires fire to open its cones and prepare the seedbed by thinning the humus layer. Fire suppression has reduced seedling recruitment and increased the understory shrubs in a number of locations. Pitch pine thrives in dry conditions on gravelly soils and rocky ridges. I have observed the species on the thin soils of granite slopes and at the edge of sphagnum bogs on granite outcrops in its few Ontario locations, near Kingston. It is found in a small region of southern Quebec, famously on the New Jersey pine barrens, and in scattered populations from New York to Georgia.

Pinus strobus, WHITE PINE. This majestic tree fueled the resource-grab mentality in North America but, just as significantly, it fed the fires of passion in artists, poets and protesters with its wind-sculpted beauty and imposing height (and the continued assault on its last uncut stands). White pine is moderately shade-intolerant; seedlings and saplings can survive up to 20 years under partial shade. It is a species of the mixed and deciduous forest and thrives in well-drained soils in a wide range of habitats, including sand dunes, sand plains, alkaline gravel ridges, swamp margins, rich forest soils and rocky ridges. Stunted trees emerging from cracks in granitic rock slopes can produce stunning natural bonsai — occasionally 200 years old at 7 feet tall — while 32 yards away, a tree from the same seed source growing in rich forest soil results in a similarly aged 82-foot forest giant.

Seed source selection may require that more attention be paid to the soil and soil moisture conditions than the exactness of climate matching. It ranges from Newfoundland across southern Quebec and north of Lake Superior, and south through Wisconsin and northeastern Iowa to Ohio and the eastern seaboard as far as Georgia.

Pinus virginiana, VIRGINIA PINE, is a small, irregularly branched tree about 33 feet in height. The needles are short, 2 to 3 inches long, and two to a cluster. The cones are only 6 inches long, shiny, reddish brown and very prickly. The open cones persist on the branches. They should not be picked until they turn from green to dark purple or reddish brown. In our region, the species is more or less restricted to the non-calcareous soils of (recently) unglaciated regions in Ohio and Pennsylvania. There it is an invader of eroded slopes and abandoned fields and frequently forms pure stands. It is shade intolerant.

The bark of older white pine trees, like most pines and oaks, is fire resistant to a point — the bark provides sufficient insulation to help prevent the sap from reaching boiling point and damaging the cambium cells. White pine tends to become hollow with age, often reaching 300 years in age, rarely 400 years. Old pines are found in various and famous "old-growth" stands such as Crow River in Algonquin Provincial Park, and Obabika Lake in the Temagami wilderness (Teme Augama First Nation) in Ontario, at Hartwick Pines State Park near Grayling, Michigan, the Joyce Kilmer tract in Bald Eagle State Forest in central Pennsylvania, and Nicolet National Forest in Wisconsin. All of these stands show evidence of fire in the blackened base of some trees. The age of the surrounding vegetation will indicate when the most recent fire took place. White pine ecosystems tend to endure a fire event at 100-year intervals.

Squirrels eat a significant amount of the cone crop of white and red pine some years. White pine disseminates its seeds in August and September, red pine in October and November, and jack and pitch pine after a fire or desiccating hot summer. Germination takes place in the first spring after dissemination.

Young female pines are receptive in late May, with pollen-bearing cones at the base of new shoots on lower branches and the female cones at the end of new shoots on stronger branches toward the top of the tree. The female cones mature in two growing seasons — you will notice them as miniature cones (upright in white pine and radiating from the terminal bud for red and jack pine). During the second summer, the cones swell and mature, with a color shift from green to yellowish and then brown in the fall. A number of insect larvae thrive on the maturing seeds inside pinecones and occasionally completely consume the seeds as the cone is ripening.

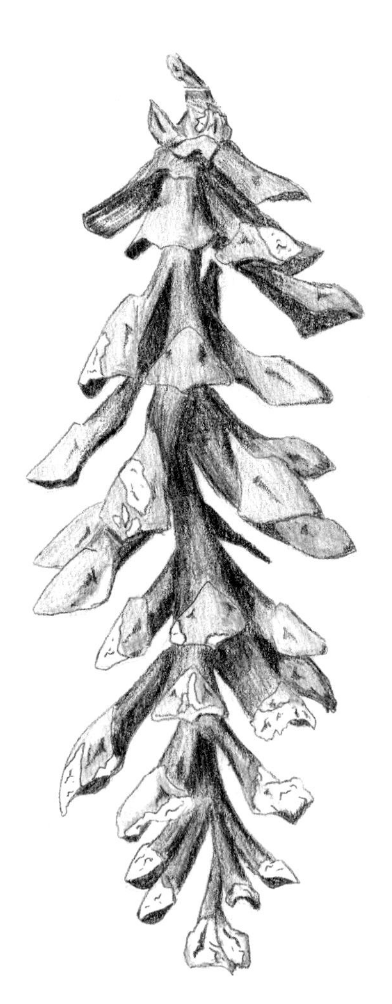

The cone of Pinus strobus *(white pine)*

Of Birds and Pines

Jack pine has a unique relationship with Kirtland's warbler — one of the rarest birds in North America. This warbler has a territorial preference for jack pine forests roughly 12 to 15 years old with a sedge and blueberry understory. The stands must be rather extensive. The State of Michigan is making a concerted effort, using controlled burns, to maintain the appropriate age and extent of jack pine woodland. Several sites on Manitoulin Island and the eastern shores of Georgian Bay in Ontario are prime jack pine ecosystems, but fire suppression has resulted in the loss of age diversity in those regions and hence the loss of preferred warbler habitat. The cones tend to release seeds only after high-temperature events such as fire or (less often) heat buildup around a branch in the full sun of a protected site.

Collect white pine cones from the ground when squirrels have dropped them in early September, even while still green (cones dropped in August are too green and will not open). Green cones are covered with extremely sticky resin. I recommend covering your hand with a plastic bag while handling these cones. Lay them out to dry immediately. Once they turn brown, the scales open and the resin solidifies. White pine cones are fully open by mid-September, release most of the approximately 80 seeds, and then drop from the tree. Cones on the ground will close once they get wet and occasionally still hold a few seeds.

Red pine cones start to open in mid-October and must be picked prior to this and dried in a rodent-proof space. Since red pines hold many of their old, empty cones for several years, be sure to pick only the current year's cones — the ones just at the base of the current season's shoot. Mature cones of jack and pitch pine (from the present season only) can be clipped from low branches at any time of the year. Commercial operations heat the cones to open the scales, but I find it easiest to place the cones in a sieve and pour boiling water over them. Once pinecones have opened and dried for a few days, the seeds can be extracted by shaking the cones in a bucket. Dry seeds of all pines can be cold-stored in an envelope for many years.

Germination of pines is enhanced with cold stratification (check the treatment guide for species-specific durations). Sow the seeds at 1/8- to 1/4-inch spacing and barely cover them with fine sand. Seedlings sprout in two weeks and are very susceptible to damping-off for the first few weeks. One-year seedlings are 3/4 to 1 1/2 inch tall. Your pot full of seedlings should be set into the warmer side of the cold frame or in a sheltered seedbed. Mice or voles may chew them if they are not protected. (*See* photo p. 251.) Two-year plants can reach 4 to 6 inches and should be planted into the nursery for another two years of growth. Out-planted trees are rarely affected by herbivores.

When planting white pine into permanent sites, the three to four-year-old plants should be planted about an inch deeper than they were in the nursery. White pine will produce strong lateral roots outward from the buried part of the stem and thus provide excellent anchoring.

White Pine Blister Rust

White pine blister rust's scientific name is *Cronartium ribicola* (*ribicola* from the genus *Ribes* — the currants and gooseberries). Like all rust fungi, two host plants are required to complete the life cycle. The "host" is always the commercially important plant and the "alternate host" is relegated to the status of a weed. Blister rust spends the fall, winter and spring on pine and the summer on the leaves of *Ribes*. Very costly campaigns took place in the early 1800s to try to eliminate *Ribes* from commercial pine forest regions when the blister rust first made its appearance. Many white pines seem to be naturally tolerant of the rust fungus, but forestry breeding programs are still underway in the pine plantation areas to produce higher numbers of rust-tolerant seedlings by crossing survivor parents.

The goals of monoculture forestry are ludicrous. The elimination of forest diversity means the loss of critical habitat that supports the diversity of predator associates. This ensures the success of insect pests such as white pine weevils, which take out the leaders and create crooks in the tree trunks. Foresters in plantation and timber-growing regions are concerned about anything that uses (attacks!) "their" trees as a host, yet the greatest threat to the pines may well be the clear-cutting and associated plantation forests.

I am often mindful of the philosophy of Merve Wilkinson of Wildwood Forest near Nanaimo, British Columbia — small roads, small cuts, small machines and small foreign profits in great forests. Using five-year-interval select cuts over 50 years, this ecoforestry pioneer has removed the equivalent of clear-cut volume from his forest yet he still has the original volume standing, without planting a single tree. The trees are between one year and 1,800 years old. Merve Wilkinson has commented that the 1,500-year-old stands need to be managed very carefully for the next 300 years to ensure that 1,800-year trees are always part of the forest. To do this, he leaves the healthiest trees standing to reseed the forest gaps with the best genotypes for the site. The land is never open to desiccating winds, rainfall erosion or baking in the sun. This wonderful forest has been purchased by the Land Conservancy of B.C. to preserve it as an education center.

Platanus occidentalis, SYCAMORE

Exotic Alert

There are four *Platanus* species native to North America and two to Asia. The Oriental plane hybridized with native sycamore when planted in England and produced the London plane tree, *Platanus* x *acerifolia*. This hybrid is occasionally found naturalized in stream-bottom lands, well away from the cities that planted this species of tree in riverside parks. It has single fruits per stem, too, but most of the fruits are in a chain of two, and occasionally up to four balls per stem. The presence or absence of multiple fruits is the only way to accurately distinguish the hybrid from the native species, which has solitary (rarely two) fruit balls. The generally acuminate leaf base and slightly khaki and yellowish bark on the hybrid are not distinctive enough, but should raise suspicion.

I once chanced to take a gravel road at night that crossed a valley dominated by sycamore, and was stunned by the bold white trunks of this striking species. Sycamore is primarily a shade-intolerant riparian species, found in floodplain forests, on river embankments, in wet woods and in decline in once-wet woods where land drainage has changed the water levels. Sycamore is a signature deciduous-forest species found in much of the eastern U.S., reaching into southern Ontario and southern Michigan and as far west as Minnesota and Texas, where Arizona sycamore replaces it in the southwest.

Fruit balls hang through most of the winter, suspended from the branch ends by a tough, fibrous stem. Each fruit has some 1,400 seeds that are dispersed by wind once the ball begins to shatter in late winter and early spring, or as finches occasion to feed in the canopy and tear open a fruit ball, but many balls fall to the ground intact to be dispersed by spring floods. Natural germination in wet ground takes place in the first spring. Natural regeneration appears to be virtually absent in relation to drier conditions of the scattered trees in non-riparian woods that have been drained.

Male and female flowers are produced separately on the same tree in April and are wind pollinated. The fruit ripens in late October but most of the fruits are produced in the upper branches, and we are often forced to wait until the snow melts to try to find an intact fruit on the ground. Because of their ability to hang in the crowns of trees through winter, the fruits are well adapted to long periods of dry, cold storage. Intact, dry fruits or dry seeds can be stored for a number of years.

The end of each seed is exposed at the outer circumference of the fruit ball. The seeds can be easily dislodged with a slight twist of the fruit, leaving a hard, round structure that was the base of the flower. Fruit balls on the ground may be dry, but if they are wet and partly exploded, the seeds must be sown right away. Isolated trees tend to produce fruit balls with a low seed

The fruit of Platanus occidentalis, *the sycamore, remains attached to the tree through the most violent of winter storms, suspended by a few tough fiber strands. The evolutionary benefit is that the fruit ball might shatter during such a storm and the tiny seeds are then carried as far as possible.*

viability count because even if only a few flowers are fertilized, the whole fruit still develops.

Separated seeds can be sown in April by scattering them densely enough that they touch (especially seeds from isolated trees) and barely covering them with soil or sand. If the seed is kept constantly moist, germination takes place after two weeks in warm conditions. The seedlings can remain together for the first growing season or be pricked out and transplanted at the three- to four-leaf stage. One-year seedlings can reach 12 inches or more and should be dug up, separated and planted into a nursery for another year. In more northerly parts of its range, the seedlings should be subjected to at least one winter to select for the hardiest seedlings, as determined by no winter injury of the shoot tip. As with all plants, for the sake of species health, do not hesitate to throw poor seedlings onto the compost heap.

Fruit of Platanus x acerifolia (London plane tree).

Sycamore, with its fibrous root system, is easy to transplant but should only be planted in the spring. The seeds are naturally adapted to germinate in the wet soils at the high-water line along creeks and rivers and the edge of swamps, but will thrive when planted on merely moist sites. They are short-lived in a dry location. Seedlings can be potted up for September planting on sites that are too wet in the spring to access. It is a lovely tree for providing shade to large ponds and along streams. The leaves shed their downy hairs during summer breezes and may cause some people to sneeze. Rabbits will devour young saplings in the winter and beavers will take these trees when all their favorites are depleted.

Populus, the POPLARS and ASPENS

Exotic Alert

Carolina poplar, *Populus x canadensis,* is a human-contrived hybrid between Eastern cottonwood and the European black poplar, *Populus nigra.* It is slightly faster growing than either parent — the nature of hybrids — and has been planted in many parks, natural areas and farm windbreaks. It is identified by the slightly wedge-shaped leaf base and the absence of lump-like glands where the leaf blade meets the leaf stalk — cottonwood has a nearly straight leaf base and leaf base glands.

White poplar, *Populus alba,* is a historically planted European species and persists as an extensive, colony-forming "weedy" tree in many rural areas. It is distinguished by the maple-like leaves, which are downy white on the underside and dark, almost glossy green above. As beautiful as it can sometimes be, a bulldozer is a useful implement for managing this tree.

I am often brought to a standstill by the subtle fragrance from the decaying leaves of cottonwood, but this may be due to the large cottonwood in my childhood backyard or my memorable autumn cycling tour through the Netherlands, where thousands of poplars grow along the famous canals.

This early-successional genus is very important, as it lays down the large amounts of organic material needed to rehabilitate eroded mineral soils and disturbed landscapes resulting from fire, clear-cutting or bulldozing — whether by glaciers or modern machines. There are roughly 40 species of poplars in the northern hemisphere, four of which are native to our area.

Female flowers (left) and male flowers (right) of Populus balsamifera *(balsam poplar)*

Populus balsamifera, BALSAM POPLAR. The long, pointed buds are coated with a sticky resin that produces a strong fragrance as the buds are unfolding in the spring — the "balsam" that is now synthetically produced for commercial use. Balsam poplar spreads by shoots arising from a vast root system in moist to wet soils near swamps and marshes, but also on shores and dunes and in the bottom of old aggregate extraction sites. The species ranges from Newfoundland to Alaska, virtually covering the Great Lakes watershed except for small areas south of lakes Erie and Ontario. (*See* photo p. 250.)

Populus deltoides, COTTONWOOD, is a majestic tree with great spreading branches. It is found in moist woodlands or along stream banks and shores. Although any poplar or willow will produce the "cotton fluffs" that carry the seed aloft in mid-June, one cottonwood tree produces an enormous amount of fluff; it is also noticed more because it has been planted close to homes. Of all the native poplars, only cottonwood grows as a non-suckering tree. It is a variable species, with the typical form occupying the watershed north and south of Lake Ontario and Lake Erie, into the southern states and down the Mississippi River.

Female flowers (left) and male flowers (right) of Populus deltoides *(cottonwood)*

Populus heterophylla, SWAMP COTTONWOOD, is an uncommon tree with us, very rare in southern Michigan and northern Ohio and only discovered recently (2002) in Ontario. It prefers even wetter conditions than eastern cottonwood and forms clonal stands in the organic soils of buttonbush swamps. The leaves are larger than our other poplar species and the pith is red rather than white. Like balsam poplar, the leaf stems are rounded rather than flattened. It flowers in late

April and May and, a few weeks later, the egg-shaped capsules on 4- to 6-inch catkins split to release cottony masses with seeds. This rarely occurs in northern clones because the sexes are isolated by distances too great for pollination to occur.

Populus tremuloides, TREMBLING ASPEN, and *Populus grandidentata*, LARGETOOTH ASPEN. The two species are very similar and both provide the sound of leaves clattering (some native peoples called this the "talking tree") that helps to drown out the sounds of industry but it is also the autumn color that excites and inspires. Aspens are colony-forming trees, with single clones often covering vast areas by way of shoots rising from wide-ranging roots. They rapidly seed into and colonize the forest ecosystems in the well-drained soils above the wet soil regime of balsam poplar. Largetooth aspen prefers the higher ground of the two aspens.

The smooth greenish white trunks and small, almost round leaves distinguish the trembling aspen from the smooth, greenish yellow trunks and the larger, coarsely toothed leaves of the largetooth aspen. The two species occasionally hybridize. Trembling aspen ranges from Newfoundland to Alaska and south from the tree line into Tennessee and Mexico. Largetooth aspen is confined to a more eastern distribution south of the boreal forest from Nova Scotia to Minnesota and south into Tennessee.

Poplar flowers appear in April as 2¾- to 4-inches long, hanging catkins (*see* photo p. 250) with the male flowers appearing bright red just before the anthers split open to release the yellow pollen. Female flowers remain somewhat greenish yellow. Individual clones can be determined at flowering time, as male and female flowers are on separate trees, or on separate clones, and each tree in the entire clone flowers at exactly the same time. The flowers are wind pollinated.

The female flowers mature as an elongated chain of small capsules that split open within a day of turning yellowish. Each capsule holds dozens of very tiny, medium green, elongated seeds, each attached to tufts of very fine cotton-like hairs. Poplar seed disperses on warm, dry days in late May when the light cotton (hence "cottonwood") bursts out of the capsules. Although it may occasionally be transported great distances on a good wind, much of it falls in the local vicinity. The cotton of poplar (and its close cousin willow) floats easily. I once found thousands of tiny two-leaved poplar seedlings emerging from their cotton mat that was floating in a pool. In the absence of soil, they all perished, but it indicates that the seeds will germinate in standing water and then come to rest and root in the soil at the edge of a natural body of water.

Playing with Poplars

With the success of the hybrid Carolina poplar, further studies of cottonwood hybrids with black poplar revealed seedlings with exceptional growth rates. These were observed, cloned and studied and much was made of the promise that forest farming of fast-growing "hybrid" poplar clones would one day provide our fiber needs and save the natural forests from being turned into newsprint and toilet paper. The predictions did not always stand the test of time, because fast growth rates were countered by increased sensitivity to pathogen invasion. Barb Boysen of the Forest Gene Conservation Association of Ontario adds that "the value of this type of wood supply could not always be justified against the cost of the necessary continuous breeding and then testing to increase the genetic base of disease resistant clones," meaning that as the diseases evolve, the clones must also evolve.

Many of the poplar clones never really got off the ground due to the ravages of canker and rust diseases. A few clones like DN 9 (*P. deltoides* x *P. nigra* #9) have performed well on certain sites and are still being grown. Cottonwood is normally found in nature as widely scattered small groups and individuals; perhaps poplars cannot tolerate the extreme monoculture requirements of clonal plantation forestry.

Present efforts to genetically modify trembling aspen trees, *P. tremuloides*, for lower lignin content — to improve efficiencies in the pulp mills — are bound to meet with similar constraints. Lignification is the secondary cell wall thickening process that ensures cell wall strength and tree trunk stability. It is unlikely that these genetically disturbed trees will have greater difficulty standing up — popularized by critics as bent-over forests. However, lower-lignin clones may have greater difficulty standing up to diseases, since fungi must penetrate cell walls in order to proliferate. The weaker cell walls are likely an open door for disease invasions in a monoculture. Even though aspen is a colony-forming species, the natural forest is made up of clone patches, each of which is genetically different — nature's way of ensuring species survival.

Collection of the chains of capsules can be done by picking them from the tree when the first hint of yellowing occurs. To obtain seeds from lofty crowns, wait for some of the catkin chains to drop to the ground — they always do — and pick up a few, even if they have started to release the seeds. If the fruits are placed inside a paper bag they will open within two days to yield a loose mound of the "cotton" with tiny green seeds throughout. Poplar seed is comprised of an embryo with a thin membrane — not a seed coat — thus making them relatively short-lived, so they are best sown right away.

The cotton can be spread thinly on pre-wetted soil and watered very lightly to avoid floating the seeds to one location. Either cover the seeds or mist them frequently, and surprise yourself with the stunning speed of germination — within 24 hours! If you do cover them, the enclosure needs to be vented for a few days prior to removal. Poplar seedlings are fairly tough, but do not overwater them. Young plants can be transplanted at the three-leaf stage and spaced about ¾ to 1 inch apart in a tray or pot and then transplanted into your nursery in late June. They will reach about 6 to 10 inches in the first year and grow another 20 inches in the second year. They have very fibrous seedling roots and are easy to transplant and establish, but cottonwood foliage, in particular, is a highly desired summer food for deer, and young trees provide a tasty bark for meadow voles in winter.

The poplars will root from winter stem cuttings but the aspens will not. Aspens and balsam poplar readily propagate by root cuttings — just keep in mind that the smallest plants establish the easiest. The horticultural industry has provided a few selections of columnar-growing poplars, most of which have fallen out of favor in preference for longer-lived columnar oaks and maples. Lombardy poplar, *P. nigra* 'Italica', was the most commonly sold and I still see massive old specimens in a few places — naturalizing extensively in some.

The tiny, light seeds are covered with short, fine hairs that aid in wind dispersion.

Potentilla fruticosa, SHRUBBY CINQUEFOIL

Only this species of the more than 200 potentillas in the northern hemisphere is truly a woody plant. *Potentilla fruticosa* is most commonly known by the numerous horticultural selections with a range of flower color from white through numerous shades of yellow to orange-red. The mounded monotony of the endless stream of new cultivars is a travesty. I recall seeing the stunning weathered multicolored trunk of an old specimen in Gros Morne National Park in Newfoundland. The wild species is a low, occasionally dense, irregular shrub growing 3 feet tall on the well-drained but moist soils of fens, shores, rocky stream edges and alvars. It has a circumboreal distribution, ranging south to New Jersey, Illinois and South Dakota and to Arizona. The North American race produces perfect flowers but those in northern Europe and Siberia are separately-sexed plants.

The striking display of five-petaled yellow flowers in June, when few other shrubs bloom, followed by sporadic bloom into September, is a welcome sight to insect pollinators. The flowers are perfect but seed set is uncommon on isolated plants or clone groups. The upright bract clusters are beautifully adapted to hold the seeds and disperse them over an extended period from late October through February. A few seeds at a time are shaken loose by wind or animals, and although local dispersal is most common, the tiny seeds with their hairy appendages may at times travel great distances with the drifting snow. Natural germination takes place in the first spring.

Each flower head produces about 20 seeds that are easily shaken from ripe, open heads. The dry seeds survive in cold storage for a number of years. They require no pretreatment and may be sown directly into a pot in the cold frame in mid to late May to germinate in about 10 days. Small seedlings can be transplanted in clumps to wider spacing or thinned out and will grow to about 3 inches tall in the first season. Potentilla has a wiry, fibrous root system and is very easily transplanted to produce 12- to 16-inch plants in the second year, after which they should go to a permanent location. Potentilla is unappetizing to all herbivores and is a fine addition to meadow gardens and stream edges.

Prunus, the PLUMS and CHERRIES

Of the more than 200 species of *Prunus* growing in the northern temperate zone, we are most familiar with the awe-inspiring flowering cherries and the fine-tasting horticultural cherries, peaches, apricots and plums. These *Prunus* cultivars all have their origins in the wild, often bitter cherries, peaches, apricots and plums of Europe and Asia. With the possible exception of plums, little effort has gone into searching for tasty selections of our own wild species. Although some are quite tart, all our native plums and cherries are edible, and I had some particularly agreeable sand cherries north of Sault Ste. Marie on the Lake Superior shores. Our wild plums are free of black knot fungus and some are delicious once the fruit begins to naturally drop.

Exotic Alert

Most of the cultivated plums were grafted onto Asian plum rootstocks, which often lived longer than the cultivar. The rootstock thickets can now be found in fence lines and around abandoned farms. Just to make sure that the invasion isn't only one way, our black cherry has now become very invasive — at least in Germany.

Sweet cherry, *Prunus avium,* is the parent species of all the popular sweet cherry varieties sold in markets. It is so named because the fruit is so relished by birds. The usual suspects, robins and small mammals, transport most of the seeds that have germinated along fence rows throughout the lower Great Lakes, but picnickers have made their contribution, too.

Nanking cherry, *Prunus tomentosa,* is a dense shrubby cherry with persistent tomentose branches and rough pubescent leaves. It flowers pink before the leaves are fully opened in mid-May and the edible, ⅓-inch wide, smooth fruits are scarlet red in late June into July.

Dispersal of cherry seeds is primarily by birds, notably robins among the thrushes and cedar waxwings. The fruit is eaten whole and the stones pass through digestion. Dispersal of plums takes place in September as they are consumed by opossums, bears, coyotes and foxes. I have found coyote scat completely filled with American plum skins and pits, which two days later were raided by small mammals for the seeds. The kernels of all *Prunus* are highly sought after by small mammals. In natural systems, if seeds are adequately covered by leaf fall or are in an abandoned cache, they will germinate in the first spring. A germinating cache may contain 30 to 40 seedlings.

Plums

Plum seeds eaten by mice

Prunus americanus *(American plum) fruit and seed.*

The plums have stones that are flattened and thorns that are actually sharp-pointed, stiff, leafy branches. Even from a considerable distance, such a lovely fragrance emanates from American plum that I am compelled to bury my face in the blossoms as the insect pollinators do. The wild plums are all very similar in their self-cloning habit, forming sometimes immense single-genome thickets up to 13 feet tall, rising from their shared root system. The plums are shade intolerant.

Prunus nigra, CANADA PLUM, is distinguished by white flowers tinged fading pink, elliptic leaves with minutely gland-tipped teeth, and 1-inch reddish orange, nearly spherical fruit without a glaucous bloom. (*See* photo p. 252.) Canada plum grows in moist, well-drained sandy or clay embankments from New Brunswick and New England states through southern Quebec and southern Ontario, scattered through Pennsylvania and Michigan, and abundant again from Wisconsin to Manitoba. In the several Ontario colonies that I have visited, the fruit was infected with an insect that causes it to abort the seed, swell to nearly twice the size that it should be and finally shrivel and dry on the stem.

Prunus americana, AMERICAN PLUM, has pure white, fragrant flowers, elliptic leaves with sharply pointed teeth, and ¾- to 1-inch red to yellow elongated fruit with a glaucous bloom. Abundant crops are produced annually where the species is abundant and all of it is consumed or stashed away by mammals. This species is found in moist to dry soils on sandy, clay and rocky embankments, fence lines and woodland edges, with a more southern range than Canada plum, from New Hampshire through southern Ontario and southern Michigan to extreme southern Manitoba and the Dakotas and south to Florida and Oklahoma.

Prunus angustifolia, CHICKASAW PLUM, can be distinguished by white flowers, lance-shaped leaves with very fine teeth and a gland near the base of each tooth, and a small ⅓-inch red to yellow glabrous fruit with a nearly round pit. It reportedly grows in dry, sandy and rocky soils from Virginia to the south shore of Lake Michigan and Kansas and from Florida to Texas. Another plum found just west of our region is HORTULAN PLUM, *Prunus hortulana*. ALLEGHENY PLUM, *Prunus alleghеniensis*, a plum of more restricted range, occurs in central Pennsylvania.

Cherries

Prunus pumila, SAND CHERRY. This is a low-growing species that is divided into varieties. V. *depressa* is a sprawling prostrate shrub found primarily on rocky shores from New Brunswick to Wisconsin and New Jersey to Pennsylvania. The variety *pumila* is a more upright, branching, several-feet-high sprawling shrub on dunes and sandy shores of the Great Lakes. Another variety, *besseyi*, is a low shrub ranging from Minnesota westward. The spherical fruit is nearly black, sweet and delicious and matures in August. After finding piles of seeds chewed open, I suspect small mammals likely get most of this seed.

Prunus virginiana, CHOKE CHERRY, is a colony-forming species, spreading by shoots arising from the roots to form dense thickets in fields and fence lines or thin colonies in the forest understory 7 to 10 feet tall, with occasional trees reaching about 20 feet in height. Choke cherry not only tolerates a wide range of light conditions, it grows in almost any well-drained soil from sand dune to swamp edge and can be found across the southern half of Canada southward from the Carolinas to California. The small fruits are held on long pendent stems and ripen from August to October, often persisting into December. They are tart and red in the typical species and nearly black and sweeter on the black-fruited variety *melanocarpa* (*see* photos pp. 252 and 253). Most plants are

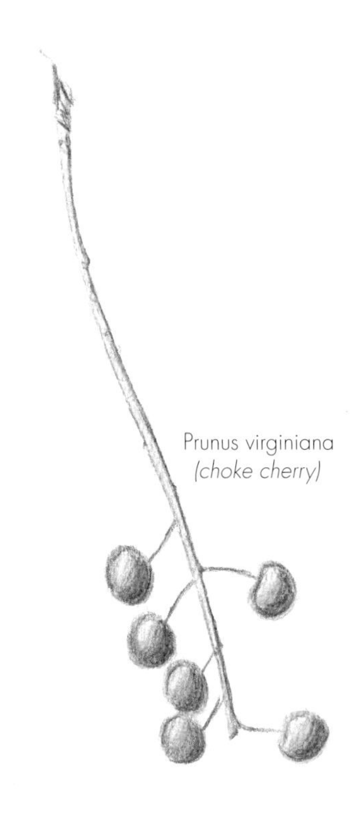
Prunus virginiana
(choke cherry)

infected to some degree with the black knot fungus that arrived in North America with the cultivated plums. (*See* photo p. 241.)

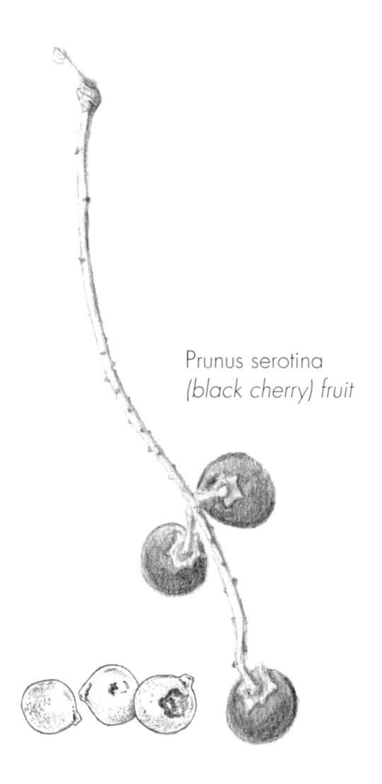

Prunus serotina *(black cherry) fruit*

Prunus pensylvanica, PIN CHERRY. The ability to sprout from the roots gives this species quite an advantage after a forest fire, and that gives rise to the less used common name of fire cherry. It occupies a wide range of soils from moist, deep-soil woodlands to rocky and gravelly barrens and sand plains from Newfoundland to British Columbia, completely surrounding the Great Lakes from Pennsylvania to Minnesota and scattered southward into Tennessee and the American Midwest. The tiny, tart red fruits are produced in clusters ripening from late July into August, and are consumed primarily by cedar waxwings. The relative absence of pin cherry in today's landscape speaks perhaps of the tragedy of fire suppression. (*See* photo p. 252.)

Prunus serotina, BLACK CHERRY, is the largest of the genus and one of the most sought-after for its premium-quality wood. Large trees generally develop only in the rich, well-drained soil of a woodland environment but, through much of the range, it is found without such protection in degraded soils, where it develops a heavily branched, rounded crown. The cherries are ripe in late August with heavy fruit crops at two- to three-year intervals. (*See* photo p. 252.) Yellow leaves are symptomatic of virus diseases brought in with the horticultural cherries; seeds from these trees should be avoided. Black cherry can be found throughout the deciduous forest, extending northward into the mixed forests from Nova Scotia through southern Quebec, southern Ontario, Michigan and Wisconsin into Minnesota, southward to Florida and Arizona and down the Sierras as far as Guatemala, and even across the equator into the Andes.

Reproduction and Propagation

Flowers are insect pollinated. The plums and pin cherry are in full bloom in early to mid-May when the leaves are just starting to expand. The other cherries bloom in early June when leaves are fully expanded. Ripe fruits are soft, the plums being at their tastiest (despite sometimes tough skins) in full ripeness, just as they begin to fall with the first frosts. Seeds (stones) are removed by eating the fruit and saving the stone or by grinding the fruit in the grit bag and floating the pulp away from the stones. The seeds must be guarded from mice and chipmunks at all times, as *Prunus* are among the most sought-after of all the seeds.

Seeds of the *Prunus* can be stored cold but are best sown in a pot or fully screened seedbed frame as soon as they are cleaned, or the seeds should be stratified in a plastic bag. Pin cherry is the earliest of the cherries to ripen and should receive 30 to 60 days of warm, moist stratification prior to the 120 to 150 days of cold stratification that all other *Prunus* require. Cherry seeds need a unique continuity of the cold-stratification period (unbroken by warming) to prevent the seed from reverting back to full dormancy. Stratified *Prunus* seeds should remain in the cold and be checked weekly toward the end of their treatment time until radicle emergence is observed through the plastic. Sprouted seeds and seeds that have cracked open can be planted at ⅓- to ¾-inch spacing (2 inches for plums) to grow for a season.

Seedlings may reach 8 to 16 inches in the first year and can be lifted, root-pruned and planted into a nursery frame or their permanent site. If they are left in the seedbed for two growing seasons, the roots tend to be heavy and unbranched. Early spring planting is essential with bareroot plums and cherries, as they rarely survive if planted when buds have begun to swell, unless cut back to near ground level. Rabbits and mice will browse young seedlings but generally leave older plants alone. The wild plums, choke cherry and pin cherry are best suited to larger yards and natural areas. Black cherry is a marvelous backyard tree. Although sand cherry is susceptible to a blossom blight that may periodically kill branches, it always resprouts from the base and is well suited to dry gardens.

Ptelea trifoliata, HOP-TREE

This shrubby, small tree of 10 to 16 feet in height grows in a range of sunny conditions from dry, sandy shores to moist, rich open woodlands and fencerows. It is found from the south shore of Lake Ontario and around Lake Erie in New York, Ohio and extreme southern Ontario, across southern Michigan and around the south shore of Lake Michigan into Wisconsin, and southward to Florida and Mexico. It meets and intergrades with the pubescent form, *P. t.* var. *mollis* (*mollis*: hairy), nearly throughout the range but more commonly farther south.

Fruits are disseminated through the winter months, traveling great distances over frozen ground, snow, ice or sand. Natural regeneration is only by seeds, which germinate in the second spring after fruit drop. Hop-trees will naturalize from planted specimens north of their range.

Hop-trees bloom in June with male and female flowers usually on separate plants. Pollination of these fragrant flowers is by insects, including flies and butterflies. Annual crops mature in September, turning from green to yellowish and ripening to brown with a wing surrounding a single (sometimes two) flattened nutlet. The fruit can be harvested any time from late September through winter. Occasional hermaphrodite plants (both sexes present in the same flower cluster) produce only a few fruits per fruiting cluster. The nutlet does not need to be removed from the wing;

Ptelea trifoliata, *the hop-tree, is an entirely North American genus and a northern member of the citrus family.*

dry fruit can be stored for a few years in a sealed container. Seed germination in the first spring is easily accomplished with a pre-germination treatment. Soak the seed in early December for 24 hours and surface-dry prior to a warm stratification for 30 days, followed by 90 days of cold stratification.

Treated seeds can be planted in early April about an inch apart and must be protected from mice. They will have enough space to grow 4 to 8 inches tall in the first year. Seedlings can be lifted and pulled apart in the following spring and planted at 4-inch spacing in full sun to partial shade for another year to reach 16 to 24 inches tall. Establish these plants in dry, sunny, open locations or at the edge of woodlands.

Roots of this easily transplanted species are deep and very fibrous. Rabbits seem to leave it alone but groundhogs go crazy for the leaves — even to the point of climbing into the small trees, often eating the bark as well and occasionally killing young plants. Meadow voles will occasionally eat the basal bark from trees up to 4 inches in diameter. Thrips infest these trees in great numbers in late summer but do not harm the plant. By September, they provide a feast for migrating kinglets and warblers. The leaves are also a principal food of giant swallowtail butterflies.

Quercus, the OAKS

Over 400 species of oaks are found through the mid-latitudes of our planet, growing in swamps, woodlands, savannahs, deserts and cloud forests in every soil type from rocky gravel to sand dunes and saturated clay. Some species are very specific, others seemingly quite adaptive to a range of soils, but even these apparent generalists may have local races that are adapted to the specific site. Consideration is needed to match the species and the seed source to your site conditions. Thirteen oaks indigenous to the Great Lakes region are described here; the ecology of oaks beyond our watershed can be referenced in other manuals — the seed handling doesn't change.

The oaks can be divided into three groups — the black, the white and the chinquapin oaks. The black oaks have sharp-pointed leaf teeth and lobes and smooth bark until the stems are at least 16 to 20 inches in diameter; the acorns take two full growing seasons to mature. The white oaks have leaves with rounded lobes, rough bark on stems as small as 4 inches in diameter, and the acorns take only one growing season to mature. The chinquapin oaks have many rounded teeth and rough bark on stems as small as 4 inches in diameter, and the acorns mature within one growing season.

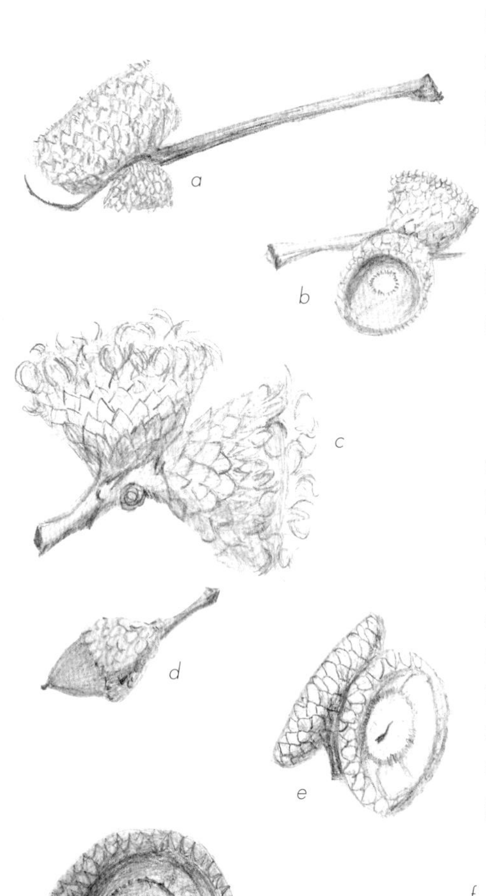

Acorn caps are a great help in identifying the different species. Shown above are the caps of: a) swamp white oak; b) chinquapin; c) bur oak; d) northern pin oak; e) red oak; f) black oak; g) Shumard oak.

A crude identification of oaks is initially based on the bark and leaf characteristics such as the presence or absence of teeth or lobes. Leaves on the same tree are variable in shape and size, depending on their exposure to sun and altitude within the tree, so other clues are needed. A more critical identification is based on acorn size, the presence or absence of longitudinal bands on the acorn, how much of the acorn is covered by the cap, the texture of the rim of the cap and the length of the cap's stalk. It is the acorn cap and stalk that hold most of the identification clues, not the leaf or acorn alone — take a few caps home, too (even those of the previous year that are still on the ground), and compare them to the illustrations here.

Quercus velutina
(*black oak*)

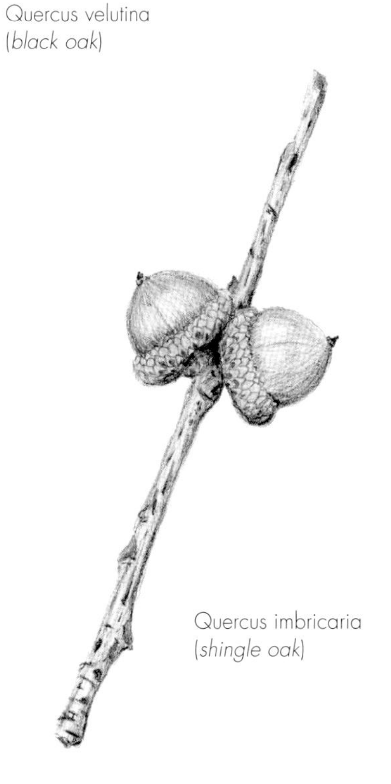

Quercus imbricaria
(*shingle oak*)

Hybrids do occur, and indeed, more than 60 are even named! Undoubtedly, back-crosses are also likely to happen, thus creating a number of species complexes that would lead one to think that naming oaks can be like naming snowflakes. Such diversity provides exciting work for systematic botanists and general confusion for the rest of us — be assured, most species are found in abundance with relatively typical characteristics.

The oaks of this region are a significant component (along with hickories) of the deciduous-forest life zone. They are all savannah and open oak-woodland species adapted to periodic ground fires (or other perturbations). A few species range northward into the mixed forest and there occupy a smaller proportion of the forest. Few genera can match the ecological importance of oaks. The abundance of woodpecker species in old-growth oak forests is a reliable indicator of wood-boring insect diversity and hence habitat diversity. Dozens of insect species feed on the foliage of oaks, from gall-forming midges to leaf miners and the giant silkworm moths. It is the acorn crop — the mast — that is perhaps most significant. This high-energy, fat-rich food supports a wide range of wildlife such as weevils, wild turkeys, deer, bears and passenger pigeons before they became extinct, not because of hunting (as brutal as it was) but because of habitat destruction and the scarcity of acorns and beechnuts associated with the land clearing that took place in the mid-1800s. Squirrels and blue jays are now the primary agents of seed dispersal.

Black Oaks

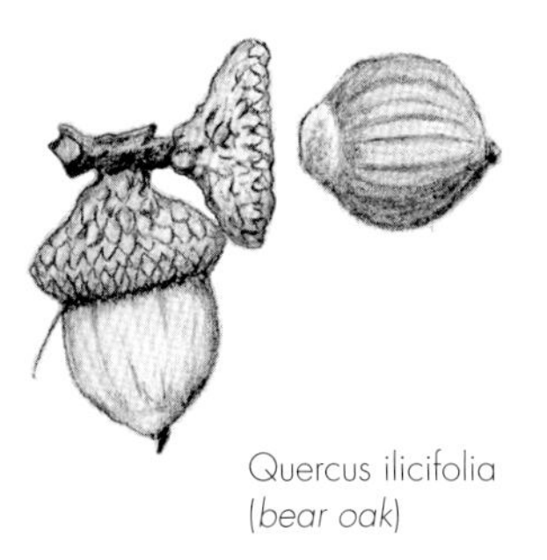

Quercus ilicifolia
(*bear oak*)

The black oaks have smooth-barked branches and produce a mature acorn over two growing seasons. The leaves have bristle-pointed teeth or entirely smooth margins in the more western species. The acorns are bitter to our taste but highly

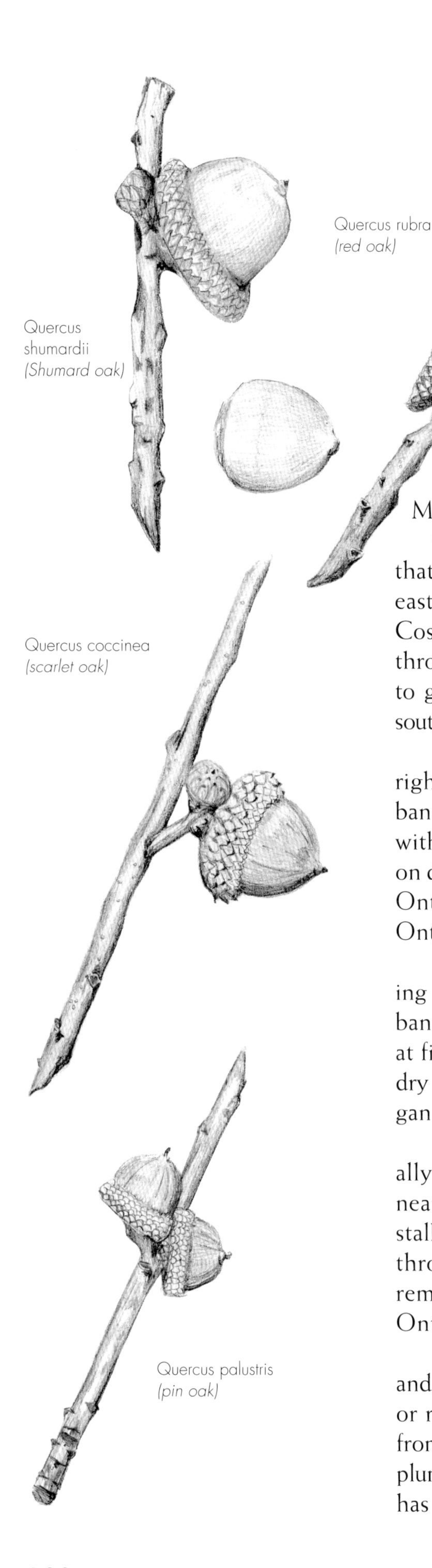

desired by wildlife. They are described here in the order of general moisture preference from dry to wet.

Quercus velutina, BLACK OAK, is a large, wide-spreading oak with stunning fall color. It is mostly associated with the dry, sandy soils of "black oak savannahs" — many of which are now unrecognizable due to fire suppression and infill planting. The acorn has dark longitudinal bands and is half-covered by a cap with a distinct fringe of projecting scales at the rim. This oak is found from Maine through southern Ontario and Michigan, into Minnesota and south to Florida and Texas.

Quercus imbricaria, SHINGLE OAK, is an oak with leaves that are entire with a bristle-pointed leaf tip (curious in the east but a common characteristic of oaks from Arizona to Costa Rica). The leaves remain tan colored and attached through much of the winter. It has a preference for dry, sandy to gravelly upland soils and is found from Pennsylvania into southern Michigan to Kansas and south to Georgia and Arkansas.

Quercus ellipsoidalis, NORTHERN PIN OAK, is a fairly upright and twiggy tree with bullet-shaped, longitudinally banded acorns that are one-third covered by an elongated cap with a thick stalk. It is found in widely scattered populations on dry to moist sandy and gravelly upland soils from southern Ontario through Michigan to Minnesota and northwestern Ontario and south to Ohio and Missouri.

Quercus coccinea, SCARLET OAK, is a large, wide-spreading tree renowned for fall color. The plump acorn is without banding and is about half-covered by a cap that is pubescent at first but becomes glossy as it matures. This species prefers dry upland soils and is found from Maine to southern Michigan and Georgia to Missouri.

Quercus ilicifolia, BEAR OAK, is a shrub oak species usually 7 to 10, rarely 16 feet high. The acorn is banded and nearly half-covered by a deep, cup-shaped cap with a short stalk. It grows in sandy soils and rocky outcrops from Maine through New York into Ohio and south to Virginia. A very remote outlying population was discovered in eastern Ontario in 1993.

Quercus rubra, RED OAK, is possibly the most common and broadly adapted oak in the region. It grows as a forest tree or relic in moist but well-drained, humus-rich soils ranging from sand to clay, gravel, loam or bedrock of any type. The plump acorn is only one-quarter covered by a shallow cap that has tightly closed scales on the rim. Red oak includes what used to be called northern red oak and is found from Nova Scotia through southern Quebec to Lake Superior and Minnesota southward into the southern states. It is distinctly absent from pockets of calcarious glacial soils. (*See* photo p. 253.)

Quercus shumardii, SHUMARD OAK, is a large, wide-spreading oak that at first gives the impression of being a red oak. The larger acorn is very plump and about one-quarter covered by the wide but shallow acorn cap. Shumard oak (which should have been named swamp red oak) grows in the slightly drier rises of low, poorly drained land — much of which is now well-drained with ditches. It is found from Pennsylvania to southern Ontario and southern Michigan to Kansas, southward into Florida and Texas.

Quercus palustris, PIN OAK, is a mid-sized species with characteristic downward-hanging low branches. The acorn is quite small, banded and covered about one-third by the cup-shaped thin cap. Pin oak thrives in moist to wet, periodically flooded soils from Massachusetts to southern Ontario, Michigan and Kansas and to the south from North Carolina to Oklahoma.

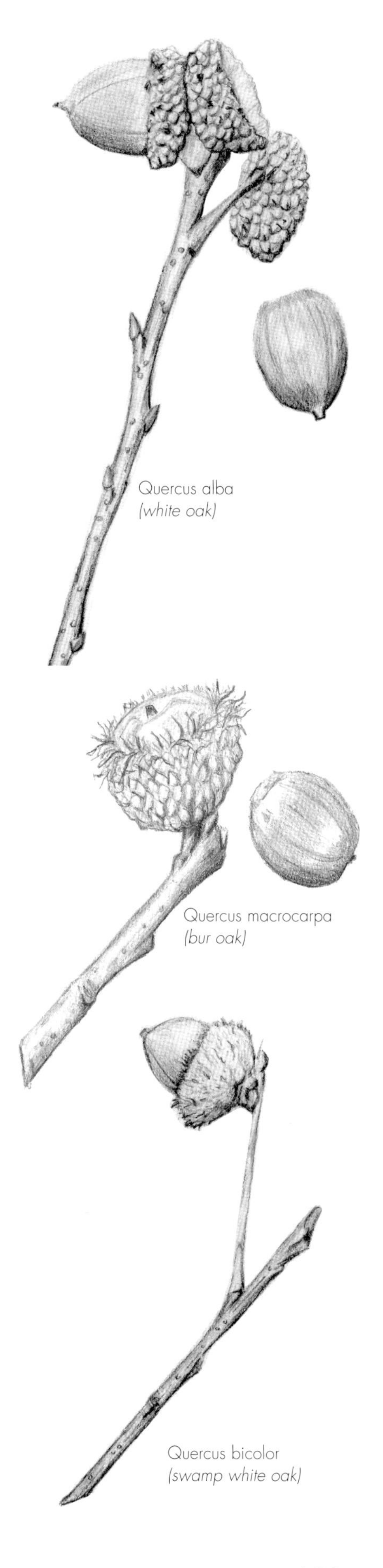
Quercus alba
(white oak)

Quercus macrocarpa
(bur oak)

Quercus bicolor
(swamp white oak)

White Oaks

The white oaks are rough-barked by the time the stems are barely 4 inches across. They produce a mature acorn within one growing season. Leaves of all the white oaks have round lobes and smooth margins. These species (and white oak in particular) are relatively sweet to eat and are highly favored by birds and mammals. The white oaks are described here in the order of general moisture preference from dry to wet.

> **Exotic Alert**
>
> English oak, *Quercus robur*, is in the white oak group and is easily recognized by the extended lobes at the base of the leaf (*see* photo p. 254) and the 1 to 1 ½ inch unusually long acorn. English oak has been planted extensively as an ornamental, ceremonial and roadside tree throughout the region and is naturalizing out from those sites. A strongly columnar form has been popular for a few decades because it fits in confined spaces. Its seeds will sometimes reproduce the rather useless columnar form but often develop as the wide-crowned species.

Quercus alba, WHITE OAK, is a stunning and attractive wide-spreading oak. The elongated acorn is about one-third covered by a warty, cup-shaped cap. This is a tree of upland ecosystems in slightly acidic soils ranging from bedrock to sand or clay, from Maine to southern Quebec and the Ottawa River, through southern Ontario and Michigan and into Minnesota and southward into the southern U.S. Acorns of white oak will germinate on the soil surface within days of falling to the ground and in wet years may germinate while still on the tree. The fall color of the leaf undersides is strikingly pinkish. (*See* photo p. 255.)

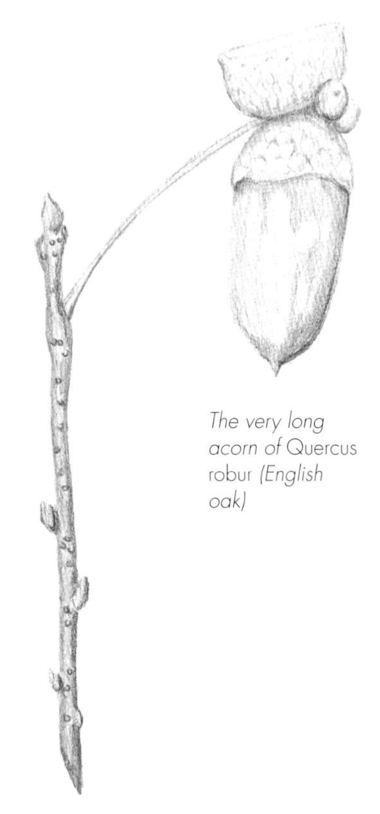

The very long acorn of Quercus robur (English oak)

English oak is the only oak with a "lobed" leaf base.

Quercus macrocarpa, BUR OAK, is occasionally a massive tree with elongated acorns that are three-quarters enclosed by a cap with a distinctly frayed rim — the bur. It grows in seasonally flooded low woods and floodplains and on high ground as well, in sand, loam, clay, calcareous gravel and limestone alvars. It is found in widely scattered populations in New Brunswick and New England and from southern Quebec to central Manitoba south to Louisiana and Texas. Bur oak has a proven tolerance to urban conditions, producing some of the largest urban trees in the Great Lakes region. (*See* photo p. 253.)

Quercus bicolor, SWAMP WHITE OAK, is an oak of medium stature whose leaves are distinctly whitish below, creating a noticeable display on a windy day. The plump acorns are about one-half covered by the cap and often paired at the end of a long stalk. This is a moisture lover found in low woods and poorly drained soils from extreme southern Quebec and the northeastern states, along the south shore of Lake Ontario, through southern Ontario, southern Michigan and Wisconsin and south to Pennsylvania, Tennessee and Arkansas.

Chinquapin Oaks

The chinquapin oaks are rough-barked by the time the stems are barely 4 inches across. The acorns mature within the growing season and they have leaves with rounded teeth and smooth margins. The nut is relatively sweet and rapidly consumed by birds and mammals. Acorns of the chinquapin oaks will germinate on the soil surface within days of falling to the ground, and once I saw radicle emergence on an entire crop of mature acorns that were still on the tree — it had been raining for a couple of days. I have occasionally used distant blue jay activity to locate sparsely distributed chinquapin oaks.

Quercus muehlenbergii, CHINQUAPIN OAK, is a rounded, usually medium-sized tree in the open and a tall tree in forests. This species is found in well-drained locations on both clay and sandy soils in dunes, moist woodlands, embankments and limestone alvars. The small acorns are almost two-thirds covered by the thin cap and produced at the shoot terminal. It is widely scattered throughout the northeastern states and around Lake Ontario and southern Ontario and more abundant from Ohio and southern Michigan to Iowa and south from Alabama to Texas. This species seems to have a considerable tolerance of cultivation and can be used as a street tree. (*See* photo p. 253.)

Quercus prinoides, DWARF CHINQUAPIN OAK, is so similar to the chinquapin oak and grows on such poor soils, one would think it to be stunted by site conditions, but it remains dwarf even when growing in rich soils. This shrub has the unique ability (for an oak) to spread by short rhizomes emanating from basal buds. The acorns are identical to those of *Q. muehlenbergii* except they are mostly axillary on the new shoots. The dwarf chinquapin grows in dry, sandy and rocky soils in barrens and savannahs from Massachusetts through southern Ontario and southern Michigan to Oklahoma and south to Florida and Texas.

Quercus prinus, CHESTNUT OAK, is a tall forest tree in upland woods and on well-drained slopes and rocky ridges. The stout acorns are about half-covered by the thin cap, which has a stalk almost as long as the acorn. The mature bark is deeply furrowed like a black oak. The swamp white oak is coarsely similar in the leaf and has a similarly stalked cap, but is a wetland species. Chestnut oak is found from southern Maine throughout New York to northeastern Ohio and south to Georgia. It was believed to occur in Ontario, but natural trees are nowhere to be found.

Quercus muehlenbergii
(chinquapin oak)

Reproduction and Propagation

The female flowers of oaks are solitary or paired in the leaf axils of the more vigorous new shoots. Although male flowers are produced on the same tree in long catkins, the timing is such that trees can rarely pollinate their own flowers. Pollen is carried on the wind through late May. Fruit development is within one season for the white and chinquapin oaks. In the black oaks, the tiny undeveloped but fertilized flowers remain unchanged through an entire year to expand and mature during the following growing season. Oaks tend to skip a year between good crops — the interim year usually without any acorns but occasionally with a small crop.

Watch the squirrels. When they start eating and dropping acorns it is time to prepare for gathering acorns. In low seed-production years, every acorn on the tree could be eaten or stashed away within a week. Green acorns that squirrels have dropped can be gathered and allowed to dry for a few days until they turn brown. They must then be soaked to rehydrate the seed before planting.

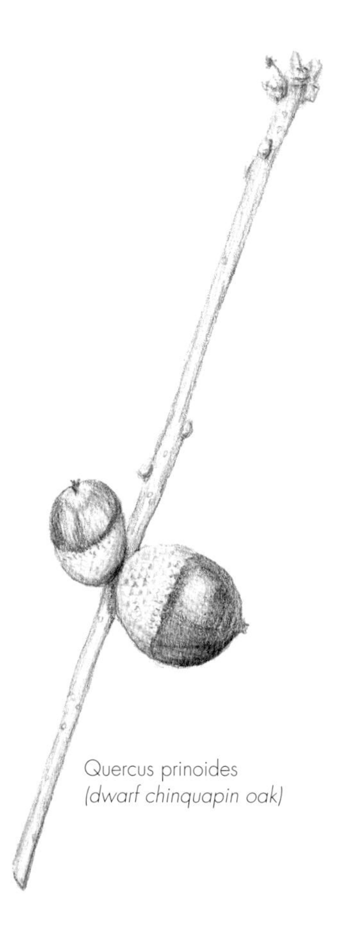

Quercus prinoides
(dwarf chinquapin oak)

When acorns are naturally falling from early September to mid-October you can pick some up for growing at home or plant some at the edge of the woods. Always break a few open to peek inside. They should be clear white or yellow throughout, and you can also look for the tiny embryo at the pointed end. Sometimes weevils eat the insides of the acorns and produce a partly or entirely brown, mealy interior. Acorns that are still on the tree when others are on the ground tend to be weevil-free seeds. (*See* photo p. 253.) Healthy seeds will sink in water whereas weevil-infested seeds tend to float.

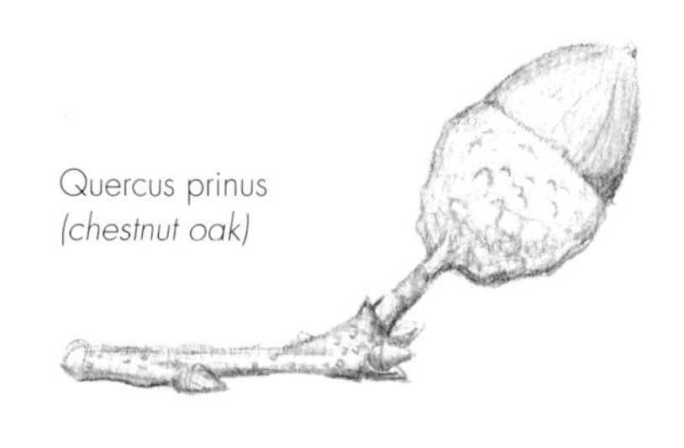

Quercus prinus
(chestnut oak)

Mature greenish red acorns on the ground among dry brown acorns are worth picking up and, if any still are in reach on the tree, picked from low branches. Acorns at this stage of maturity can be soaked and planted without delay. Once acorns have turned brown and are lying on the ground, become more suspicious of quality, although during bumper crop years, good acorns lie on the ground into late October.

Freshly gathered acorns can be kept in a plastic bag for no longer than a day or two at room temperature. Remove the caps but do not remove the seed coat. The float test is an excellent way to determine the soundness of seeds. If you still have doubts, cut some of the floaters open. Acorns are very prone to drying out and can be destroyed within a week if they are not protected in the fridge; acorns from the black oak group can be kept damp all winter in the fridge and will germinate by early spring.

Seedlings emerge in May and will grow to about 6 inches high the first season. After two to three years the trees will be large enough (12 to 20 inches) to plant in a permanent place. Oak seedlings have a deep root and must be dug with care to keep at least a 12-inch length of root intact. Seeds can be planted right where you want a tree to grow; just make sure the animals can't get at them.

When collecting acorns, look for the older trees. Oaks, in particular, are potentially long-lived trees and we haven't a clue which gene pool is the best to survive climate change and diseases. The older oaks in the area provide some evidence of important genetic characteristics that distinguish trees that are able to tolerate drought episodes, pests and diseases. If you are unable to identify a particularly healthy old oak, don't let this prevent you from growing its seeds. Just make sure to note its soil preference. The oaks should be planted wherever opportunity allows, since little natural recruitment is evident in many locations.

Rhus (including *Toxicodendron*), the SUMACS

There are around 150 species of sumac in subtropical and temperate regions around the world. I have retained the name *Rhus* for all species although some authors place the toxic species in the genus *Toxicodendron*. The poisonous species have either yellowish white or grayish white fruit. All other species with reddish fruit are not toxic and, more interesting still, the fresh dried fruit is used for making a vitamin-C–rich sumac "lemonade" by steeping a small handful of berries in boiled water.

The irritating rash or internal infection that comes when their smoke is inhaled makes the few poisonous members of the genus fascinating, but the sumacs are also interesting in their ecological role as pioneers in fire-induced succession, either from root sprouts or a flush of dormant seed germination, and as a survival food, too: some 20 bird species are reported to feed on sumac berries through the winter — reportedly including the pine warbler. I have watched crows increasingly depend on staghorn sumac for the survival of their huge overwintering flocks. But it is returning bluebirds that I watch for in March, waiting out a spring snowstorm and eating sumac fruit with the earliest robins.

The seed is disseminated locally and occasionally for long distances by birds migrating in either direction. The hard-coated seed is separated from the dry fruit in the crop and passed intact in the droppings, and may lay dormant in the soil for a number of years. Natural germination occurs soon after low-intensity fire or solar heating. I have observed sporadic germination in shaded nursery areas and it is likely that they germinate after a few years of microbial decomposition of the hard seed coat.

Rhus aromatica, FRAGRANT SUMAC, is the only non-colonizing sumac in our area. It is a sprawling, mounding plant from a single crown of stems, occasionally reaching 7 feet tall and 16 feet across but often half that size. Fragrant sumac thrives in dry to mesic sites in full sun to partial shade in sand dunes, oak savannahs, rocky soils, limestone alvars and clay embankments. In the oak savannah of Pinery Provincial Park, Ontario, it came in by the thousands after the first controlled burn back in the early 1990s. This yellow-flowering species is found from Vermont to Quebec, in southern Ontario (into Manitoulin Island) and southern Michigan to Illinois, and from Louisiana to Kansas. Fragrant sumac is variable in habit, tending to a more upright form in Missouri and Illinois, and strikingly similar to the ill-scented sumac *Rhus trilobata* — a western, upright shrub that may reach 7 feet tall, with greenish yellow flowers and smaller leaves. *Rhus trilobata* may show up in natural gardens in the east, a result of the importation of plants to meet a burdensome demand for mass planting. Look for its greenish flowers and small leaf size of less than 1 inch.

Rhus copallina, SHINING SUMAC. "Winged sumac" is a more fitting name from the distinctively "winged" leaf petiole. This is a relatively slow, but still extensive, colony-forming shrub, spreading by shoots rising from the shallow spreading roots. It reaches 16 to 20 feet in height in the southern part of its range and barely 7 feet in the north. It thrives in dry conditions on sandy soils and rocky outcrops in the deciduous region from Maine, New York, a few isolated areas in Ontario, much of southern Michigan and Illinois, into the southern states.

Rhus glabra *(smooth sumac) leaf (above) and fruit (left), shown at approximately half their original size*

Rhus glabra *seeds at four times their size*

Rhus glabra *berries, enlarged*

Rhus typhina, STAGHORN SUMAC. The distinctive felt-like hairs on the branches are discernable from quite a distance in good sunlight. As colonies reach about 25 years old, they begin to die out in the center to give way to other species — the true sign of an early-succession plant. It spreads rapidly by shoots rising from the spreading root system to form colonies often over 100 feet across. Staghorn sumac grows in dry, gravelly soils and rocky cliffs, embankments, forest edges and old fields from Nova Scotia to Minnesota and in the south from the Carolinas to Iowa. (*See* photo p. 254.)

Rhus glabra, SMOOTH SUMAC. Other than the near-complete lack of felt-like hairs on the stems and the taller, more pyramidal and brighter red fruit cluster, this shrub is virtually identical to the closely related staghorn sumac. It has a similar range in the east but extends farther west to British Columbia and Mexico. Indeed, the two are known to hybridize, forming a species complex known as *R.* x *borealis*. From almost every seed lot of *R. glabra* that I grow, nearly half the seedlings are intermediates, indicating the possibility that smooth sumac is naturally being overtaken in the east.

Rhus (Toxicodendron) radicans, POISON IVY. Everyone should know this widespread and wide-spreading plant in the wild. It is generally divided into a groundcover race (var. *rydbergii*), found in the mixed forest southward, and a climbing race with aerial roots (var. *radicans*), found primarily in the deciduous forests. Poison ivy creates natural "No Entry" zones and maybe there should actually be more of it for that reason. It is reported to be a very important plant for wildlife, providing fruit for woodpeckers and many other birds, and it can be exquisitely tinted in the fall. Its range covers much of North America, primarily south of the boreal forest. Poison oak is a very similar species (with more indented leaves) in the west, likely with intermediates where ranges overlap.

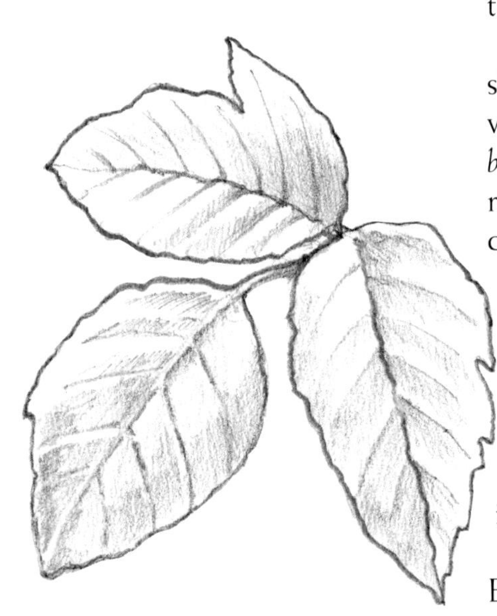

Poison ivy leaves

Rhus (Toxicodendron) vernix, POISON SUMAC, or POISON ELDER. This is a rarely encountered species due to its affinity for growing in wet places. It is a sparsely colonizing small tree in swamps but can form dense colonies on wet slopes. Poison sumac is most noticeable in autumn when in orange fall color. It grows throughout the deciduous-forest life zone of the eastern U.S. with isolated populations northward in Quebec and Ontario.

Sumacs are dioecious and the colony-forming species can easily be studied for male and female clone characteristics. The flowers are bee pollinated in June. The fruit is a single roundish "berry," fuzzy and red in nontoxic species, smooth and yellowish white on the poisonous species. The cone-shaped clusters of fruit persist into the winter and even into a second year. The seed is a pill-shaped disc.

Collection of fruits can be done from September to April and the dry fruits can be kept in a paper bag until you are able to clean them. The seeds are very hard and I use a blender to extract the seeds — a few minutes at a moderate speed is all that is required. The filled seeds sink and empty seeds float with the pulp. I have found entire stands of staghorn and smooth sumac where all of the seeds were empty. This means a fruit crop can be produced with no viable seeds. Cleaned seed is dried at room temperature for a week prior to dry storage.

Seeds will swell to twice their dried size during the hot water–treatment soak and can be planted at ½- to ¾-inch spacing. Germination is readily obtained in 10 days and one-year seedlings will reach 4 to 12 inches, depending on the species. They can be transplanted in the spring and are relatively easy to establish in full sun, despite the cord-like, nearly hairless roots. There is little need to plant the clone-forming species at 6-feet spacing by the hundreds, as I have too often seen done. They colonize quickly enough and, in the interim, the herbaceous cover is important insect habitat.

Although seed acquisition and germination is easy enough, there will be some interest in the distinctive characteristics of individual clones, especially of smooth sumac. The colonizing species of *Rhus* are easily increased by division as long as you cut the shoot off just above ground level once it is planted. If predators are diminished, rabbits will chew the one-year stems of red-fruited sumacs down to the snow line in most winters. Sumac is a survival food but may lure rabbits away from important fruit crops such as raspberries and apples or other desirable plants in natural landscapes.

Ribes, the CURRANTS and GOOSEBERRIES

For my taste, black currant juice and gooseberry jam place near the top of the comestible list, but then, my ancestry is also from northern Europe, where the cultivated fruits originate. There are over 100 species in the Americas, Eurasia and North Africa. The fruit is a berry that contains several seeds and, as an important food, is widely dispersed by many birds and mammals, from catbirds and robins to chipmunks, and foxes and humans. The common prickly gooseberry produces its flowers early enough in May to be a critical nectar source for hummingbirds after their strenuous migration north. The berries of all currants and gooseberries are ripe from late July through August and are edible and mostly of excellent flavor.

Ribes flowers are perfect, with most species pollinated by insects and ruby-throated hummingbirds from late May into June.

The currants produce several flowers (hence fruits) in chains and, with the exception of bristly black currant, all of the currants in our area have thornless stems. The gooseberries produce solitary fruits (rarely two) at the leaf axils and the branches are armed with sharp spines. The identification of gooseberries is determined by the distribution or absence of glandular hairs or soft prickles on the fruit, as well as its color. Of about ten native species found in the Great Lakes region, I describe eight after familiarizing you with the exotics.

Gooseberry fruit

Exotic Alert

Ribes rubrum (*R. sativum*), the European red currant, can be found in many wild places. It is a more upright shrub with larger berries in longer chains and in drier sites than the native red currant, *R. triste*. The European black currant, *R. nigrum*, of gardens is also a more strongly upright plant and often grows on drier sites than its native counterpart.

Ribes americanum, WILD BLACK CURRANT, is a loose shrub growing up to 3 feet with branches usually reclining and layering. The delicious fruits are black and smooth, in chains of several berries. This is a common shrub of moist meadow edges, woodland clearings and stream edges, growing in mixed and deciduous regions from Nova Scotia to Alberta and southward from Delaware to Colorado.

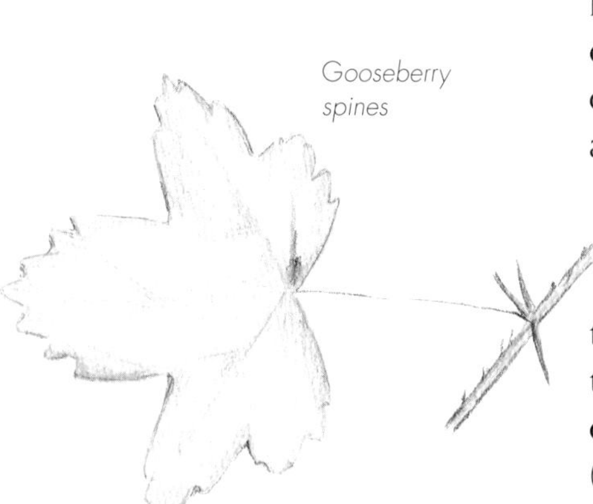
Gooseberry spines

Ribes cynosbati, PRICKLY GOOSEBERRY, is a low, sprawling shrub with upright branches rarely reaching 3 feet tall before reclining again and rooting. The branches have one to three spines located at each node. Flowers are born from early May into June and the wine-colored solitary edible fruits (despite the brownish, soft prickles) mature from mid-July into August. This is a common plant of moist to dry woods, rock outcrops and stumps in mixed and deciduous regions ranging from New Brunswick through southern Quebec and south-central Ontario, Michigan and North Dakota, south to Georgia and Oklahoma.

Ribes glandulosum, SKUNK CURRANT. Easily distinguished by the skunk-like odor when any part of the plant is bruised, this ground-hugging shrub is common but rarely noticed. The long, trailing stems will root where they touch ground. Upright flower clusters open in late spring. This species might be dioecious since some plants are without fruit. The small, spherical red berries are covered with glandular hairs and are quite unappetizing to us. This shrub is found in moist woods, swamp edges, rocky ground and clearings across North America in boreal and mixed forests from Newfoundland to British Columbia, south to Vermont and Minnesota and through the Appalachian Mountains to Tennessee.

Red currant

Ribes hirtellum, WILD GOOSEBERRY, is a loose, upright shrub that may reach 3 feet in height, with few spines (often none) on the stems. The fruit is delicious, bluish black or dark purple and smooth, produced solitary in the leaf axils from mid-July into August. A common shrub on moist soils in clearings and on rocky embankments and shores, it grows from Newfoundland to James Bay and Alberta, and south to Pennsylvania and Nebraska. Wild gooseberry is very similar to *R. oxyacanthoides*, the bristly wild gooseberry that is found from Michigan and north of Lake Superior westward to Alberta and Montana, and the two likely intergrade where their ranges overlap.

Ribes missouriense, MISSOURI GOOSEBERRY. This species looks like *Ribes cynosbati* but without the prickly fruit. It too is a small bush that will grow under many conditions from moist to dry and shady to sunny, including old fields and pastures on all soil types. It has great value for wildlife. The species ranges from the west side of Lake Michigan right across the tallgrass prairie region.

Ribes odoratum, CLOVE CURRANT. This unarmed, upright shrub grows 7 feet tall. The leaves are small, fan-shaped and generally three-lobed. Spicy-fragrant, tubular yellow flowers really set it apart. It has a small but tasty black fruit. The species is not fussy as to soils or exposure although it suffers in deep shade and wet sites. It thrives in prairies, savannahs and disturbed sites. Although native west of Lake Michigan, it has escaped from cultivation all through the Great Lakes area and can be found in old fields, fencerows and roadsides.

Ribes triste, SWAMP RED CURRANT, is a scraggly shrub with layering low branches. The small red berries are produced in small chains of only half a dozen berries as compared to twice that number on the garden varieties. It grows in moist ground in woodlands, bogs and swamps and along streams across North America as far south as Virginia and through the northern states to Oregon.

Fruits can be gathered when they are mature to ripe in late summer and the very hard seeds removed with the grit bag, giving them an extra-firm scrubbing to scratch the very tough seed coat. Commercial operators place the seeds in a sulfuric acid bath for 10 minutes. Each berry contains several seeds and they should be planted in a clay pot that is plunged into your cold frame, mulched and forgotten until early April, when the mulch should be removed. Germination is usually in the first spring and one-year seedlings should reach 6 to 8 inches. The fibrous roots can be teased apart in the spring and the seedlings planted into pockets of your garden or a natural area. I have not seen evidence of herbivore interest in this genus. Compromised white pines in your area may become infected with blister rust and be killed — fair warning — an important part of evolutionary selection.

Ribes and White Pine Disease

All *Ribes* species are alternate hosts for white pine blister rust. In the early 1900s, millions of dollars were spent in a failed attempt (fortunately) to eradicate all currants and gooseberries from white pine growing areas. The rust fungus killed thousands of trees, but stands of pine within the range of blister rust now seem to tolerate the disease. The forester's strategy has shifted to identifying rust-resistant pines for seed orchard establishment. There is still an out-dated and prevailing recommendation that *Ribes* be eradicated — but to what end ... has the white pine not already naturally selected for rust tolerance on its own?

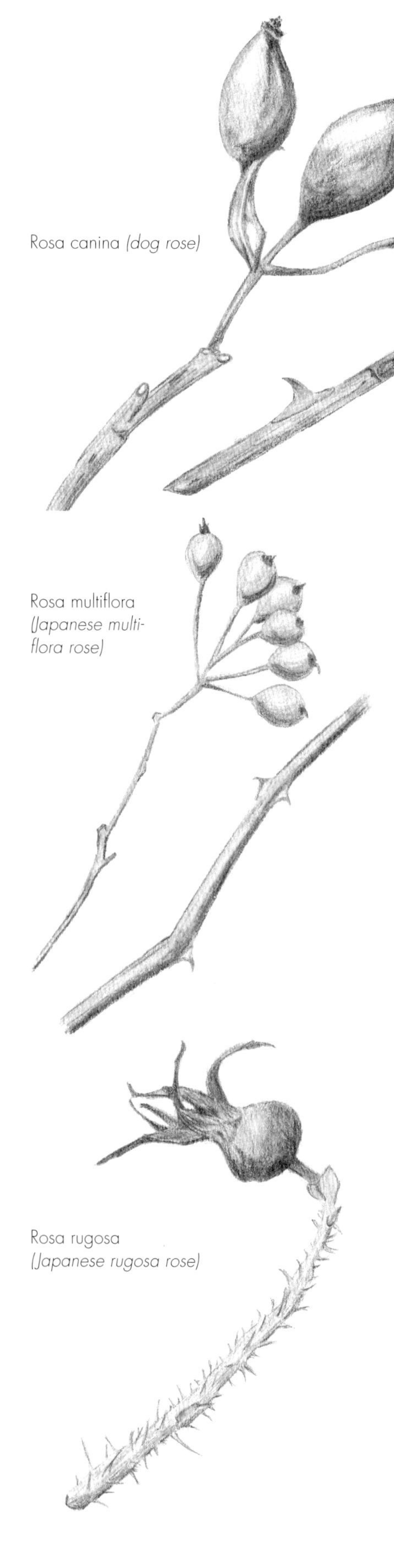
Rosa canina (*dog rose*)

Rosa multiflora (*Japanese multiflora rose*)

Rosa rugosa (*Japanese rugosa rose*)

Rosa, the ROSES

Few other genera have had so much horticultural attention as roses. Although historical interest was focused on flower color and form, current interest is being placed on returning the fragrance, hardiness and disease tolerance to garden roses. The petal-filled flowers of the horticultural hybrids (not to be confused with natural wild hybrids) are rarely accessible to pollinators and are no match for the simple, ephemeral beauty of wild rose blooms, let alone the autumn display of red rose hips and leaf colors. The wild roses grow in a wide range of sometimes extreme conditions — from swamps to deserts and often in some of the poorest soils — so different from the intensive-care ward required for most horticultural roses.

Exotic Alert

Several garden and "wildlife" planted roses are found in the wild but can be readily distinguished from the indigenous species. Most of the introduced species tend to be upright-growing shrubs, with even small colony formation rare as compared to the low colony-forming nature of our wild roses. The four most common escapees are described here.

Dog rose, *Rosa canina*, is an upright European shrub with loosely spaced basal stems reaching 7 to 10 feet tall. The outward-arching branches are armed with stout, randomly spaced, hooked prickles. The red fruits are up to ¾ inch long and often quite tasty by midwinter if you learn to pinch off the seed-free basal end of the hip with your teeth — like a small packet of rose hip jam. (*See* photo p. 255.)

Sweet briar, *Rosa eglantaria* (*Rosa rubiginosa*), is cultivated for the delicately fragrant foliage (reminding me of green apples). This is a distinctly upright European shrub with tightly crowded basal stems reaching 7 to 10 feet tall and arching outwards, and round, orange-red fruit. The vigorous stems are armed with vicious randomly spaced, back-curved prickles.

Japanese multiflora rose, *Rosa multiflora*, was formerly planted for wildlife cover and in some places now covers the landscape with 7- to 10-feet shrubs that may arch up around 10 feet into the canopy of tall pasture shrubs. Large clusters of small, 2- to 3-inch, white, flowers are borne from mid-June into July. The thick, curved prickles are absent on some plants. The fruits are very small and consumed by mockingbirds, the chief agent of secondary dispersal.

Japanese rugosa rose, *Rosa rugosa*, is a commonly planted ornamental shrub occasionally escaping cultivation, especially in coastal areas where it forms dense colonies up to 7 feet high from spreading roots. The leaves are deeply veined above (rugose), and the flowers usually have more than one series (semi-double) of white to pink petals.

Of the more than 100 species of roses in the northern hemisphere, the half-dozen native roses of this region are mostly relatively low growing and form colonial thickets either by underground stems or, in the case of prairie rose, by layering when the long, outward-arching branches touch the ground. These low thickets provide nesting habitat for many species of birds. The annual fruit crop (rose hips) is persistent through much of the winter, obviously more a survival food (though an important one) than a preference. Natural dispersal takes place throughout the winter as mammals and birds like ruffed grouse, mockingbirds and bobwhite take the fruits. Mice also harvest, cache and consume a considerable amount of the seed. Natural germination is usually delayed to the second spring, with the seed lying dormant through an entire year after late winter dispersal.

Rosa acicularis, PRICKLY ROSE. Acicularis is a term for needle-like and accurately describes the slender, straight prickles that densely cover the stems right to ground level. The 1½- to 2¾-inch flowers of this northern species are solitary or occasionally two or three in terminal clusters through June, producing a ¾-inch bright red, smooth fruit that matures in September. It is a low-growing, rapidly spreading species of dry, open slopes and open woodlands on almost any soil type, throughout Eurasia and from Newfoundland to Alaska and south to Virginia through Michigan to New Mexico.

Rosa blanda, SMOOTH ROSE. The botanical name was obviously not intended for promotion but refers to the near absence of prickles in the upper parts of vigorous shoots. The 1½- to 2½-inch diameter pink flowers are usually solitary and are produced from late May to July. The small red fruits are glabrous and are mature in September. This is a vigorous, colonizing shrub to 3 feet in dry, rocky, gravelly and sandy soils of clearings, embankments and dunes from New Brunswick to Saskatchewan and south from Pennsylvania to Nebraska.

Rosa blanda *(smooth rose)*

Rosa carolina *(pasture rose)*

Rosa carolina, PASTURE ROSE. Perhaps "upland rose" would have better suited this species. The stems are armed with a pair of straight prickles at the leaf node and numerous shorter, straight bristles between the nodes. The solitary flowers are produced from mid-June through July. The small, ⅓-inch red fruits are covered with glandular hairs and ripen through September and October. (*See* photo p. 256.) This is a slow-colonizing rose on dry, sandy soils in clearings, upland woods and dunes primarily in the deciduous forest region, from Maine through extreme southern Ontario, scattered throughout Michigan and Minnesota and south into the Gulf States.

Rosa palustris, SWAMP ROSE. As the name implies, this is a shrub of marsh edges, stream banks, bogs and wet, woodland clearings. The stems are seriously armed with a pair of sturdy, downward-curved prickles at each leaf node. Flowers are produced in May through June and the gland-covered orange red fruits mature in late September, followed by a stunning autumn leaf display of orange and red to purple. It forms extensive colonies from Nova Scotia to Minnesota and south to the Gulf States. Seedlings should be grown in pots for the summer in order to plant them in late August when wetlands dry up a bit. The introduced cinnamon rose, *R. majalis*, is similar in habit and thorn characteristics and is occasionally found naturalized on dry sites, but it has a glabrous fruit rather than the glandular fruit of swamp rose.

Rosa setigera, PRAIRIE ROSE. The stems are well armed with scattered, very sturdy, downward-curved prickles that allow one to reach into the shrub but not to extricate oneself easily, hardly bristly as the name *setigera* suggests. The spherical red fruits are ⅓ to ½ inch across and mature in October. This is indeed a rose of the prairies, growing in moist to dry conditions in open woods and prairie edges. It is found from southwestern Ontario and Ohio to Kansas and into the southern states from Georgia to Texas. This is the only native climbing rose in our region, sometimes forming thickets 8 feet or more in height. It is similar in habit to the introduced multiflora rose, which has smaller fruit only ¼ to ⅓ inch wide.

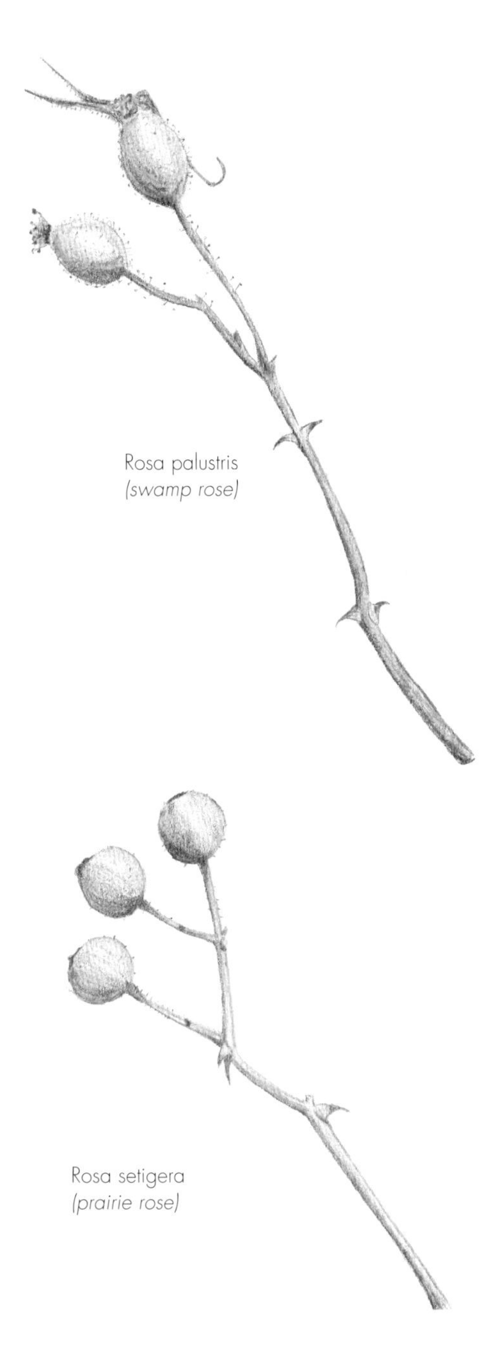

Most of our wild roses have five-petaled pink flowers, so it is better to take note of the thorn placement, fruit characteristics and habitat.

The roses mostly bloom in midsummer and are insect pollinated, providing both pollen and nectar rewards. The fruits contain 20 to 30 seeds and are ripe when fully colored and beginning to shrivel in mid-October. When collecting rose hips, take a stem sample, too, or at least sketch the prickles and their distribution as well as taking note of the moisture regime. Accurate field identification of roses relies on the presence or absence of glandular hairs on the fruit, but more so on the distribution and nature or absence of thorns on the stem and the habitat. The species are found with some variation and occasional hybridization, so grow the regional variants as they are found — species and hybrid alike — but do check that your find is not one of the numerous exotic roses.

Whole fruits are dried for storage and seed extraction. The seeds are easily separated from dry fruits by crushing the hips to dislodge the seeds from the dry chamber they are housed in. The dry pulp can be blown off and the seeds placed in water for 24 hours. Germination requires a long cold stratification, so mid- to late winter fruit collections are best stored until October. Rose seeds will germinate at cold temperatures during the later stages of stratification. Plant the seeds as soon as there is evidence of root emergence. Roses do not transplant well when in leaf; therefore, thin the seedlings to about ⅓-inch apart to allow room for development. They are very hardy and can be plunged into a nursery bed to grow for the first season. In early spring the seedlings can be teased apart and set out into a nursery bed.

Protection of rose seed from mice is very important, and seedlings should be protected from mice and rabbits for the first two years. Rabbits will browse on twigs in winter but this is normal and I have not observed feeding to the point of decline. Still, I wonder about the relative scarcity of native roses in the landscape as compared to the exotics. No natural garden or naturalization project can be final without a representative of this lovely genus. The ephemeral bloom period makes them as precious as spring wildflowers.

Rubus, the BLACKBERRIES, RASPBERRIES and THIMBLEBERRIES

I occasionally watch my neighbors' children eat the berries from *Rubus* canes that have spread to the other side of the fence. It is little wonder that at least one of these fabulous fruits is cultivated in almost every garden. I rank them as my favorite. It is easy to gorge on both wild and cultivated black and red raspberries. With over 400 species in the temperate regions of the northern hemisphere, the genus provides dense cover for all manner of wildlife; due to the extended, sequential ripening over three to four weeks, this is likely one of the most important groups of summer fruit for birds and many mammals. No landscape is complete without some members of this genus and, since they are so prolific, it is inevitable. The *Rubus* are significant colonizers, proliferating from underground roots after fire or seeding in after clear-cutting or any other disturbance.

The tangles created by the plants themselves hardly compare to the botanical complexity of this genus. There are likely several hundred forms of *Rubus* in the temperate world, a number of which are quite distinct species, but many are varieties, hybrids and hybrid complexes — especially among the blackberries. Since they are all clone-forming from seedlings, variation and hybridization give rise to isolated patches with unusual characteristics that may be perpetuated because blackberries are also able to set seed without pollination (apomictic) — all of this means endless joy for the botanists! The clear-cutting of eastern North America in the 1800s clearly left many open spaces for such an expression of diversity — only slightly complicated further by introductions of European fruit selections, which will inevitably hybridize with the native species.

The distinct species groups are frequently encountered and I have described the ecology and distribution of only a half-dozen species here. Important identification features include the color and dryness of the fruit and whether or not it slips free of the base of the flower (thimbleberries and raspberries) or includes the flower base when picked (blackberries). Other features include the density or absence of bristles or prickles and if they are straight or curved back, the number and shape of leaflets, and the height and method of clone spreading — by robust canes rising from the roots, by arching branches that root at the tip, or by ground-hugging stems that root where the branch touches the ground. Regional diversity will baffle even the botanical world's finest, but don't get stuck on trying to name them so much as capturing a desired flavor or colony characteristic.

Rubus idaeus *(red raspberry)*

Raspberries and Thimbleberries

The fruit of raspberries and thimbleberries is thimble-like, slipping off the base of the flower.

Rubus idaeus, RED RASPBERRY. Biennial canes (vegetative shoot in the first year and fruiting in the second) rise 3 to 5 feet from a shallow, spreading root system. Extensive colonies are found in clearings on moist, well-drained soils and rocky ground across Eurasia and, on this continent, from Newfoundland to Alaska, south to the Carolinas and west to Arizona and California. This is a variable species throughout its range, with the European races primarily generating the larger-fruited garden varieties. Hikers enjoy the smaller wild fruit in July and early August if the wildlife has left any.

Rubus odoratus, PURPLE FLOWERING RASPBERRY, a thornless, colorful, raspberry that spreads rapidly from stems rising to 3 feet or more from shallow underground stolons. The purpleflowering raspberry grows in partial shade, in moist to mesic soils on embankments and in clearings and forest edges from Nova Scotia to Michigan and to the south from Georgia to Tennessee. It is commonly cultivated as an ornamental in European gardens. Sadly, its fruit is withered, dry and inedible unlike the tasty fruit of its western white-flowered relative *R. parviflorus*, THIMBLEBERRY. That species name implies small flowers, but they are the largest flowers of any raspberry. Thimbleberry is identical in habit to the purple flowering raspberry but is found in drier, rocky sites from northern Ontario, northern Michigan and Minnesota westward to Alaska, Arizona and Mexico.

Rubus occidentalis, BLACK RASPBERRY, is unique in habit, with the long, arching stems touching down to the ground and rooting at the tip to produce a new plant from the "upside-down" stem. This, along with the bluish purple bloom on the canes in winter, makes it easily identified. The fruit is ripe in July. It thrives in dry to moist gravelly or rocky soils in clearings, fence lines and forest edges from Quebec to Colorado and south to Georgia and Arkansas. It occasionally hybridizes with *R. idaeus*, producing *R.* x *neglectus*, from which the cultivated purple raspberries are selected.

Blackberries

The base of the flower remains with the fruit of blackberries when it is picked.

> **Exotic Alert**
> The European blackberry, *Rubus fruticosus*, and its hybrid complex provide the larger-fruiting horticultural selections that have 10 to 13 feet long, arching canes that are often thornless. These have naturalized in North America and have become overwhelming in some regions.

Rubus allegheniensis, COMMON BLACKBERRY. Biennial, high-arching canes rise 7 feet or more from a deep, spreading rootstock on dry soils in clearings, forest edges and roadsides. It is distinguished by the pubescent leaf stems and prickles on the leaflet midveins. It is found from Nova Scotia through southern Quebec, southern Ontario, Michigan and Minnesota and south from the Carolinas to Missouri. The fruit ripens to black in late July and early August and is variably delicious, from sweet to tart. Common blackberry, and dewberries in particular, form more than 30 regional hybrids. The nearly identical *R. canadensis*, smooth blackberry (without pubescent leaf stems or prickles on the leaflet veins), grows in much the same habitat and range.

Rubus allegheniensis
(common blackberry)

Rubus flagellaris, NORTHERN DEWBERRY. One of the several dewberries, this one has been found to be apomictic — producing seeds from ovaries without fertilization by pollen. The dewberries are prostrate or low arching shrubs with long, trailing stems that tend to root at nodes. The northern dewberry thrives on poor, dry soils of sandy shores, old fields and rock outcrops, producing large, tasty fruits in late summer. *R. hispidus*, swamp dewberry, as can be expected by the name, thrives in wet soils of bogs and gravelly shores or woods, producing small, dry fruits in late summer. Dewberries range from Nova Scotia to Wisconsin and into the southern states.

Rubus flagellaris
(northern dewberry)

Rubus pensylvanicus, PENNSYLVANIA BLACKBERRY, is a vigorous, formidably armed, reclining shrub providing both food and near-impenetrable cover. This is the one to plant where a serious barrier is needed. The leaves are pubescent beneath. The black aggregate berries ripen over an extended period from July on. It is common in Ohio, less so in Michigan and rare in southern Ontario except for the Lake Erie islands. The entire range encompasses Maine to Minnesota, south to the Virginias, Indiana, Illinois and Missouri.

Blackberries and northern dewberries are among the finest natural treats there are.

Raspberry flowers are borne in many-flowered racemes in May and June, the individual flowers in succession over two to three weeks, depending on the species, thus providing a continuous pollen and nectar reward for the bees. The fruits are dispersed by bears, foxes and, reportedly, nearly 70 species of birds, with catbirds being, perhaps, the most significant. Natural germination is likely in the first spring. The seeds are only about 1⁄16 inch long, with beautiful, vein-patterned, hard seed coats that can take a firm scrubbing in the grit bag. Germination percentages are usually relatively low. The scarified seeds should be planted in a clay pot or seed tray as soon as they are cleaned. A warm period of 30 to 60 days preceding a long cold stratification will provide modest results. The few seedlings that emerge can be left for one season or transplanted out as the seedlings reach the three-leaf stage.

All species of *Rubus* are so easily propagated by a division taken from a desirable clone that seed propagation is hardly worth the effort for a few plants. Two or three clones established as single plants in any natural area are all that is required for their local establishment. Naturalization projects that specify hundreds of *Rubus* plants installed at 3-foot spacing are obviously overfunded. Rabbits enjoy the one-year stems of most members in this genus, especially the red raspberries.

Salix, the WILLOWS

The pussy willow is an irresistible botanical sign of spring — there are several species, with different flower sizes, that bloom before the leaves emerge; others bloom at the end of new shoots, after mid-May. Willows are a very diverse and important group of trees and shrubs, with over 300 species ranging from low carpets on the tundra to the massive European trees that have naturalized here. They grow in swamps, marsh edges, stream banks and ditches, moist meadows and shores and in rocky woodlands and dry savannahs, depending on the species, of course — some are specialists.

Even though our eyes tend to glaze over when we think about attempting it, identifying willows is relatively easy once you start examining them more closely, even with the natural variation. Hybrids are suggested by a number of authors but only a few species seem to be involved. The introduced willow species in particular are commonly misidentified and cause much of the confusion. All willows support significant insect activity, which is noticeable by the abundance of neotropical birds in willows during the migrations.

The flowering of willow species is almost sequential from late March to early June, as species after species comes into bloom. The male catkins, white at first and yellow as pollen is released (*see* photo p. 257), and female catkins, with green capsules, occur on separately-sexed plants. Both male and female catkins produce nectar that attracts swarms of several species of bees, flies and butterflies that overwinter as adults. On a warm, sunny day in early May there is a lot of commotion around willows in full bloom. The female catkins produce mature fruit capsules in about three to four weeks.

Midge gall on Salix erio-cephala *(woolly-headed willow)*

Exotic Alert

The classic "pussy willow" is the male plant of *Salix caprea*, with huge catkins that are silver white before the yellow stamens protrude. All of the large, massive-trunked trees belong to the European white and crack willow hybrid complex. They have been planted in parks and along creeks for nearly a century and, along with a few shrubby species, have naturalized extensively. The exotic willows don't seem to be too disruptive in the natural landscape but they restrict the amount of wetland space for the important diversity of native willows. The French pussy willow along with European weeping willow hold an all too prominent place in our idea of what a willow is.

(French) pussy willow (goat willow), *Salix caprea*, is a tall, upright shrub 6 to 20 feet high with several trunks. It is a male plant that is only propagated by cuttings. This shrub is inevitably sold as *Salix discolor* in nurseries due to huge confusion because they share the same common name. *Salix caprea* can now be found in almost any naturalization site and embankment stabilization or "bioengineering" project. The leaves have pronounced raised veins on the woolly leaf underside and a wavy leaf margin, very different from the native pussy willow, *Salix discolor*.

(European) crack willow, *Salix fragilis*, is a bold-statured 50- to 80-foot tree, usually with a single, often low-branching trunk. It is usually naturalized along rivers. The twigs are brownish green in winter and snap off easily (like the native black willow) but the leaves are distinctly lance-shaped instead of long-tapered and bent. Foresters for decades confused this tree with the small-statured native black willow even in books and, to this day, the nursery industry may still provide *Salix fragilis* named as *Salix nigra*.

(European) white willow, *Salix alba*, is a tall, narrow tree to 50 feet. The normally single trunk has a prominent taper. It is found naturalized in moist soils of abandoned quarries and along ditches and edges of swamps. The leaves are covered in fine, closely pressed hairs that reflect light and make the leaf surfaces look white. The 'Austree' sold through farm catalogs is a hybrid of *S. alba* and *S. matsudana*.

(European) reddish willow, *Salix* x *rubens*, is a large tree, 65 to 100 feet tall, with few to several large, upright trunks. The stems are often leaning or splayed apart to produce massive sprawling specimens. It has naturalized extensively along marsh edges and in floodplain forests and wet woods. The twigs are often bright yellow, orangish yellow or occasionally reddish in late winter. The weeping willows have two or three parent origins, the names of which are often interchanged (*babylonica, tristis*). The cultivar with long downward-hanging yellow branches (*S. alba* var. *pendula*) is likely a mutation of *Salix* x *rubens*, not the white willow as the name implies.

(European) bay-leaved willow, *Salix pentandra*, has very similar leaves to the sprawling native shining willow but grows upright to become a small tree up to a height of 10 feet or more, and is usually planted along ditches and in yards — it is rarely naturalized.

(European) basket willow, *Salix purpurea*, is a large, many-stemmed shrub to 10 feet high and wider, with very smooth, smallish leaves that are uniquely (for willows) arranged sub-opposite on the slender stems. It is commonly found naturalized in ditches and swamp

Salix discolor
(pussy willow)

Exotic Alert (continued)

edges. It would be great if more people made willow baskets and decimated this space-consuming species.

(European) silky osier, *Salix viminalis*, is a large, few-stemmed shrub to 30 feet that has long-tapered, narrow leaves that have a yellow midrib beneath. It is occasionally escaped from old homesteads, as it was commonly used for basketry.

This outline covers the 11 common native willows found in the southern Great Lakes area and surrounds. There are three others: *S. myricoides*, blue-leaved willow; *S. pedicellaris*, bog willow; and the fairly common *S. petiolaris*, slender willow, in the area, and *S. caroliniana*, Carolina willow to the south. Five other species touch the north shore of Lake Superior including the beautiful *S. pyrifolia*, balsam willow. There are about a dozen more of the arctic willows that at low elevations extend no farther south than James Bay in Ontario and Quebec but also are found on the tundra-like peaks of the highest mountains in the Adirondacks.

Salix discolor, PUSSY WILLOW, is the first of the native willows to bloom. This is a large, upright, almost tree-like shrub with only a few trunks 7 to 20 feet high. It grows in marshes, ditches, wet meadows and the margins of swamps and is found across Canada and south from Kentucky to Idaho. The leaf is dark green above and glaucous and slightly hairy below. The fruit is mature around the third week of May.

Salix eriocephala, WOOLLY-HEADED WILLOW. The common name is a literal translation of the botanical name and is useful for identification. This is the most common host species of the woolly "pinecone galls" that are produced on the ends of twigs by a tiny insect called a midge (the slender-twigged upright, *S. petiolaris* is the other significant host). Woolly-headed willow is a large, sprawling shrub with several stems growing 7 to13 feet high and broad. It is a typical swamp-edge species, growing also in ditches and wet meadows and on shoreline dunes in the reach of high waves. It grows from Nova Scotia to Yukon, south across the northern states and into Arkansas. The foliage is highly variable but the unfolding leaves are consistently tinged reddish and downy and the stipules are large and persistent. The fruits are mature in the last week of May.

Salix humilis, UPLAND WILLOW or PRAIRIE WILLOW. It is fascinating that a willow has evolved to thrive in dry, sandy soils in black oak savannah and dunes. I have found it in moist ground at the edge of wet woods but also on limestone where it is very dry in the summer. Upland willow is usually a small, sprawling shrub with smooth branches to 3 feet, rarely to 10 feet high. The leaves are green with impressed veins above and glaucous with prominent net veins below. The fruit matures in mid-June.

Salix bebbiana, BEAKED WILLOW, is a multiple-stemmed, very upright, large shrub reaching 7 to 20 feet and occasionally tree-like. It grows in wet meadows, ditches, moist woods and marsh edges. This is one of the most common willows and is found from Newfoundland to Alaska and south across the U.S. from Maryland to California. The leaves are distinctively net-veined and downy on the underside. The fruit is mature in the first week of June.

Salix cordata, DUNE WILLOW, is generally upright in growth but reaches only 3 to 10 feet tall. It is a dryland willow in the dunes and upper shorelines beyond the wave and flood reach of lakes and rivers, from Newfoundland to Maine and inland around the Great Lakes. The leaves are very downy when unfolding and the woolliness persists on the upper surface. The fruit matures in the first week of June.

Salix candida, SAGE-LEAVED WILLOW, is a gorgeous willow. It is an upright, multiple-stemmed shrub attaining 7 to 10 feet in height. The male flowers have red anthers that open to expose the yellow pollen. It grows in calcareous wet sand, marsh edges, fens and rocky shores from Labrador to Alaska and New Jersey to Colorado. The leaves are an exquisite blue gray, with deeply indented veins on the upper surface and usually have persistent white, woolly hairs covering both surfaces. Fruits mature in the first week of June. (*See* photo p. 257.)

Mature fruit of *Salix nigra* (black willow)

Salix nigra, BLACK WILLOW, is the "obscure" willow. The twigs snap off at their branch junction with very little resistance — in the same manner as the European crack willow, which is often mistakenly identified as black willow. With us, black willow is a small, sprawling, tree-like shrub with a few trunks or occasionally a single, partly reclining trunk. It attains a height of 10 to 40 feet, often forming pure stands along the edge of quiet water, creeks and streams. It is found from New Brunswick to Minnesota and into the southern states. The foliage has a rich green, fine texture and the leaves are narrow, long-pointed and often curving to one side. Fruits mature in the second week of June.

Salix exigua, SANDBAR WILLOW, is the only clone-forming willow in our region. Thickets are dense and extensive — sometimes covering 160-foot swaths. It grows generally 3 to 10 feet tall along shores, beaches, wet meadows, ditch banks and marsh edges. It is perhaps the most widespread of any species and is found throughout North America. Its leaves are very linear, with small teeth remotely spaced along the edges. The fruits are mature in the third week of June.

Salix nigra (black willow) with empty capsules

Salix lucida, SHINING WILLOW, is a multiple-stemmed, sprawling shrub with orangish yellow twigs and very shiny leaves. It grows 3 to 13 feet tall in ditches and the edges of swamps, with its base below the high-water line. Shining willow is found from Newfoundland to Saskatchewan and from Delaware to Nebraska. It is distinctive with its very shiny dark green leaves (compare with the introduced bay-leaved willow). The fruits are mature in the third week of June.

Salix amygdaloides, PEACH-LEAVED WILLOW, our tallest native willow, occasionally reaches 60 feet. It has single, often slightly leaning, sparsely branched trunks but a densely branched crown. The species grows in wet woods, sloughs, swamps and cattail marshes. It is found from Vermont through southern Ontario and Michigan to British Columbia, and from Pennsylvania to Arizona. The leaves are broad at the base and long-tapered, like the leaves of a peach tree. The fruits mature in the third week of June.

Salix serissima, AUTUMN WILLOW, is the only willow fruiting in the fall (mid-October). It is an upright shrub reaching 7 to 10 feet and grows on wet ground in cattail marshes and edges of bogs from Newfoundland to Alberta and south of the Great Lakes from New York to Colorado. The leaves are similar to shining willow but have yellowish brown twigs and upright habit; the two are easy to tell apart. I have kept the seeds sealed in a jar in a fridge and sown them in spring, with good results.

Natural seed dispersal of the native willows takes place from late May through June, with one species in October (autumn willow). Each species has a dispersal period of only a few days. The capsules are perfectly designed to open within a few warm, windy days and shed the seeds just before a rain. Under these favorable conditions, the seeds, with their light, silky appendages, will float and can be carried by the slightest breeze great distances to a soil/water interface where they may germinate.

Gather the chains of capsules when they turn yellow or just start to split. Seeds will be released with their light "cotton" within a day as they dry. Place the drying catkins in a quiet location or in a bowl or paper bag, since the seed will waft into the air with the slightest disturbance. Willow seeds, like those of their relatives the poplars, comprise the embryo with just a membrane cover and no seed coat. The seeds are relatively short-lived and should be sown by spreading the loose cotton thinly on premoistened soil and then wetting it down with a quick pass of the watering can. Germination occurs within 24 hours and seedlings 2 to 6 inches tall can be expected in the first year, depending on how closely spaced they are. Willows have a very fibrous root system and are easily transplanted. Beavers may take down some fine specimens but they invariably sprout again.

There is a place in every garden for at least one willow and significant numbers of them can be planted in wetland (or upland) restorations. I have found that most of the native willows will grow from dormant cuttings placed in water in March. Curiously though, *S. discolor, S. bebbiana* and many clones of *S. eriocephala* will not root easily. At least for these earliest-flowering species, try your hand at growing seedlings. For all applications to natural areas, use seedlings to obtain the important genetic diversity and mix of sexes.

Sambucus, the ELDERBERRIES

I made elderberry wine once, from fruit picked along the Eramosa River in Ontario, and it was quite highly regarded by critics — delicious, smooth, full-bodied and dry — very nice. More famous for pie and jelly, the dark purple fruit of common elderberry commands a premium price now. Summer-fruiting red elderberry is reported to be inedible for humans but not poisonous.

Native elderberries are critical wildlife plants and some 30 species of birds are reported to devour the fruit of both. The very small seeds pass through their digestive system to germinate in favorable conditions in the first spring.

Exotic Alert

There are 30 to 40 *Sambucus* species worldwide — ours are North American representatives of what may eventually be accepted as two circumpolar species complexes. The small differences rarely merit the division into species, but that doesn't mean we should abdicate our responsibility to grow the local, native genotypes.

European black elderberry, *Sambucus nigra*, is often found in the wild, having escaped from gardens and elderberry farms. It is very similar to *S. canadensis* with the exception that it grows on high, drier ground and is a multiple-stemmed shrub that does not form extensive colonies by underground shoots. Several times I have found its golden-leaved form, *Sambucus nigra* 'Aurea', in naturalization planting sites due to supplier misidentification. These offensive yellow shrubs do not "fit" in the wild and should be returned and refunded. I have twice found colony-forming plants on dry embankments and these could be a hybrid of *Sambucus nigra* x *Sambucus canadensis*.

European red elderberry, *Sambucus racemosa*, differs from the native *Sambucus pubens* only in attaining half the size and having glabrous young shoots and leaves, where *Sambucus pubens* has pubescent new growth — ultimately becoming glabrous or nearly so.

Sambucus canadensis, COMMON or CANADA ELDERBERRY. Large, flat or slightly domed clusters of white flowers in July and intricately black spaced berries in August and September immediately catch our attention. (*See* photo p. 258.) This is a 7- to 10-feet high, colony-forming shrub that spreads by underground stems to produce somewhat open clones

Sambucus canadensis
(common or Canada elderberry)

usually 10 to16 feet and sometimes 100 feet across. It is restricted to wetlands and moist ground of swamp edges, ditches and wet meadows. It is found from Nova Scotia through Quebec, along the north shore of Lake Huron to Manitoba and south to Georgia in the east and west to Oklahoma.

Sambucus pubens, RED ELDERBERRY. You will know this plant as the only bush in the forest with bright green leaves and purple flower buds when the trilliums are blooming. The flowers open around mid-May and are insect pollinated. It is adapted to capture sunlight before the forest canopy leafs out. This is a bushy, multiple-stemmed shrub, rarely to 15 feet in dry to moist soils of open woods, clearings, forest margins and embankments. The huge berry crop ripens in July and early August and is among the first ripe fruit available to birds. Red elderberry is commonly found from Newfoundland to Manitoba and in the south from Georgia to Missouri.

Elderberry flowers have both sexes (perfect flowers) and are pollinated by numerous insects including bees and butterflies. Individual berries are ¼ inch across with three to five seeds in each. The seeds are easily extracted in the grit bag and have an immature embryo that requires a warm stratification of 60 days followed by a cold period of 120 days to break dormancy. Seeds are scattered at roughly ¼-inch spacing. Young seedlings can be transplanted out or thinned to produce 4- to 8-inch plants in the first year. I have occasionally found hundreds of seedlings growing in the disturbed ground around red elderberries in gardens.

Red elderberry seedlings can be planted into dry to moist sites in spring and Canada elderberry can be potted up for planting at the edge of wet sites in the fall. I have not seen herbivore feeding on either species, although elderberry stem borer will take out stems of older plants that are stressed by not being matched well to the soil moisture. The borers are a favorite of downy woodpeckers.

Sassafras albidum, SASSAFRAS

The delicate colors of the fruit, the odd mitten-shaped leaves, the fragrance of the root and its association with root beer, the yellow flowers on upright shoots and the awesome fall color ensure that no one can forget their first encounter with the sassafras tree. Sassafras is occasionally a tall tree of the forest canopy but is most often found as a bushy, many-stemmed thicket, growing up to a height of 10 feet or more at the forest edge or along fence lines, with branches that can be reached easily. The stems rise from a deep, spreading root system and readily resprout after fire. This species is fussy as to soils, preferring slightly acidic, drier locations in rich loam woodlands, sand dunes, prairie and savannah ecosystems and sand plains. It occasionally composes the relic vegetation of fence lines and ditch embankments. Sassafras is essentially a tree of the deciduous forest region from Maine through southern Ontario, south-central Michigan to Missouri and southward into the southern states. A pubescent form found in much of the eastern states intergrades with the northern, smooth-leaved race.

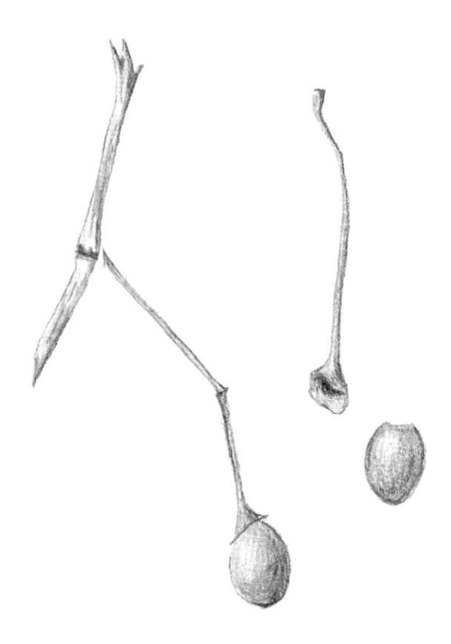

Sassafras albidum (sassafras) fruit

The greenish yellow, six-petaled flowers of sassafras open in early to mid-May with the unfolding leaves. (*See* photo p. 258.) This is a dioecious species, with male and female flowers on separate trees. The female colonies produce a modest annual crop of beautiful fruits that mature in August. The regular winged gang arrives to consume the fruit: robins and the mimics (brown thrashers, catbirds and mockingbirds). Reportedly, a few primarily insectivorous birds such as the kingbird, crested flycatcher and red-eyed vireo take this fruit as well — possibly it is an important food for building up stored fats to give them energy for their long migration to South America. The canopy feeding results in many fruits falling to the ground where turkeys, bobwhite quails and mice take their share. When searching the ground for fruits, I can find whole fruits but no cleaned-off seeds — indicating that whole fruits are eaten and the seed is passed through birds' digestive systems. Natural germination takes place in the first spring but new recruitment seems to be a rare occurrence in uncultivated sites.

The ovoid fruits ripen with a change from green to dark blue in succession over a few weeks. The single-seeded berry is held by a swollen red cup at the end of a long red peduncle and can be picked from low branches or gleaned from the ground while birds are feeding. When the berry is plucked free, the base is bright yellow where it was attached to the red peduncle. Extract the seeds soon after collection to avoid fermentation but take into account that the brown seed coat is relatively thin and cannot be subjected to a firm grinding in the grit bag. The skin and pulp are so soft that a firm scrubbing is not required. Cleaned seeds should be refrigerated for a month and planted in a clay pot in the cold frame and mulched for the fall and winter. Alternatively, they are stored cold and dry for a few months before presoaking and cold-stratifying for five months in order to plant the seeds in early spring.

The seeds will germinate in warm conditions and produce a 4- to 8-inch seedling in the first year. These can be grown out for a year or two longer in a nursery before establishing in suitable habitat. The fragrant roots are fibrous but the species is relatively difficult to transplant bare-root. Few plants need be grown, so you can take the time to handle seedlings carefully and keep the soil intact around the roots. Attempts to propagate sassafras by digging up root sprouts will usually fail unless a section of the horizontal root is obtained and the sprout severed at ground level to force a new shoot from basal buds. Seeds need protection from mice and chipmunks; deer and spice-bush swallowtail larvae may browse on the foliage.

Few gardens have room for the natural spread of this tree unless you use a mower. It is best suited for the wilder parts of a yard or natural area, taking into account its preference for slightly acidic soils. Trees grown from southern seed sources do not produce the fall color of the northern counterparts.

Shepherdia canadensis, CANADA SOAPBERRY

Exotic Alert

Autumn olive, *Elaeagnus umbellata*, has been planted extensively for wildlife food and cover in most watersheds. Cedar waxwings disperse the seed in huge quantities and now the shrub is becoming frighteningly invasive. Autumn olive produces fragrant greenish white flowers in June when *Shepherdia* has ripe fruit. The red fruit of both is similar in shape and could be confused, but autumn olive fruits ripen in September and October.

Canada soapberry is the only species of this genus in the world (a monotypic genus). It is a small, relatively short-lived sprawling shrub reaching 3 feet tall and twice as wide — one of many shrubs that evades notice until you are right beside it or see the ripe red fruit a short distance away. The current year's shoots and the underside of the grayish leaves are distinctively flecked with copper-brown scales. (*See* photo p. 258.) It is a shrub of partly shaded dry environments at the edge of woodlands, along lakes and rivers and in open woodlands and alvars, growing in a range of well-drained sandy and gravelly soils. The species is most abundant on calcareous soils but otherwise widely scattered through many life zones, from tundra to mixed forest to prairie and desert. It ranges across much of Canada from Newfoundland to Alaska and south from New York through Ohio to Arizona.

The tiny yellowish flowers are pollinated by bees and flies in early May, before the leaves expand, with male and female flowers on separate plants. The single-seeded fruits change from green to bright red in mid-June to July, and any effort to gather the seeds may be foiled by the rate at which they are devoured by cedar waxwings, catbirds and reportedly many other birds and mammals, including grouse and black bears. The randomly dispersed seeds lie dormant through the summer and germinate in the early spring.

Elaeagnus umbellata
(Japanese autumn olive)

One must be opportunistic to gather the red fruits of this species when it is encountered on a hike, not for consumption

— they are quite bitter — but to grow this lovely shrub. The hard-coated seed is covered by a very soft pulp and skin and is easily cleaned in the grit bag. The seeds can be planted immediately in a rodent-proof frame or dried and stored in a fridge until February, when they should get a soak and 90-day stratification treatment prior to spring planting.

Plant the seeds about ¼ to ⅓ inch apart and watch the watering carefully once they germinate — this is a dry ecosystem species and very susceptible to damping-off fungi, but the seedlings are very hardy. They can be grown in a nursery for the first year, where 8 inches of growth should be obtained without difficulty. Transplant them to 4-inch spacing the following spring to produce a healthy young plant for establishing in natural gardens a year later. This shrub is easily transplanted and the fibrous roots have nitrogen-fixing nodules on them that should be left in place. Aside from keeping small mammals out of the seeds, little protection is required for this species. Several individuals should be planted in dry landscapes and natural areas to ensure that both sexes are present and fruits are produced.

Smilax, the CATBRIERS or GREENBRIERS

There are about 350 *Smilax* species worldwide — mostly in the tropics and subtropics. Our catbriers are mainly herbaceous woodland plants that die to the ground at the end of the growing season, but three are woody vines. Generally they grow in deciduous woods and thickets. All three climb by tendrils and are armed with spines, some of which resemble a cat's claws. When dense, they make walking difficult. The flowers are green-yellow and hardly of notice during the May-June blooming period. The sexes are on different plants. The long-lasting fruits are an important winter food for birds.

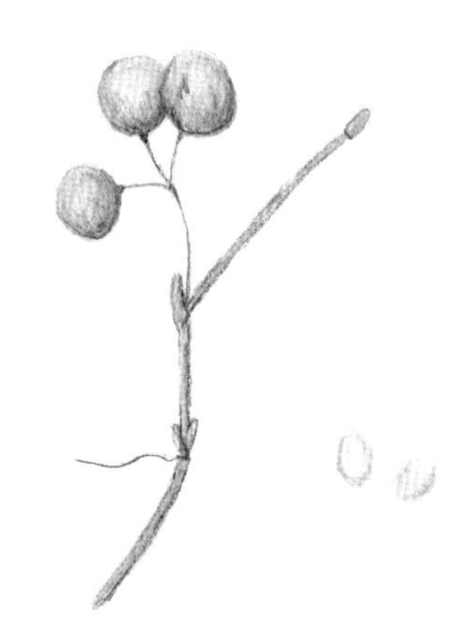

Round-leaved catbrier fruit and seeds

Smilax glauca, SAWBRIER, forms low, dense, shrubby tangles of tough, wiry stems armed with stout prickles. The leaves are whitened beneath and the fruit is bloomy like those of riverbank grape. They occur in small globular clusters that often hang on through the winter. The leathery leaves may be retained in mild winters and take on a reddish hue. Sawbrier grows on slightly acid soils of all textures and moisture levels in forest clearings and second-growth thickets. The range includes the south shore of Lake Erie to the Atlantic and Gulf coasts.

Smilax hispida, BRISTLY CATBRIER, is densely armed with black slender prickles on the older stems; the upper branches are generally unarmed. It is high-climbing to 20 feet rather than thicket forming, and the fruits are black (more of a dark green on close examination) without a waxy bloom. They occur in round clusters of five to twenty. The ovate leaves are evergreen where winters are mild. Bristly catbrier grows in slightly acid to neutral soils of thickets, open woods, shores and riverbanks from the Great Lakes to the Atlantic coast and south to the Gulf of Mexico.

Smilax rotundifolia, ROUND-LEAVED CATBRIER, is one of the most viciously armed of native plants — right up there with devil's-club, *Oplopanax horridus*. The prickles are broad-based and very sharp. It is a common and wide-ranging species of acid, often sandy soils of varying moisture. In thin woodlands, it forms dense tangles and climbs 20 feet or more. The flowers are more bronze-colored than yellow-green. The bloom-covered berries are in small clusters of half a dozen or fewer. Round-leaved greenbrier is most common south of the Great Lakes but is found in a few places in southern Ontario and in the southwest corner of Michigan. The continental range is from Nova Scotia to Oklahoma south to Florida and Louisiana. (*See* photo p. 258.)

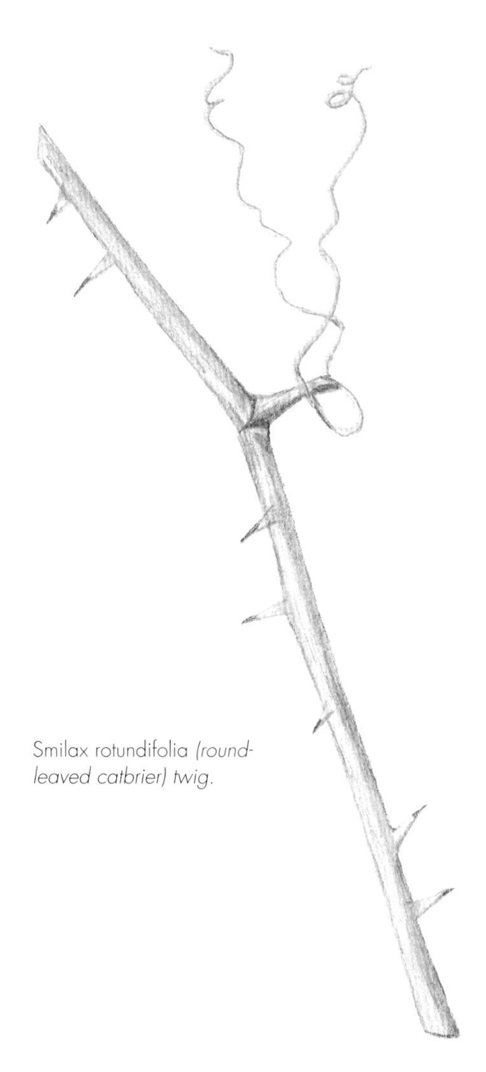

Smilax rotundifolia *(round-leaved catbrier)* twig.

Catbrier berries ripen in September but hang on into midwinter — you might still find some in February and March. Each ¼- to ⅓-inch berry contains one, two or even three seeds. Not many people would want to propagate these species; they are unsuitable to home grounds and a couple would be plenty in most restoration and wildlife plantings. They do provide winter food and excellent cover for wildlife, and could perhaps be planted for those benefits in an out-of-the-way location away from pedestrian traffic or used to direct such traffic around sensitive areas. Gather the berries in September and remove the seeds from the berry flesh either in the grit bag or, if only a few are required, with your fingernails. Plant immediately and place the pot in a cold frame. Label them well because they are unlikely to germinate until the second spring. The woody *Smilax* species propagate more quickly from cuttings and layers; indeed, you may find naturally layered vines that can be dug and transplanted — landowners should be consulted of course but are unlikely to object, not even to a truckload.

Sorbus, the MOUNTAIN ASHES

The mountain ashes are not ash trees of the genus *Fraxinus*; they are, in fact, members of the rose family. The North American species are so named for the compound leaf and their affinity for slopes, cliffs and rocky mountaintops — places where browsers like moose don't easily reach them. Only on high ground are they found in solid stands or as trees 25 to 30 feet high; otherwise, the native *Sorbus* are more likely to be seen as isolated large shrubs. Fruit of the native species is reported to be dispersed primarily by cedar waxwings, sharp-tailed grouse and grosbeaks. Natural germination takes place in cool, moist soils and I see many seedlings in the wild (especially along the disturbed soils of portage trails), but few survive for more than a few years, due to shade or browsing. There are possibly as many as 80 species of *Sorbus* in the northern hemisphere, primarily in Asia, with hybrids often appearing where several species are planted together in botanical institutions.

Exotic Alert

The European mountain ash or rowan, *Sorbus aucuparia*, is the species most of us are going to find on a regular basis, especially south of the 43rd parallel and away from higher elevations in the Adirondacks. It has naturalized extensively throughout our area — the seeds were dispersed primarily by robins, which now thrive on many exotic fruits. The rowan tree has also been planted in gardens since early settlement. It was believed to offer protection against witches, among other things, and juice from its berries was used as a folk medicine. Rowan is most readily distinguished from the two indigenous species by its relatively small leaves (*see* photo p. 260) and the downy (non-resinous) winter buds. Hybrids with the indigenous species are inevitable and a hybrid with *Sorbus americana* has been known since 1850 as *Sorbus* x *splendida*.

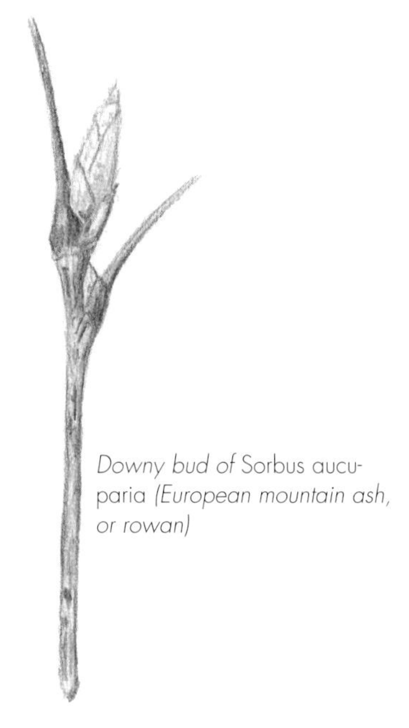

Downy bud of Sorbus aucuparia *(European mountain ash, or rowan)*

Sorbus americana, AMERICAN MOUNTAIN ASH, is usually a tree with one to several trunks, most often found in woodlands and the thicket-like edges of clearings and occasionally on rock outcrops. It grows in moist, humus-rich, acidic soils of mixed forests from Newfoundland through Quebec, Ontario and Minnesota, is more widely scattered in the south from Pennsylvania to Illinois and is abundant again in the mountains south to Georgia.

Sorbus decora, SHOWY MOUNTAIN ASH, is a similar species to the preceding but with a more northern range and a more robust and shrub-like habit associated with twice the chromosome count. Showy mountain ash tends to grow in more open, drier conditions of rocky embankments along rivers and rocky islands in the Great Lakes, often where little humus is present. It is essentially a boreal species, found in exposed conditions (including nearly barren islands) from Labrador and Quebec through northern Ontario and northern Michigan, and sparsely from the Adirondacks in New York to northern Ohio and from northern Illinois through Wisconsin to Minnesota.

The mountain ashes are in full bloom in June and insect pollinators swarm the clusters of tiny white pungent flowers. The fruits mature in September with a shift from green to orange and orange-red and contain usually one to three filled seeds out of a possible ten. The fruits of *S. americana* are relatively small at ⅛ to ¼ inch and those of *S. decora* relatively large at about ⅓ inch in diameter. The fruits cannot be highly desired by wildlife, as they persist through much of the

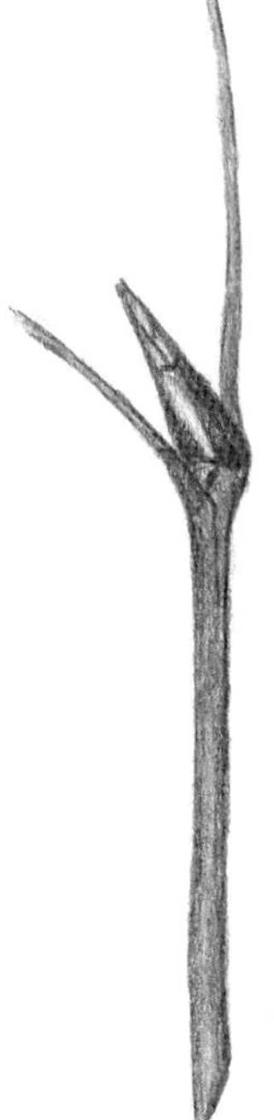

Resinous bud of Sorbus decora *(showy mountain ash)*

Winter buds of Sorbus aucuparia, *the European mountain ash, and* Sorbus decora, *showy mountain ash, help distinguish them.*

Fruit and seeds of Sorbus decora *(showy mountain ash)*

winter, making seed gathering relatively easy. The fruits can be picked from low branches at any time, dried and stored for a number of years.

The seeds are relatively small and their extraction from the fruit, as with *Amelanchier,* is a little tricky. Presoak the berries for a day to soften the fruit, and because the seeds do not have a tough seed coat, place the fruits in a tough plastic bag and burst them open with a rubber mallet or the palm of your hand before grinding them moderately in the grit bag. Good seeds (sometimes very few) will sink and thin, empty seeds will float and can be decanted with the pulp. I have found that the northern seed sources of *Sorbus* species require a longer cold stratification than recommended in most books.

Because the seeds are so small and cold-temperature radicle emergence occurs toward the end of stratification, it is best to sow them in late fall in a clay pot that is enclosed in plastic and held in the fridge or in a humus-rich outdoor frame so as to avoid handling small germinating seedlings. Early spring germination is very uniform and young seedlings can be thinned or transplanted at the three-leaf stage if they have germinated too thickly. Two-year seedlings of native species should reach 4 to 6 inches and benefit from close spacing and a well-protected environment. These can be transplanted in the spring to a spacing of a few inches. The roots of *Sorbus* are often quite elongated, with little fibrous root, and should be clipped each time they are transplanted.

In cultivation, the native species are fine garden plants with a far richer fall color than the European mountain ash (rowan). They are suited to natural areas if protected from browsing deer and meadow voles. Try to establish a few plants so that good pollination is obtained — be sure they are not grafted clones. For purposes of identification, a grafted plant has a crook at the base and the lighter-colored (gray-brown) bark of rowan is just below the crook; grafting is a common practice among commercial propagators, who do not make the effort to obtain and germinate the wild seeds. The columnar selection called *S. aucuparia* 'fastigiata' has all of the characteristics of *S. decora* (except for an open crown) and may be incorrectly named.

Spiraea, the MEADOWSWEETS

The mid- to late summer extended bloom of these shrubs is a delight for many insects, including butterflies. Of some 70 species in the northern hemisphere, only two are found in the Great Lakes region and northeastern U.S. These are thicket-forming shrubs, expanding by underground stems. It is possible that the tiny, almost dust-like seeds are carried as hitchhikers on the wet fur of mammals or stick to the feet of birds as they travel between or within watersheds. I can't imagine any other method of dispersal except perhaps wind.

Spiraea alba, MEADOWSWEET. Dense, single-clonal thickets up to 3 to 5 feet deep can blanket a significant area. The heavily branched underground stems travel about 20 inches each year, and send up dozens of shoots. It thrives in moist, gravelly to sandy and organic soils on shores, in wet meadows and along swamp edges. Meadowsweet is found from Newfoundland to the Carolinas and west to Alberta and the Dakotas. The species is variable, with a wide-leaved and broad inflorescence form (var. *latifolia*) in the east and a narrow-leaved, narrow inflorescence form (var. *alba*) growing in the west. (*See* photo p. 259.) The two overlap so extensively and intergrade so frequently that it seems pointless to separate them. Once again, grow the local expression.

Spiraea tomentosa, STEEPLE BUSH or HARDHACK. The tight, spire-like terminal clusters of small pink flowers can draw attention from considerable distances and give rise to the common name. This is a botanical jewel — a lovely small shrub with a few tightly clustered stems and attractive downy foliage. It thrives in acidic, sunny conditions poking out of the gravel in a rocky shoreline crevice or a wet sedge meadow. It is found from Nova Scotia to Minnesota and south from North Carolina and Georgia to Arkansas. This plant does not adapt well to normal garden cultivation due to its requirement for moist, acidic soils.

Hundreds of tiny five-petaled perfect flowers are produced in terminal inflorescences on the current season's growth. The fruit matures with a shift from green to brown in September and the upright, five-segmented capsules (each with dozens of seeds) open in dry conditions from October through the winter, to scatter the seed as the wind or wildlife weaves through or alongside the plant. The seeds have no dormancy characteristics and germinate in moist soil in early spring — which places plants just above the high-water level under natural conditions.

The seeds are so small that they need only be scattered onto a moist soil surface and covered for a week with Plexiglas before they germinate. Seedlings can be thinned or pricked out and transplanted and should reach 2 to 4 inches in the first year. A second year of growth will produce usable plants of meadowsweet but steeple bush is slower growing. Rabbits will browse on the twigs in winter, but the new growth is always stronger. Meadowsweet is also easily grown by division or softwood cuttings. The white-flowered species is suited to the coarser, wilder landscape and the pink-flowered species for more refined acidic soil locations at the edge of pond gardens and streams.

Staphylea trifolia, BLADDERNUT

As a child I called bladdernut bag-tree because of the fascinating papery inflated fruits. They hung on all winter and rattled when I bumped the branches as I walked through a stand near my home in Bright's Grove, Ontario. This is a colony-forming species from shoots rising generally 10 feet but up to 16 feet from a slowly spreading root system. In the woodland understory, the stems are well spaced through the colony but the species also grows in the open, where it forms a rather tight, nearly impenetrable thicket. Bladdernut is a healthy shrub, seemingly affected by nothing. It grows in moist conditions on clay and rocky embankments, on alluvial flats, in moist, sandy woodlands and, occasionally, on drier embankments and shallow soils over limestone. It ranges just north of and through the deciduous forest region, from southern Quebec through southern Ontario and Michigan to Minnesota and into the southern states.

The white, bell-shaped flowers are perfect, occur in drooping clusters and are insect pollinated in May, as the leaves are expanding. As the fruits mature, they change in color from green to yellow and finally tan brown. I haven't a clue how the seeds are dispersed over the range they cover — surely not all by wind and water — other than the fact that they are so hard that some seeds could survive a feeding by mastodons — pure interpretive conjecture. Perhaps grackles pick up some of the shiny seeds as they forage ravenously through the forest floor — again, conjecture. Mice will disperse the seed locally. Natural germination may take two to three years.

Fruits can be plucked from a few shrubs as they turn yellow in September and the seeds peeled free and planted right away to obtain some germination in the first spring. Once the fruits have turned brown a deep dormancy sets in and delays germination to the second spring and even the third. Seed of this species requires patience and protection, since mice will empty a seedbed. One-year seedlings are 8 inches tall and reach 16 to 20 inches in the second season, after which they need to be planted into the landscape. The dense fibrous root of bladdernut is very easy to transplant. However, I have seen meadow voles debarking the stems. If this happens, cut back recently planted stock to sprout from below ground. Established plants can recover if a few stems are not debarked — otherwise, cut them off, too.

This species is essential at the edge of woodlands, in larger backyards and in naturalization projects. Divisions may occasionally provide the few plants required for establishing the species. Plantings in dry sites will perish during droughts. Seedlings grown from open, drier locations should be further studied for drought tolerance in comparison to those from shaded, moist woodlands.

Depending on pollination, usually only two to five seeds form of a possible 12 in the unmistakable three-chambered inflated fruit of the bladdernut (Staphylea trifolia).

Symphoricarpos, SNOWBERRY and CORALBERRY

Symphoricarpos albus *(snowberry) berries. The white berries are spongy, air-filled sacs.*

Symphoricarpos colonies produce a dense, low thicket, spreading by shallow underground stems in clearings and partly wooded landscapes. The spongy fruits persist well into winter but slowly disappear as cardinals and thrushes occasionally take the fruit. Dispersal over the extensive northward distribution of the white-fruited species is likely the work of mammals such as bears, foxes or wolves (as the common name of the western species, wolfberry, suggests).

There are about 15 species in North America and one in China, only two of which are natives in the Great Lakes area. The tall plants with large white fruits were introduced from the west coast.

Exotic Alert

Wolfberry, *Symphoricarpos occidentalis*, and western snowberry, *S. albus* var. *laevigatus*, grow 3 to 7 feet high and produce fruits that are up to ⅓ inch across, in large terminal clusters. They are both west coast species that have been planted extensively in eastern gardens and are now found naturalized in widely scattered locations throughout the region. The highly visible fruits remain on the plant well into winter.

Symphoricarpos albus, SNOWBERRY, is a dense, colony-forming low shrub rising from underground stems to a height of 20 inches and rarely 40 inches with fruits that are up to ⅓ inch across. It grows in dry to mesic sand, clay or gravelly soils in open woodlands and on partially shaded embankments from Quebec to Virginia and west to British Columbia and Colorado.

Symphoricarpos orbiculatus, CORALBERRY, is a vigorous spreading shrub to 5 feet tall with long, arching stems rising from underground stolons. The small, tightly clustered deep pink fruit persist well through winter. It grows in dry and stony soils in clearings and open woods in the deciduous forest and prairies from Connecticut west to Colorado and south to Louisiana.

Symphoricarpos orbiculatus *(coralberry) berries*

Symphoricarpos are in bloom from July into August and the tiny clusters of pink and white flowers in the upper leaf axils are beepollinated. The fruits are spongy, with two to four hard-coated seeds in each berry. They can be picked when fully ripe from late August into winter, as long as they persist. The seed coat requires a vigorous scrubbing in the grit bag to tear away at the outer protein layers, or a one-hour soak in sulfuric acid prior to a 45-day warm stratification and 150 to 180 days of cold stratification. Seeds should be planted at ¼-inch depth and kept relatively moist until they germinate.

Symphoricarpos orbiculatus *(coralberry) seeds, enlarged*

First year plants may reach 4 inches but are usually smaller. They transplant and establish easily and are not damaged by herbivores. These species are easily propagated by division. The snowberry spreads rather slowly but the coralberry can be quite challenging to contain.

Taxus canadensis, CANADA YEW or GROUND HEMLOCK

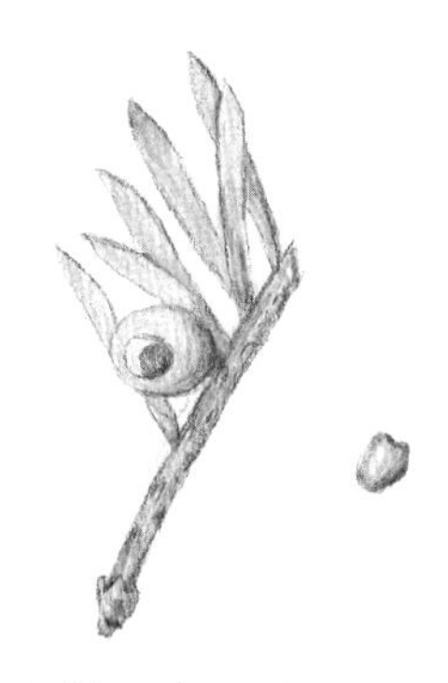

Taxus canadensis *(Canada yew)*

Interest in the in the 10 species of yew found in the northern hemisphere soared in the mid-1980s after the discovery that a promising antitumor drug, taxol, could be isolated from the bark of Pacific yew. The Pacific yew is now rare, placing some harvesting pressure on Canada yew even though it has a lower taxol content. The fruit contains a single seed surrounded by red fleshy tissue. It is a bit slim yet sweet and edible; the Japanese produce jelly from their native yew. The seeds and foliage are highly poisonous.

Exotic Alert

The pyramidal tree known as Japanese yew, *Taxus cuspidata*, is commonly sold in nurseries along with cultivated forms of the hybrid with English yew, *Taxus baccata*. These horticultural yews are occasionally found naturalized in the near-urban woodland, the seed transported by wildlife from a local garden.

Canada yew is a sprawling, shade-loving shrub that provides important cover for woodland birds. It has ground-level upward-arching trunks radiating symmetrically outward to produce a flat-topped, spreading shrub from 3 to 6 feet tall and 10 feet across or more. It thrives in moist but well-drained, humus-rich soils in deciduous, mixed and conifer forests. Fruits are consumed by a number of bird species and the hard seeds dispersed by passing intact through digestion. I have watched thrushes and cedar waxwings take the fruit. I once noticed several very active yellow warblers in a thick stand of heavily fruited yew on the Mink Islands in Georgian Bay but do not know if they were eating the yew berries, as insects were numerous as well. Little of the crop is left by September and Canada yew is locally abundant where found, indicating frequent local seed dispersal. It ranges south of the boreal forest from Newfoundland to Manitoba, southward to western Virginia and Kentucky.

The yews are dioecious plants; the male plants produce conspicuous small cones that turn bright yellow as pollen is released in May. The females bear solitary organs in the leaf axils, so small they are rarely noticed. Fruits are produced annually but some years provide bumper crops. The berry-like fruits are bright green through early summer and mature with a rapid change to bright red in late July. Each fruit contains a single sound seed that is visible through an opening at the terminal end of the soft red flesh. The fruits can be gathered through August into autumn.

To avoid fermentation, the seeds should be separated from the fruit soon after collection and planted before they are able to dry out. Germination is always delayed until after the second winter. Few plants are needed in new landscapes and two dozen seeds can be sown in a clay pot that is plunged into the shadiest end of a cold frame. Germinated seeds stop growing after the seed leaves expand and should not be disturbed for a couple of years, until they have made a couple of inches growth. The seedlings have well-branched roots with white growing tips and are easily transplanted in spring. It will take a few more years before they are large enough to plant into a shade garden or woodland restoration. Heavy browsing by rabbits and deer over a period of a few years can kill the plant.

The Canada yew occasionally self-layers, and cuttings can sometimes be obtained with roots already formed. These divisions should be nurtured for a year in the protection of a cold frame. This is a fairly large shrub for yards; it is more suited to rural properties and woodland restoration. Several specimens should be planted to ensure both male and female plants.

Thuja occidentalis, WHITE CEDAR or ARBORVITAE

Old split-rail fences provide a clue as to the abundance of cedar and human industry in the historical landscape. A cedar woodland can be magical with its occasional leaning trunks and uplifted roots, some of which provide rich reddish brown colors among the usual moss carpets. The current use for straight-line hedging is an insult to the beauty of this species, especially since they age with such informal grace. In stressful environments, trees can reach very great age — over 1,000 years — often with a dead terminal stem and sections of trunk without bark, weathered to a smooth grayish white. Some of the oldest trees in the region are cedars found as stunted plants in the limestone and dolomite cliffs of the Niagara Escarpment, sometimes suspended by only a twisted root lodged in a narrow crack.

The white cedar is common where found, often in pure stands, but also in mixed forests with hemlock, yellow birch, red maple and black ash. It has a range of habitat from moist to wet woodlands, hummocks in swamps, rocky stream banks and shores to steep rock cliffs and the seams in limestone outcrops and occasionally on gravelly slopes. It is absent in dry soils. The species name *occidentalis* refers to this being the western cedar (west of Europe) as differentiated from the Oriental (eastern) cedar of Asia — *Thuja orientalis*. White cedar is found from southern Nova Scotia and Prince Edward Island to the south shore of James Bay, into Manitoba and south from Maine through New York, most of Michigan and into Minnesota, with a few, widely scattered populations into the adjacent states and down the Appalachian Mountains.

Separate male and female cones are produced on the same tree and are windpollinated in May. The small green cones mature in one growing season, turning a pleasant yellowish color in late August and ripening in September when they turn brown and start to open on dry days. The flattened, winged seeds are found in twos and occasionally threes under each of the mid-cone scales, for an average of eight sound seeds per cone. The cones are held upright but seeds are easily dislodged by sudden branch movements caused by wind or the activities of squirrels and chickadees. Flocks of pine siskins descend upon white cedar forests during winter and pull seeds out of the open cones with their well-adapted long-pointed beaks. Dispersed seeds are consumed by ground-feeding birds such as juncos, ruffed grouse and tree sparrows. A very high percentage of seeds falling on cool, moist soils will germinate in the first spring.

The cones can be picked from September through winter but with fewer and fewer seeds remaining in the cones as the months go by. The seeds require no pregermination treatment and are best shaken from dry cones and stored in a paper envelope until March. Sow the seeds on a damp, humus-rich, soil surface in a clay pot with a Plexiglas cover or in soil with the seeds barely covered with a sprinkling of sand or soil. Germination takes a few weeks and the seedlings produce juvenile, needle-like leaves at first. The first true (flat spray) leaves appear toward midsummer when seedlings reach about 2 inches. Plunge the whole lot together into a cold frame for the first two winters. Two-year seedlings can be teased apart and planted in the spring at a few inches spacing to grow for a couple more years. Cedars are easily transplanted due to a compact, fibrous root system and can be set into the landscape when they are 6 to 8 inches tall. Rabbits and deer will browse on the foliage; in fact, cedar is a primary winter food for deer, as evidenced by the browse line in cedar woodlands and along the shores of northern lakes. Deer herds will typically "yard" for the winter in dense cedar woods.

The cones of the white cedar, Thuja occidentalis, *are held upright and persist for several months.*

Natural variation in *Thuja occidentalis* is significant but most variants are shaded out in the wild because they are slower growing. By propagating these variants, the horticultural industry provides many cultivars with dark green to yellow leaves, from the very dwarf 'Tiny Tim' and the rare, permanently juvenile foliage form called 'Ohlendorfii' to globose, pyramidal and columnar selections. The cultivated Oriental cedar, *Thuja orientalis*, has distinctively upright, fan-shaped sprays of foliage, the cone scales are fleshy and the seeds are without wings. Untrimmed wild cedar is an important addition to natural landscape in moist areas, providing easy cover for songbirds when predators such as sharp-shinned or Cooper's hawks are on the hunt.

Tilia americana, BASSWOOD

Basswood is often ignored. The wood is not highly valued today except by carvers and it was hardly known horticulturally due to the difficulty in germinating the seed. The creamy yellow, fragrant flowers in July are eagerly sought by insects, especially bees — both the native and introduced honeybee. The nectar yields a prized honey. Basswood grows in rich, moist soils as a shade-intolerant canopy tree in mixed forests and a wide-crowned tree at the forest edge and along fence rows and stream banks. The occasional trees on dry embankments need to be studied for drought tolerance.

Exotic Alert

Little leaf linden, *Tilia cordata*, and numerous other Eurasian species, hybrids and cultivars are raised extensively in the nursery industry and planted throughout cities and parks and even for roadside "beautification." Lindens can now be found naturalizing extensively. Undoubtedly, our native species will hybridize with the exotics, so make sure your basswood seed is collected in an area that is remote from the European species or remove the lindens. I suspect we will see *Tilia platyphyllos* naturalizing in the future. Its leaves are about two-thirds the size of basswood but it is more readily identified by its distinctively five-ribbed fruit. I have found sound fruits dispersed across snow and some are bound to germinate.

There are about 30 species of *Tilia* in the northern hemisphere, with three species in eastern North America. Basswood has a distribution through the deciduous and mixed-forest life zones into the prairies, from New Brunswick to Manitoba and south to Florida and Texas. The species is variable. The white basswood, *Tilia heterophylla*, has white hairs on the leaf undersides and reddish winter buds. It ranges south from southern Pennsylvania, southern Ohio and southern Illinois. The intermediate, *Tilia neglecta*, is found throughout the overlapped ranges, suggesting a species complex of considerable extent. Such variability need not always be given species status; perhaps habitat differences are more significant.

Basswood flowers are insect pollinated in July. The sweetly fragrant perfect flowers hang in dense pendulous clusters but

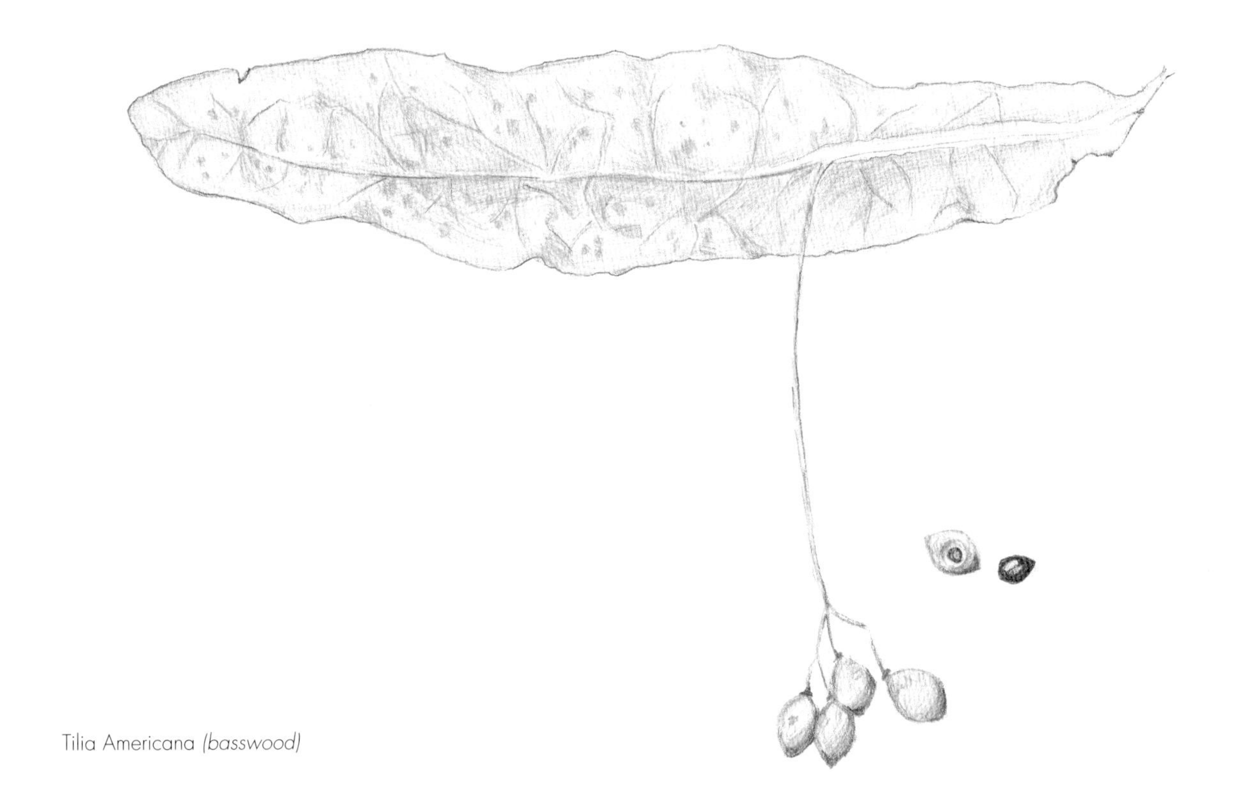

Tilia Americana *(basswood)*

only five to ten of the flowers produce a fruit. The fruit matures in September with a shift in color from green to yellow, ripening to tan-colored in October. (*See* photo p. 260.) Dispersal takes place from November into winter, when fruits may blow across crusted snow during storms. Mice and chipmunks gather and cache the seeds a short distance farther, but these animals can hardly explain the tremendous range of a tree with such a heavy seed and a relatively hard seed coat. I am left to ponder the role of the mastodon and mammoth again ... just what did they eat, or what did they not eat? I have often noticed dozens of young basswood trees in close proximity to older, fruit-bearing trees, indicating that the digestive tract is not involved in seed germination. Natural germination is sporadic. Why does only a low percentage of apparently viable seed readily germinate? The fruit has a hard seed covering that is thought to interfere with germination, but even when the heavy seed coat is opened or removed, germination percentage is hardly altered.

Growers have tried various ways to improve germination and have made few gains. The best success is still obtained if fruits are collected as they turn yellow and are sown in an outdoor frame immediately, but don't let this critical timing stop you. A deep dormancy sets in when the fruit is fully ripe and the seeds will then sit in the ground for two winters to break the more complex dormancy. In this case, the most critical factors for growing basswood are patience and enclosure. Mice and chipmunks will overturn the entire seedbed in search of this tasty food.

So, at best, you can expect only a small percentage of the seed you plant to sprout, but a dozen seedlings will provide even large sites with all that is required in basswood; it is not a pure-stand species. Basswood, with its prolific fibrous root system, is easily transplanted but much effort will be required to protect the trees from deer and rabbits. I have seen 10-foot saplings with all their branches broken by deer that have reared up to pull branches into reach. Rabbits will tear most of the bark off young trees. Saplings that are slow to break out of transplant shock or are damaged by browsing should be cut off just above ground to force a vigorous new shoot from the dormant basal buds. Seeds should be gathered from older healthy trees in drought-stressed locations. Basswood is a tree for large yards and natural areas.

Tsuga canadensis, HEMLOCK

There are about a dozen species of hemlock in Asia and North America. Hemlocks have a lovely ability to produce a straight trunk from a continuously bent-over terminal — the cells on the inside of the curve are able to elongate during secondary growth to straighten the stem as the shoot tip continues to grow and lean to the side. Perhaps the Japanese word *tsuga* is related to this characteristic.

Hemlocks are majestic trees by any measure, reaching great size in rich woods of old-growth forests, where the needle carpet can be so thick and the shade so great that little that is photosynthetic can be found beneath. The eastern hemlock thrives on a considerable range of moist, well-drained soils from granitic rock and sand to clay and calcareous glacial till or limestone bedrock, as long as the temperature is cool. It is found from Nova Scotia through southern Quebec to Ontario north of Lake Huron, Michigan and Wisconsin, with a curious outlier stretching west from Thunder Bay to Rainy Lake. The southern range is from Pennsylvania and Ohio and south in the mountains to Georgia. In the mountains it overlaps the range of Carolina hemlock, *Tsuga caroliniana*, which grows at higher elevations and differs in having darker green foliage and cones twice the size of eastern hemlock.

Male and female hemlock cones are produced on the same tree and are wind pollinated in May. The seed cones develop in one growing season and are mature when still green in early October. Seed crops are at two- to three-year intervals; a large part of the cone crop is ripped into by red squirrels and crossbills. The cones change to brown in late October and remain on the tree into the following summer. Once the cones open,

Hemlock Woolly Adelgid

Eastern hemlock, by pollen records, has already survived one population crash and recovery. Hemlock's natural requirement for cool conditions will certainly reduce survival in the southern reaches of the species' range. The generally hotter and drier episodes over the last few decades may stress the trees enough to induce natural decline. This, in turn, may predispose them to the current threat in the south, where thousands of trees are dying from an insect infestation called hemlock woolly adelgid. The adelgid arrived in North Carolina in the 1950s on hemlock nursery plants imported from Japan. Of course it was announced that this insect has no known predators — but they just hadn't been noticed yet. An Asian ladybug that has naturalized in North America has gravitated to the infested trees and shows signs of bringing the adelgid numbers down, doing something our native predators would have done as well. Recommendations that gardeners simply substitute exotic hemlocks more resistant to the adelgid are no longer acceptable.

Hemlock cones hang downward; their scales open only during dry weather.

pine siskins and chickadees will hang upside down to pull out the seeds. Seeds are dispersed locally by wind and consumed on the ground by mice, juncos and sparrows, which feed on the seeds from late fall through to spring migration in April, when the cones open on warm days and release the last seeds. Is it possible that the northward migration of hemlocks is brought about by these small birds ingesting and passing uncracked seeds intact? Natural germination occurs in the decomposition layer of the forest floor, typically with yellow birch, in rotting stumps and decaying logs (often called nurse logs).

One can access cones from canopy trees by gathering branchlets filled with green cones from the ground as red squirrels drop them. The cones will open when they are pulled from the branchlets. Cones can be plucked from low branches from the time they start to change color into midwinter; remember that there are fewer seeds left in the cone as winter progresses. Mature cones will open fully when dried indoors and the tiny winged black seeds can be placed in an envelope to store until early January. Eastern hemlock requires a cold stratification and I have obtained the highest rates of germination with durations of 90 to 120 days after a 24-hour soak in water.

The seeds must be germinated in a relatively cool and shaded location, as they are not tolerant of the heat buildup associated with direct sunlight. Seedlings benefit from a slightly acidic soil and you can even try germinating them on a piece of decayed log. Scatter the seeds at ⅛-inch spacing in a clay pot and cover them with a small amount of organic material. Plunge the pot into a cold frame in March to complete the stratification with natural diurnally alternating temperatures. Germination is usually quite uniform, but seedling development usually stops when the cotyledons expand; this makes them particularly susceptible to frost heaving. Seedlings can remain in the pot but plunged into the cold frame for a year. The following spring, the clump of seedlings can be unpotted and set into the cold-frame soil to grow undisturbed for a of couple of years.

Hemlocks benefit from close spacing. When your seedlings reach 2 to 4 inches they can be lifted, teased apart, root-pruned and replanted to 1½- to 2-inch spacing to grow for two more years to reach 8 inches. After this stage they can take a bit more sunlight. Hemlock transplants easily due to its compact, fibrous root system. Protect the germinating seeds or sparrows will pluck the seedlings while trying to get the seeds. Rabbits devour hemlock foliage and bark from trees even ¾ to 1¼ inch in diameter. Such heavy feeding accounts for low recruitment of hemlock.

Eastern hemlock makes a fine yard tree provided it is in an area of low soil compaction, but it is best suited for mixed woodlands. Avoid planting it in hot sites such as south or west slopes. The hotter the summers become, the farther north this species will be growing.

Ulmus, the ELMS

Although relatively abundant as young trees throughout the region (young trees can generally resist disease), the old elms are almost legendary now. White elm, in particular, once sheltered our towns. The Dutch elm disease (DED) brought about their near-demise, shattering the shelter our elders grew up with. City life has yet to recover the grace and beauty it had under the canopy of elms. Elms are a tough, early-succession, shade-intolerant genus of about 20 species in the northern hemisphere. Large survivors of DED are out there and those with trunks greater than 10 feet in circumference are worth watching and protecting from the chainsaw. Those that are greater than 13 feet in circumference are worth reporting to a state or provincial forestry representative or elm recovery program. This effort is most appropriate for large trees in more remote areas that may not have been found by researchers yet.

Exotic Alert

Three species have been widely planted and are occasionally found as large trees in and around cities, especially in pavement areas. None of the exotics has the classic umbrella shape of the white elm.

Siberian elm, *Ulmus pumila*, presents the greatest concern due to its ability to successfully encroach upon fields and ravine forests. Siberian elm has very thin twigs and was planted mostly to create fast-growing hedges — many of which grew out of control. To this day, remnant rows of old Siberian elm are still found. *U. pumila* is often, mistakenly, called Chinese elm, which is another Asian species entirely, and notable as an autumn-flowering, small-statured tree with beautiful bark and glossy, tiny leaves as implied by its name, *U. parvifolia*. Despite its beauty, it can be as invasive as its homelier relative.

Scotch elm, *Ulmus glabra*, is upright in growth like its close relative the slippery elm. It differs in not having red hairs on the ends of buds, and in that leaves on vigorous shoots occasionally display an extra leaf "tip" or two. Otherwise the leaves of both species have prominent vein branching visible on the leaf underside. It tends to be the preferred host of elm leaf beetle, which causes periodic episodes of extreme defoliation. Scotch elm can show significant tolerance to DED, and very large trees should be considered for germ plasm export back to Europe. It is usually found naturalizing in hedges of urban pavement areas and rarely in the natural landscape. The rather attractive (because it is not overplanted) broad, arching Camperdown elm is a selection of *Ulmus glabra*.

English elm, *Ulmus procera*, is upright in growth and colony-forming, and has very corky low branches and twigs. It is closely related to the native rock elm, which also produces a corky bark on lower branches and twigs. Rock elm has very distinctive seeds and lacks the vein branching that is so visible on the underside of the leaves of English elm. *U. procera* is occasionally found in the wild, where it tends to form a large clonal colony in the absence of mowing.

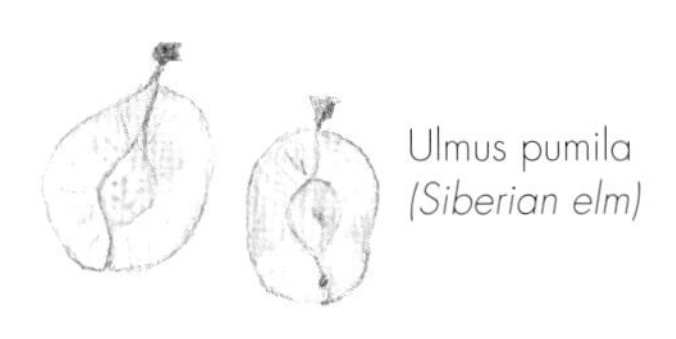

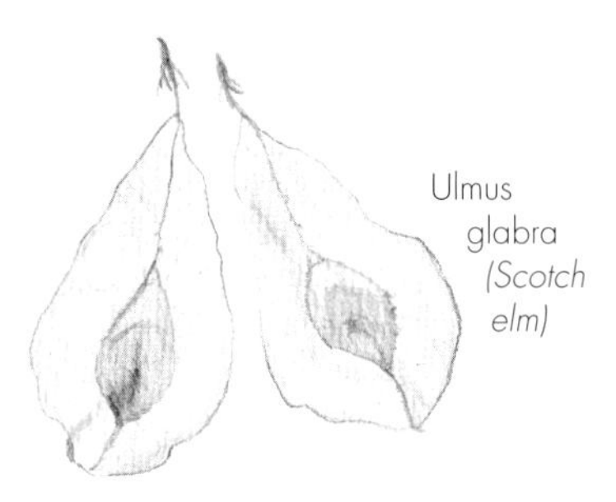

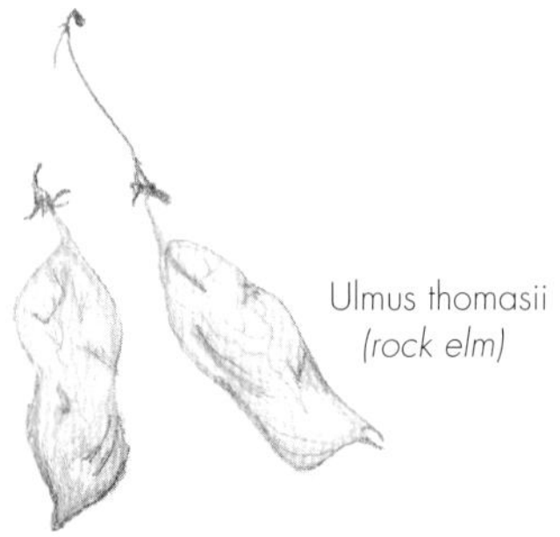

Positive identification of the elms requires fruit.

Ulmus americana, WHITE ELM or AMERICAN ELM. In spite of DED, majestic and massive trees can be seen even to this day; DED survivors in the landscape at 13 to16 feet in circumference are occasionally found. The U.S. national champion is located in Buckley, Michigan, standing about 121 feet tall and nearly 25 feet in circumference in 1997. White elm grows in the wet soils of swamps and moist soils along ditch banks and creeks but equally often on dry ridges and in field fencerows and urban streets, with the roots covered in pavement and concrete. Trees on wet sites seem shorter-lived. It is native from Nova Scotia through southern Quebec and Ontario to southeastern Saskatchewan and the Dakotas, and into the southern states as far as eastern Texas. The name white elm is associated with the white layers visible in a cross-section of the bark.

Ulmus rubra, RED ELM or SLIPPERY ELM. Red comes from the solid reddish brown color in a cross-section of the bark, and slippery from the mucilaginous inner bark that is used for throat protection during severe coughs. So few of these trees remain in the landscape, with only occasional trees of 7 feet or, rarely, 13 feet circumference that are potentially tolerant of DED, that it should be considered for endangered status so that the best elms are not cut down before they can contribute to a recovery program. Red elm grows in moist to dry conditions in a range from rich woodland soils to hard clay and gravelly soils or shallow soils over limestone. This species is found from Maine through southern Quebec and southern Ontario, throughout Michigan to Minnesota and into the southern states.

Ulmus thomasii, ROCK ELM is a colony-forming species, growing from shoots rising from a shallow root system. Colonies are found with stems reaching 8 inches in diameter; individual trees up to 3 feet in circumference are rare. Perhaps, rather than giving up on this species, it too should be provided the possible protection associated with endangered status. It is found in mesic to dry gravelly and clay soils and shallow soils over limestone, widely scattered through New England and New York, extreme southern Quebec and Ontario and most of Michigan into Minnesota, and in the south from Ohio to Arkansas.

The elms are windpollinated and flower in early April — swelling flower buds can be observed during a warm spell in late February. The single-seeded winged fruits are falling by late May to early June, especially on dry, windy days. A high percentage of the seeds are viable and can be stored dry for a few years. Natural germination takes place soon after dispersal wherever seeds fall in appropriate conditions (often seeds with other seeds on top acting as a mulch). These are rapidly growing trees and it is not unusual for seedlings to grow 3 feet high in one summer.

The dry fruit are sown with the wings intact and can be scattered densely enough to be touching under a thin soil covering. Keep the soil moist during this potentially dry time of the year by spreading a thin mulch of wood shavings, or even by placing a burlap or plywood cover over the seedbed (watch for mice moving in). Germination takes seven to ten days and solid coverings should be removed as soon as you see the cotyledon hook at the soil surface. Seedlings can be thinned to a spacing of ¾ to 1 inch, and 20-inch and higher saplings are ready to plant out the following spring. They have deep but fibrous roots and transplant with a high rate of success, but deer and groundhogs will browse heavily on the foliage and rabbits may strip the sapling bark during harsh winters, while meadow voles will nibble at the base of young plants. By the time elms are a few yards tall, they are generally left alone.

The elms are important in early succession on many sites and there is simply no excuse for not employing them in naturalization and restoration projects. Seed obtained from the largest surviving elms may carry genes for disease tolerance, and perhaps one in a hundred could reach 70 to 80 years. In the meantime, most of the seedlings will live for about 20 to 30 years, providing cover for other hardwoods; some will remain cropped by browsing deer, effectively providing a preferred food to reduce impact on longer-lived species. When trees in a natural area do finally die, they will release the canopy for other species and provide excellent habitat for woodpeckers and the various parasitic organisms that prey on wood-boring insects. This is a contribution to the landscape that is so desperately needed that there should be an injunction against cutting down dead trees when they are not in a hazardous location.

As to planting elms in yards? Work as a neighborhood and purchase a variety of clones to plant in the local area but, more importantly, produce several of your own if you can by chip budding the largest surviving trees in your area. Any efforts of this nature can only complement the work of elm recovery projects.

Renewing the White Elm

Survivor selection is not new. 'Liberty' is a six-clone mix of *Ulmus americana* that was released around 1986. 'Valley Forge', 'New Harmony' and 'Princeton' are each single-tree clones now sold by mail order in the U.S. These are energetic efforts to put the white elm back in the landscape, but reliance on fewer than a dozen clones is biologically unsound. The white elm is not a clone-producing species, and the lack of genetic diversity in the relatively few clones cannot stand the test of time. The good news is that the mass clone release being carried out in the U.S. can be salvaged from potential disaster by planting several clones of the best local old survivors in close proximity to the 'Liberty' and 'Princeton' group in order to produce hundreds of mini breeding orchards — for tolerant x tolerant seed production. Where species recovery is required, we must progress beyond hybridizing with exotics or relying on single clones and move toward survivor mating and seedling release to provide the genetic diversity required to stand the test of time.

Vaccinium, BLUEBERRY and CRANBERRY

Vaccinium is a complex genus of about 400 often hybridizing species. The flowers are beepollinated and the fruits are dispersed by many species of mammals (including humans), and bears in particular, as well as birds such as the wood thrush and catbird. The fruit of some is so prolific as to have generated a significant industry. Commercial operations in the east rejuvenate and invigorate "wild" low-bush blueberry with controlled burns and, at the same time, remove woody and exotic invasive plants (what a concept). Highbush blueberry selection is so advanced it has tended to sink to the grocery store standard of larger, tasteless "display" fruit. Cranberries are commercially harvested by flooding bogs so that the fruits float to the surface and can be raked up. *Vaccinium* is one of about 125 genera in the immense heath family (Ericaceae), in which are numerous dry-fruiting genera, the native ones of which are described under *Kalmia*.

Vaccinium angustifolium, LOW-BUSH BLUEBERRY, is the very common "wild" blueberry that is picked by hikers or purchased at roadside stands throughout the region. The delicious fruit ripens sequentially, peaking from mid-July to mid-August. It is a species that recovers easily from underground stems after ground fires and can form vast colonies. Low-bush blueberry grows to about 12 inches high in acidic soils in open and partly shaded rock outcrops, sand plains, woodlands and the borders of sphagnum bogs. It is found from Newfoundland to North Carolina and west to Manitoba and Minnesota. Usually growing in the same patches are *V. myrtilloides*, the velvet-leaved blueberry, and *V. uliginosum*, bog-bilberry, with nearly black fruit.

Vaccinium corymbosum, HIGHBUSH BLUEBERRY, is the parent of the many cultivated varieties whose berries are sold in grocery stores. The blue fruit from wild plants is delicious and just under ⅓ inch in diameter, ripening in July. The species forms small colonies of loose shrubs up to 8 feet high in open, moist, acidic woodlands around bogs and ponds and occasionally in sandy openings. *V. corymbosum* is a complex and variable species found from coastal Maine, southern Quebec, southern Ontario, and much of Michigan south to Florida and Texas.

Vaccinum macrocarpon
(cranberry)

Vaccinium macrocarpon, CRANBERRY. Who said a blueberry had to be blue? And what makes this a *Vaccinium* when the flowers are not similar? This just adds another layer of confusion that makes one wonder if the blackflies were too voracious and prevented botanists from further studying this genus. Cranberry is a sun-loving species and only found in sphagnum bogs — sometimes very small bogs in the hollows of rock outcrops — from Newfoundland to Manitoba and Virginia to Illinois. This is a true specialist.

Vaccinium stamineum, DEERBERRY, is a low, slender shrub usually 3 feet high but taller in the south. The leaves are deciduous. Although the white flowers are less than ⅓ inch in diameter, they make a show in late May and June. The fruit is similarly sized and yellow-green to bluish when ripe in July and early August. At maturity the fruit falls to the ground — it is not particularly palatable to humans but is esteemed by wildlife. Deerberry is not fussy as to soil as long as it is well drained and acidic. It will grow in rocky woods and floodplains, on ridges and barrens, and in old fields. The species ranges from eastern Ohio through Pennsylvania and New York State, south to Florida and the Gulf Coast. It barely enters Ontario along the Niagara River (apparently extirpated) and at St. Lawrence Islands National Park.

The fruits of all *Vaccinium* species begin to mature around mid-July and ripen through August. (*See* photo p. 261.) Fruits should be dried and stored until January and then either be partly crushed to separate the small brown seeds, or soaked so that seeds can be separated in the grit bag. *Vaccinium* seeds require a 90-day cold stratification after sowing in peat. Growth is slow for the first year and the seedlings should be handled as described under *Kalmia*.

One of several fascinating blueberry relatives is worth mention. Bearberry, *Arctostaphylos uva-ursi*, is a mat-forming shrub with long branches that root into the humus layer. It grows in sandy soils and rock crevices in both acidic and calcareous parent material. Bearberry is a circumboreal species growing as far south as Virginia and New Mexico. Bees pollinate the flowers and the inedible fruits ripen red in August and are dispersed by birds and mammals. There are about six very small seeds in each fruit and they require a 90-day cold stratification to germinate. The seedlings are minute for the first year and some care is needed to ensure that they are not upended during weeding or by rodents.

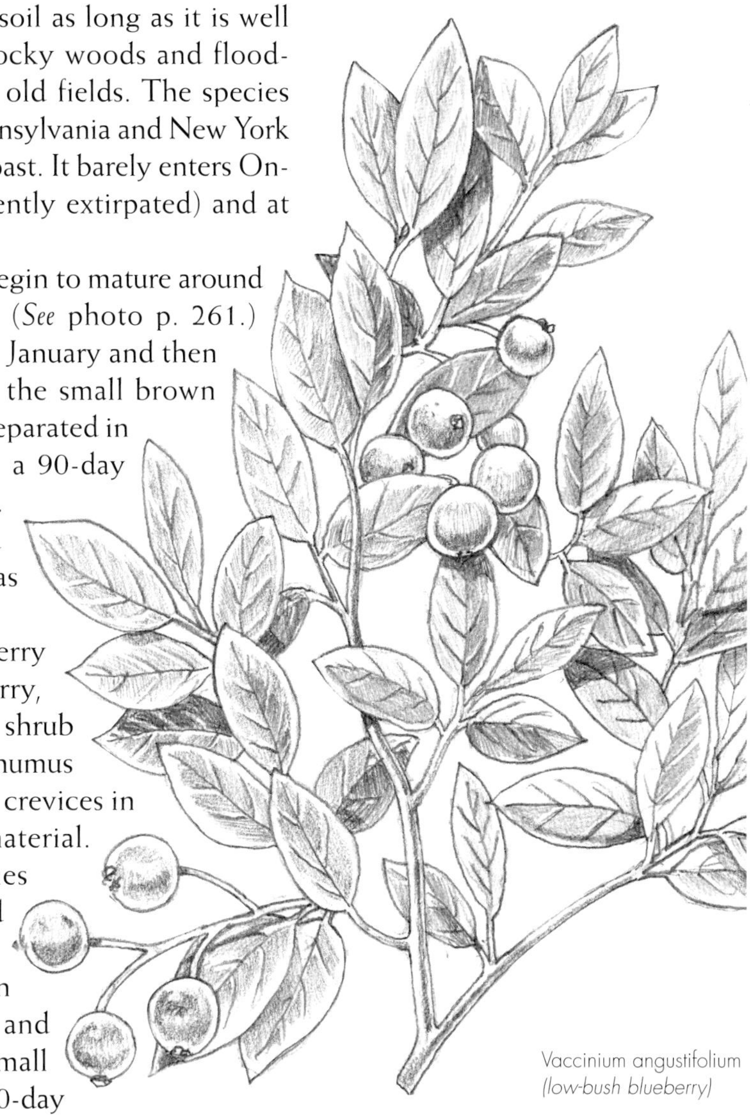

Vaccinium angustifolium
(low-bush blueberry)

All of the heath family plants are quite fussy as to soil moisture and acidity and can be used in making beautiful "downspout" and bog gardens that rely on natural rainfall. Blueberries and their kin are generally free of herbivore pressure except for rabbits.

Viburnum, the VIBURNUMS

The viburnums are very handsome shrubs, important as habitat and food. Over 250 species grow in the northern hemisphere, with the greatest diversity in Asia. At least nine indigenous species and two exotics are found in our area.

The fruits of wild raisin and native highbush cranberry are quite tasty but I know of one instance where jelly was made from the nearly identical European highbush cranberry and

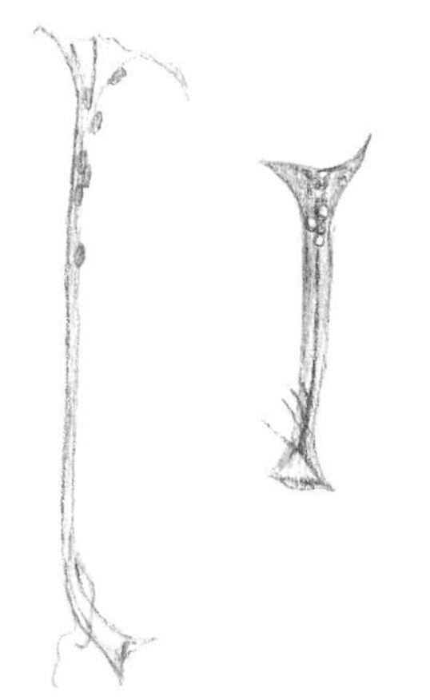

The petiole glands of Viburnum trilobum *(highbush cranberry) right, and* Viburnum opulus *(European highbush cranberry), left. (See photo on p. 262.)*

given to friends for Christmas, only to find out later that the fruit of the European species is simply wretched. If the jelly had been made from the native species it would have been quite a gift!

The *Viburnum* species grow in a range of ecosystems from swamps and old-growth wooded slopes to dry, open spaces. Fruits, which are technically known as drupes, mature in September and October and both mammals such as foxes, opossums, chipmunks and white-footed mice, as well as birds including thrushes, cedar waxwings, great crested flycatchers and pileated woodpeckers disperse the seeds. A two-stage germination strategy in most species results in radicle emergence in late summer to early fall, nearly a year after dispersal, and shoot development in the following spring.

The identification of viburnums is not that difficult, as each species is fairly distinct with only minor variation within some species. Significant confusion between the North American and European highbush cranberries can best be resolved by tasting the fruit or with careful examination of the petiole glands. There is no question that hybrids of these two species (or subspecies) have established in many areas.

Exotic Alert

European highbush cranberry, *Viburnum opulus*, is also known as guelder rose. (*See* photo p. 263.) This and *Viburnum sargentii* from northeast Asia are very similar to the native highbush cranberry. One should immediately suspect the Eurasian race when plants are found on high ground or in shaded woodlands and the fruit tastes foul. The fruit tastes so awful that normally it is not consumed by wildlife until it has frozen and thawed a few times. Red squirrels peel off the fruit to get at the seed. The differentiation of the petiole gland characteristic is useful until hybrids are encountered. Above all, taste the fruit!

Wayfaring tree, *Viburnum lantana*, is a Eurasian species that has naturalized extensively throughout the region. It is a tight, multiple-stemmed, mounding shrub that was formerly a staple in ornamental landscapes for its leathery leaves and attractive red fruit that turns black (not to be confused with hobble-bush, which has large heart-shaped leaves). Germination (typical of species adapted to human disturbance regimes) is very high, with hundreds of seedlings found underneath and in close proximity to naturalized plants. (*See* photo p. 263.)

I have arranged the native species that follow according to similarity of appearance and describe the distinctive characteristics of the plant, fruit, ecology, range and use.

Viburnum alnifolium, HOBBLE-BUSH. This sprawling, 3- to 10-feet tall understory shrub has attractive, large, distinctly two-ranked, heart-shaped leaves and naked winter buds on upward-pointing branches. The name comes from the numerous hoops created by its branch-layering habit, which, when you are trying to walk through the bushes, can trip you up as though you were hobbled around the ankles. The fruits mature

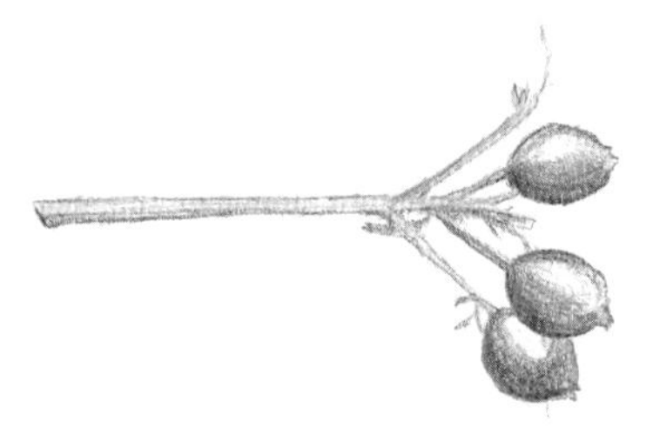

Viburnum alnifolium *(hobble-bush)*

red and are mostly consumed before all the fruit have fully ripened and turned purple-black. The stage when fruit clusters have both red and black fruits present, in early to mid September, is quite striking. (*See* photo p. 262.) This species is a habitat specialist, requiring moist to wet conditions in deep, acidic humus soils over granitic bedrock, although I have observed it on one site in a glacial soil with little humus. It is found in low woods and moist wooded slopes from Nova Scotia to Georgian Bay in Ontario and two locations in west-central Michigan, south through the Appalachian Mountains to North Carolina. Hobble-bush requires moist, acidic soil and rainwater in cultivation. Attempts to cultivate this species in alkaline soils result in chlorotic leaves and rapid decline. Rabbits, deer and moose browse on the winter twigs quite heavily.

Viburnum cassinoides, WILD RAISIN. The common name is appropriate for this delicious fruit. The fruits are in upright clusters and change from greenish white through bright pink to blue (when they taste best) to dark and shriveled. Often, two to three color phases of the fruit are evident in early to mid-September when the leaves are also beginning to change color. (*See* photo p. 262.) It is a tight, multiple-stemmed shrub from 3 to 7 feet tall in moist to wet, organic and sandy, acidic soils at the borders of streams, swamps and bogs. This is a rarely cultivated shrub, likely due to the habitat specialization, but I have found that some seed sources will tolerate moist alkaline soils in cultivation. Wild raisin is found from Newfoundland to Manitoba, from the northeastern U.S. to Illinois and south in the mountains to Alabama. It is worth the effort to grow this well-behaved shrub in gardens bordering backyard streams and ponds. A southern shrub called possum haw, *V. nudum*, is barely distinguishable beyond the shinier leaves, and the literature suggests that the two form a hybrid complex where their ranges overlap. Possum haw is the easier of the two to cultivate.

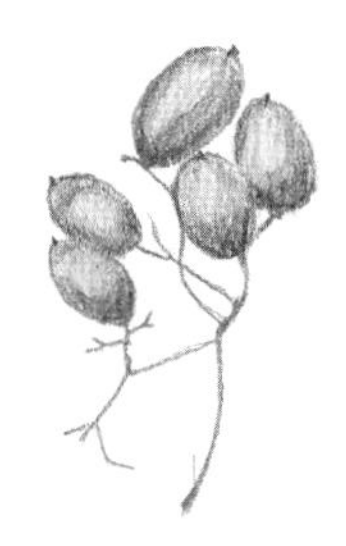
Viburnum cassinoides *(wild raisin)*

Viburnum lentago, NANNYBERRY. The fruits are quite large (⅓ to ½ inch), in drooping clusters, turning from green to shades of pink, then dark bluish purple in September and persisting through winter. (*See* photo p. 262.) Shoots rise from a shallow, spreading root system to produce a dense, colony-forming shrub usually about 10 to 13 feet high and over 10 feet across. It is commonly found in full sun to partial shade in a wide range of moisture conditions from swamps to dry embankments, from Quebec through southern Ontario, Michigan and southern Manitoba to Wyoming and south from New Jersey to Colorado. This is a commonly cultivated viburnum, and is essential in native plant gardens and restoration sites.

Viburnum lentago *(nannyberry)*

Viburnum prunifolium, BLACK HAW. I only know this plant from cultivated specimens. It is a small tree-like shrub with only a few shoots rising from the roots. The fruits are edible

Viburnum Leaf Beetle

A few years ago, viburnum leaf beetle (VLB) provided ample challenge for gardeners in some regions. Both the larva and the adult consume foliage and they left lovely plants of highbush cranberry and arrowwood leafless and declining. Spraying ensued, of course, but enough plants were left untreated that predator populations were able to build up, and now several species, including assassin bugs, are able to keep the VLB at bay. It is once again "safe" to plant the native highbush cranberry and arrowwood, but pay close attention to the soil moisture requirements so as not to stress the plants.

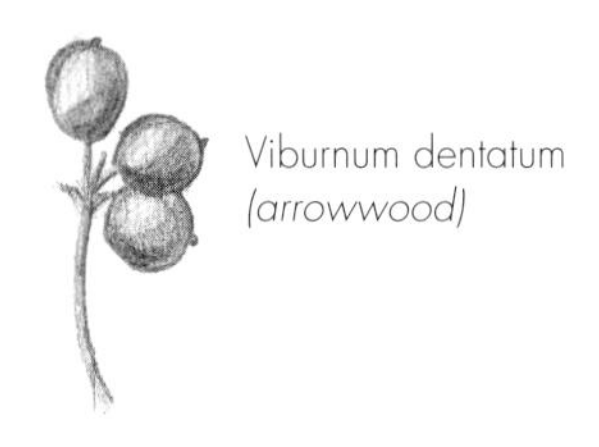
Viburnum dentatum *(arrowwood)*

Viburnum rafinesquianum *(downy arrowwood)*

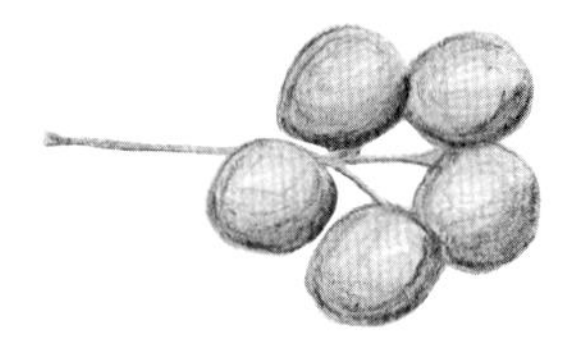
Viburnum trilobum *(highbush cranberry)*

and turn to a bluish purple. It is reported to grow to heights of several yards and have a preference for dry soils. It can be found from Connecticut to extreme southern Michigan and Kansas, south to Georgia and Texas. (*See* photo p. 263.)

Viburnum recognitum (*Viburnum dentatum* var. *lucidum*), ARROWWOOD, is a highly variable and intergrading species complex. Arrowwood is best distinguished by the lack of leaf stipules, a smooth lower leaf surface and ovoid fruits that change from green to dark blue. It is a tight, multiple-stemmed shrub spreading slowly by underground stems with outward-arching branches reaching 7 to 10 feet high. It grows in moist woods and swampy forest margins from Maine to the extreme southern parts of both Ontario and Michigan, and from New York through Ohio to Illinois. This is an uncommonly available plant that is worthy of much more attention by both tidy and natural gardeners. A southern race variety, *V. dentatum*, with a pubescent lower leaf surface, is primarily coastal. The interior varieties have leaf stipules with leaves either smooth or pubescent beneath and are found primarily in the regions of the Ohio River and its confluence with the Mississippi.

Viburnum rafinesquianum, DOWNY ARROWWOOD. The ovoid, flattened fruits are in upright clusters, ripening to a purple-black in September and drying and shriveling on the stem if not eaten. Shoots rise 1 foot to rarely 7 feet from the roots to produce an open, loose colony up to 13 feet across in dry to mesic rocky, gravelly or clay soils. It is found in the open or partial shade as a relic in fencerows and as widely scattered clones at the edges of woods, on embankments and on shallow soils over limestone, ranging from southern Quebec to Manitoba and in the south from New Jersey to Arkansas. This rarely cultivated plant should be used more in dry gardens and naturalization sites.

Viburnum acerifolium, MAPLE-LEAVED VIBURNUM. The nearly spherical dark purple-black fruits ripen in early October and are held in upright clusters, persisting into winter if not eaten. Shoots rise, quite dispersed, from a wide-ranging root system, producing open colonies usually 3 feet but occasionally up to 7 feet tall. It grows in moist but well-drained humus-rich and slightly acidic soils in gently hilly woodlands and on the slopes of old-growth ravine forest. It ranges from New Brunswick and southern Quebec through southern Ontario and Michigan to Minnesota and south from Georgia to Louisiana. This apparent habitat specialist is rarely found in cultivation but is a very beautifully structured plant and a prime candidate for woodland gardens.

Viburnum edule, MOOSE BERRY, or LOW-BUSH CRANBERRY. The fruits and seeds are very similar to *V. trilobum* but in smaller clusters. This is a low-sprawling shrub to about 3 feet high, growing in cool, wet meadows and bogs and along streams. It is encountered from Newfoundland to Alaska and south and less commonly from Pennsylvania to Michigan's upper peninsula, northern Minnesota and Oregon. This plant is for diehard plant collectors only, but it could provide "garden-friendly" selections.

Viburnum trilobum, HIGHBUSH CRANBERRY. Botanically, *V. trilobum* will likely become described as *V. opulus* var. *trilobum*, as part of a large circumboreal species complex. It is distinct only in the flavor of the fruit, the habitat and the leaf petiole glands — important enough. The spherical fruits hang in heavy, pendulous clusters and change from green through yellow and finally bright red in early September, becoming very soft and juicy once they fully ripen by mid-October. This fruit is tart at first taste but distinctly cranberry flavored and worth exploring. The fruit is normally consumed before winter by birds such as grouse, which will perch in the upper branches to reach the fruit. This is a multistemmed, clump-forming shrub of 7 to13 feet or, rarely, 16 feet tall, occasionally with branches bent to the ground and rooting. I have only found this species growing in cool, wet soils, close to the edge of water in marshes, creeks and swamps. It is native from Newfoundland south to Pennsylvania and west to British Columbia and Washington.

Reproduction and Propagation

Viburnums flower from late May through June with dozens of small perfect flowers in tight, terminal clusters. *V. opulus, trilobum* and *alnifolium* produce a single outer series of sterile flowers with largish white petals. (The gardener's snowball viburnum is *V. opulus* 'Rosea', which has all sterile flowers with large white petals.) The single-seeded fruits, or drupes, mature in late summer to early fall with every seed fertile. These fruits can be gathered when they are ripe and beginning to soften or when shriveled and dry. Fruits can be dried for storage or soaked and cleaned for immediate treatment.

The tough viburnum seeds are easily extracted in the grit bag using a good firm grinding. The seed-coat scratching may be important for quick germination. Cleaned seeds are immediately planted in a warm rodent-proof location or warm-stratified until radicle emergence is evident. If no evidence of germination occurs in three months, the seeds are placed in cold conditions until May and set into the cold frame for the summer. If radicle emergence is evident, the germinating seeds are planted into trays at ½ inch spacing and kept warm until the cotyledon stems begin to show at the soil surface before

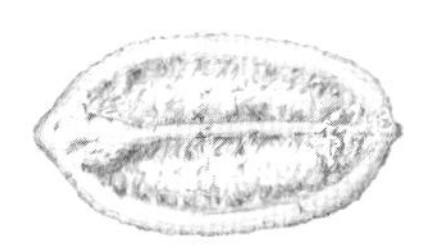

Critical identification of viburnum can be made by the presence and number of or absence of grooves on the two sides of the seed and the shape of the seed.

Seeds (enlarged) from top to bottom:
Viburnum alnifolium *(hobble-bush)*
Viburnum cassinoides *(wild raisin)*
Viburnum dentatum *(arrowwood)*
Viburnum rafinesquianum *(downy arrowwood)*

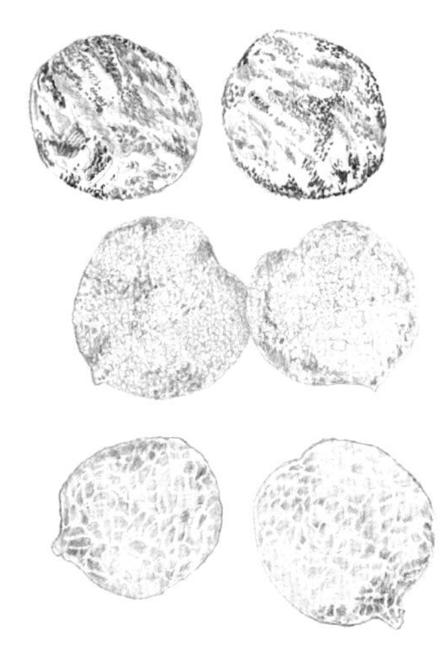

Seeds (enlarged) from top to bottom (continued):
Viburnum lentago *(nannyberry)*
Viburnum trilobum *(highbush cranberry)*
Viburnum acerifolium
(maple-leaved viburnum)

placing these half-germinated seeds into cold conditions (refrigeration or an unheated garage) for three months before placing them in the cold frame in the spring. Seedlings can be left undisturbed for the first year and then transplanted into suitable growing conditions for one to two more years of growth before outplanting.

Extreme care is required to prevent mice from gaining access to your seeds, all of which they will devour in short order. Even one-year seedlings of most species will become mouse food. Two- to three-year-old plants are generally left untouched by mice. Viburnums have very fibrous root systems and are easily transplanted in spring or fall. No garden, natural area or restoration is complete without at least one of these lovely species.

Vitis, the GRAPES

The wild grapes and escaped cultivated grapes are turning forest edges, thickets and hedgerows into a vine-tangled paradise for insects, birds and foxes. I am often asked if the grapes should be removed to protect trees. One has to step back for a moment to figure out why a myopic focus on trees has so pervaded our sensibilities. The trees and thickets are perfectly designed to support the vine tangles, and yes, may occasionally be smothered, after which the whole works may crash to the ground. So what?

There are some 60 species of grapes in the northern hemisphere, nearly half of them in North America, but only four species grow as natives in the Great Lakes watershed. Identification of the indigenous grapes is a little confusing at first due to the variability of the leaf in all species, which is dependent on growth rate. The presence or absence of hairs and the color of the mature leaf underside provide the important clues. Our native grapes are all high-climbing vines that will form dense mounds over any support or they will trail and root along the ground where no vertical support is available. Wild grapes are consumed by at least 40 species of birds. Major consumers include the ruffed grouse, mockingbird, cedar waxwing, cardinal and fox sparrow.

Vitis aestivalis, SUMMER GRAPE. The young leaves are covered in fuzzy, rusty hairs and the mature leaves are glaucous (bluish or whitish) below (especially in the poorly defined variety *argentifolia*), with the rusty pubescence remaining on the veins but usually more or less covering the lower surface. This species grows in dry to mesic soils from Massachusetts through southern Ontario and southern Michigan to Minnesota and into the southern states. The ¼- to ½-inch blue-black, sour berries ripen from late September into

October (the common name misleads many budding botanists).

Vitis riparia, RIVERBANK GRAPE. The leaves are pale green below and thinly pubescent. It is found in moist soils in woodlands and on embankments from Nova Scotia across the north shore of Lake Huron to southern Manitoba, and south to Virginia and west to the American southwest. The ¼- to ½-inch dark blue-black berry is sweet when fully ripe in late September and early October.

Vitis labrusca, FOX GRAPE. The leaf underside has a firm covering of rusty hairs. It grows in moist to mesic soils from Maine through extreme southern Ontario and southern Michigan and south from Georgia to Kentucky. The large ⅓- to ¾ -inch reddish to dark purple fruits ripen in September-October and are sweet to sharp tasting. The popular cultivars such as 'Niagara' and 'Concord' are selections of this species.

Vitis vulpina, FROST GRAPE, is similar to riverbank grape except the grapes have no waxy bloom and the leaves are mostly unlobed; they look somewhat like basswood leaves. The species has the same high-climbing habit as riverbank grape, engulfing small trees and shrubs. Probably most people walk past it without thinking it is anything different. The shiny black grapes supposedly ripen after frost. It is commonly found south of the Great Lakes (west to Nebraska) and south to Texas and Florida.

Grape flowers are small, numerous and borne in dense to open clusters in late May and early June. They are perfect and are insect pollinated. The berries mature unevenly in the cluster but are persistent and ultimately dry or become moldy on the vine. Each grape contains two to four sound seeds and they can be picked at any time from maturity to fully dried.

Gibberellic acid reportedly improves germination in grapes but I suspect that is true more for the cultivated grapes, which evolved by selection of seedlings that likely germinated in piles of fermenting waste after the grapes were stomped. Gibberellin treatment is inevitably the product of neglect as the fruits usually ferment for a day or two in a bag after collection. The tough seeds are readily extracted from soft, ripe or presoaked dry fruit using the grit bag. A firm grinding will scratch the seed coat surface and prepare the seeds for uniform germination after five to six months of cold treatment. Seeds can be fall planted at ⅓-inch spacing in a frame or stratified and planted in the spring. One-year seedlings are 8 to 16 inches long and can be planted out the following spring. Grapes are easy to establish but may be browsed on lightly by deer. The grapes are useful for improving the habitat of chain-link fences, piles of debris and abandoned cars.

Vitis aestivalis *(summer grape)*

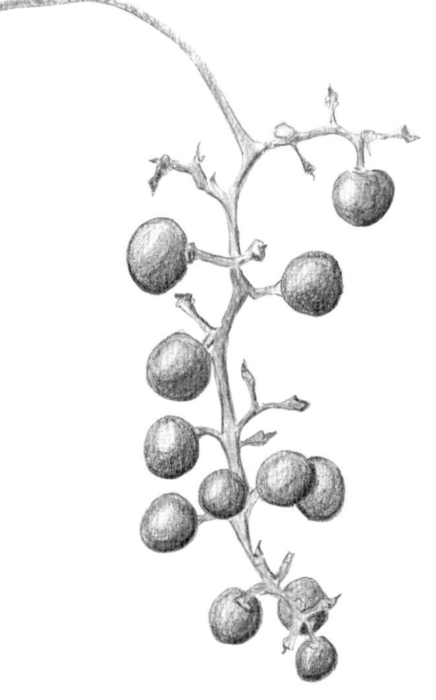

Summer grape is a gardener's delight — before and after fermentation.

Zanthoxylum americanum, PRICKLY-ASH or TOOTHACHE TREE

Xanthoxylum *(prickly-ash)*

The delight of seeing the giant swallowtail butterfly in flight makes it worthwhile living in an area where both food plant and butterfly are found. This is not a true ash species; the folly of common names will lead you almost anywhere. It is known as toothache tree for the mild localized freezing effect obtained by chewing on a small piece of bark or fruit. It is the northernmost member of the citrus family and the beautifully aromatic fruit makes it one of the most interesting species from which to gather seed.

Prickly-ash is a colony-forming species with shoots rising from a slowly spreading root system. In moist woodlands, it produces a somewhat open colony of well-spaced prickly stems reaching up to 13 feet to create barely penetrable regions. More often, it is found in the open as impenetrable thickets on moist to dry soils along fence lines, embankments, forest edges and clearings. It ranges from southern Quebec, through southern Ontario and Michigan into the Dakotas and south from Georgia to Oklahoma.

The reddish flower buds expand into tight clusters of greenish yellow flowers in April and May in separate male and female colonies. The flowers are insect pollinated and fruit is produced annually. The elegant fruits mature in August, the small capsule turning orange-red and splitting in two halves to expose one and usually two jet-black seeds, which ultimately fall to the ground. I have not observed seed dissemination for this species but would expect that ground-feeding finches would pick the seeds and some would pass uncracked through digestion. Natural germination is in the first spring.

Fruits should be gathered in early August as the capsules are just starting to open, or seeds can be picked from ripe capsules later in August until early September. The seeds require a five-month cold stratification before they will germinate. One-year seedlings are 2 to 4 inches tall and should be planted into the nursery for a couple years, after which the now 16- to 24-inch plants can be set out into the landscape. The densely fibrous roots make it easy to transplant in spring or fall. Since it is a clone-forming dioecious shrub, it can be grown by division once the sexes are identified.

This is an excellent plant for naturalization, but be mindful of where it is planted. More than almost any other plant, the impenetrable nature of prickly-ash creates what I call sacred ground ... no one can get through it without contemplating an altered course and destination.

Enjoy growing your woodlands from seed.

Planting nuts requires a vision for a future that goes beyond one's mortal reach. If we envision ourselves as participants in the same grand, complex web of interactions as the forest, then planting acorns is like planting part of ourselves.

— Bernd Heinrich —

Abies balsamea *(balsam fir) twig*

Pseudotsuga menziesii *(Douglas fir) twig*

Acer negundo *(Manitoba maple) leaves and fruit*

Acer spicatum *(mountain maple) leaves*

Acer rubrum *(red maple) leaves*

Acer pensylvanicum *(striped maple) leaves*

Acer saccharum *(sugar maple) leaves*

Acer nigrum *(black maple) leaves*

Aronia melanocarpa *(black chokeberry) fruit and foliage*

Aesculus flava *(yellow buckeye) flowers and leaves*

Betula pendula *(European white birch) bark*

Betula lenta *(cherry birch) bark*

Betula nigra *(river birch) bark*

Betula papyrifera *(paper birch) bark*

Betula alleghaniensis *(yellow birch) bark*

Carya cordiformis *(bitternut) twig and bud*

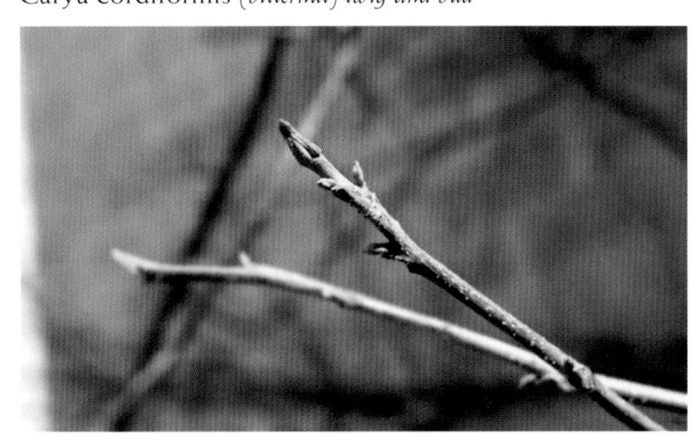

Carya illinoensis *(pecan) twig and bud*

Carya laciniosa *(shellbark hickory) twig and bud*

Carya ovata *(shagbark hickory) twig and bud*

Betula populifolia *(gray birch) bark*

Catalpa speciosa *(northern catalpa) flowers*

Cephalanthus occidentalis *(buttonbush) maturing fruit*

Cornus alternifolia *(alternate-leaved dogwood) berries and leaves*

Black knot fungus is commonly found on native cherries.

Castanea dentata *(American chestnut) young leaves, fruit and male catkin*

Cornus alternifolia *(alternate-leaved dogwood) in the open*

Cornus canadensis *(bunchberry) flowers and leaves*

Cornus drummondii *(rough-leafed dogwood) flowers*

Cornus florida *(eastern flowering dogwood) flowers*

Cornus amomum *ssp.* obliqua *(silky dogwood) flowers*

Cornus foemina *ssp.* racemosa *(gray dogwood) flowers and leaves*

Cornus rugosa *(round-leaved dogwood) flowers and leaves*

Corylus americana *(American hazel) fruit, leaves and next year's male catkins*

Cornus stolonifera *(red osier dogwood) flowers and leaves*

Crataegus *(hawthorn) field*

Dirca palustris *(leatherwood) flowers*

Euonymus atropurpurea *(wahoo) fruits*

Dieback of Fraxinus *(ash) due to the emerald ash borer*

Hamamelis mollis *(Chinese witch hazel) flowers*

Gymnocladus dioica *(Kentucky coffee tree) fruit*

Ilex verticillata *(winterberry) female flowers*

Ilex verticillata *(winterberry) male flowers*

Kalmia latifolia *(mountain laurel) flowers*

Rhododendron maximum *(rosebay rhododendron) flower and leaves*

Lonicera japonica *(Japanese honeysuckle) fruit and foliage*

Lonicera maackii *(Amur honeysuckle) fruit and foliage*

Lonicera *(honeysuckle) flower*

Magnolia acuminata *(cucumber tree) fruit*

Malus baccata *(Siberian crabapple) flowers and foliage*

Malus floribunda *(Japanese crabapple) fruit and foliage*

Nemopanthus mucronata *(mountain holly) fruit and foliage*

Physocarpus opulifolius *(ninebark) foliage and flowers*

Populus balsamifera *(balsam poplar) catkins*

Populus tremuloides *(trembling aspen) emerging catkins*

The closed young cones of Picea glauca *(white spruce)*

Pinus resinosa *(red pine), girdled by voles or mice*

Prunus nigra *(Canada plum) fruit and foliage*

Prunus pensylvanica *(pin cherry) flowers and leaves*

Prunus serotina *(black cherry) foliage and fruit*

Prunus virginiana *(chokecherry) fruit*

Prunus virginiana *(chokecherry) fruit and leaf*

Acorns of Quercus *(oak).*
Many are destroyed by rodents and insects.

Quercus rubra *(red oak) leaf damage*

Quercus muehlenbergii *(chinquapin oak) leaf*

Quercus macrocarpa *(bur oak) seedlings*

Quercus robur *(English oak) leaf*

Rhus typhina *(staghorn sumac) fruit*

Quercus alba *(white oak) leaves*

Quercus *(oak) seedlings protected from rodents during their first year*

Rosa canina *(dog rose) fruit*

Rosa carolina *(pasture rose) fruit*

Salix *sp.* *(willow) male flowers in catkins*

Salix candida *(sage-leaved willow) mature fruit of the female plant*

Sambucus canadensis *(common elderberry) flower and foliage*

Sassafras albidum *(sassafras) flowers*

Shepherdia canadensis *(Canada soapberry) leaf surfaces*

Smilax rotundifolia *(round-leaved catbrier) leaves*

Spiraea alba *var.* latifolia *(wide-leaved meadowsweet) fruit clusters*

Sorbus aucuparia *(European mountain ash or rowan) fruit and foliage*

Tilia americana *(basswood) fruit*

Close-up of wilted leaves caused by Dutch elm disease on a host elm

Vaccinium *(blueberry) fruit and foliage*

Dutch elm disease on a host elm

Viburnum alnifolium *(hobble-bush) fruit and foliage*

Viburnum lentago *(nannyberry) fruit and foliage*

Viburnum cassinoides *(wild raisin) fruit and foliage*

Petiole glands of Viburnum opulus *(European highbush cranberry)*

Viburnum opulus *(European highbush cranberry) fruit and foliage*

Viburnum lantana *(wayfaring tree) flowers and foliage*

Viburnum prunifolium *(black haw) flowers and foliage*

Viburnum lantana *(wayfaring tree) fruit and foliage*

Aesculus hippocastanum *(horse chestnut) fruit and foliage*

Appendix I – Invasive Species

Abies, the FIRS
Douglas fir, *Pseudotsuga menziesii*

Acer, the MAPLES
Norway maple, *Acer platanoides*
Amur maple, *Acer ginnala*
Asian striped maple, *Acer rufinerve*

Aesculus, the BUCKEYES
European horse chestnut, *Aesculus hippocastanum*

Alnus, the ALDERS
Green alder, *Alnus viridis*
European alder, *Alnus glutinosa*

Amelanchier, JUNEBERRY and SERVICEBERRY
Saskatoonberry, *Amelanchier alnifolia*

Aronia melanocarpa, BLACK CHOKEBERRY
Red chokeberry, *Aronia arbutifolia*

Betula, the BIRCHES
European white birch, *Betula pendula*

Carpinus caroliniana, MUSCLEWOOD or BLUE BEECH
European hornbeam, *Carpinus Betulus*

Castanea dentata, AMERICAN CHESTNUT
Chinese chestnut, *Castanea mollissima*

Celastrus scandens, CLIMBING BITTERSWEET
Oriental bittersweet, *Celastrus orbiculatus*

Cornus, the DOGWOODS
Tartarian dogwood, *Cornus alba*

Crataegus, the HAWTHORNS
English hawthorn, *Crataegus monogyna*

Dirca palustris, LEATHERWOOD
Paradise plant, *Daphne mezereum*

Euonymus, the BURNING BUSHES or SPINDLE TREES
European spindle tree, *Euonymus europaea*
Winged burning bush, *Euonymus alata*

Fagus grandifolia, AMERICAN BEECH
European beech, *Fagus sylvatica*

Fraxinus, the ASHES
European ash, *Fraxinus excelsior*

Juglans, the WALNUTS
Heartnut, *Juglans ailanthifolia* var. *cordiformis*

Juniperus, the JUNIPERS
Chinese juniper, *J. chinensis*

Larix laricina, TAMARACK
European larch, *Larix decidua*

Malus coronaria, WILD CRABAPPLE
Siberian crabapple, *Malus baccata*
Japanese crabapple, *Malus floribunda*
Chinese crabapple, *Malus prunifolia*

Morus rubra, RED MULBERRY
(Asian) white mulberry, *Morus alba*

Picea, the SPRUCES
Norway spruce, *Picea abies*
Colorado spruce, *Picea pungens*

Pinus, the PINES
Scots (or Scotch) pine, *Pinus sylvestris*
Austrian pine, *Pinus nigra*

Platanus occidentalis, SYCAMORE
London plane, *Platanus* x *acerifolia*

Populus, the POPLARS and ASPENS
Carolina poplar, *Populus* x *canadensis*
White poplar, *Populus alba*

Prunus, the PLUMS AND CHERRIES
Sweet cherry, *Prunus avium*
Nanking cherry, *Prunus tomentosa*

Quercus, the OAKS
English oak, *Quercus robur*

Ribes (and *Grosularia*), the CURRANTS and GOOSEBERRIES
European red currant, *Ribes rubrum* (*Ribes sativum*),

Rosa, the ROSES
Dog rose, *Rosa canina*
Sweet briar, *Rosa eglanteria* (*Rosa rubiginosa*)
Japanese multiflora rose, *Rosa multiflora*
Japanese rugosa rose, *Rosa rugosa*

Rubus, the BLACKBERRIES, RASPBERRIES and THIMBLEBERRIES
European blackberry, *R. fruticosus*

Salix, the WILLOWS
(French) pussy willow (goat willow), *Salix caprea*
(European) crack willow, *Salix fragilis*
(European) white willow, *Salix alba*
(European) bay-leaved willow, *Salix pentandra*
(European) basket willow, *Salix purpurea*
(European) silky osier, *Salix viminalis*

Sambucus, the ELDERBERRIES
European black elderberry, *Sambucus nigra*
European red elderberry, *Sambucus racemosa*

Shepherdia canadensis, CANADA SOAPBERRY
Autumn olive, *Elaeagnus umbellata*

Sorbus, the MOUNTAIN ASHES
European mountain ash or rowan, *Sorbus aucuparia*

Symphoricarpos, SNOWBERRY and CORALBERRY
Wolfberry, *Symphoricarpos occidentalis*
Western snowberry, *Symphoricarpos albus* var. *laevigatus*

Taxus canadensis, CANADA YEW or GROUND HEMLOCK
Japanese yew, *Taxus cuspidata*

Tilia americana, BASSWOOD
Linden or lime, *Tilia cordata*

Ulmus, the ELMS
Siberian elm, *Ulmus pumila*
Scotch elm, *Ulmus glabra*
English elm, *Ulmus procera*

Viburnum, the VIBURNUMS
European highbush cranberry, *Viburnum opulus*
Sargent viburnumn, *Viburnum sargentii*
Wayfaring tree, *Viburnum lantana*

Appendix II – Seed Dispersal Calendar

Late May	*Aspens*, elms, red maple, silver maple, *willows*: pussy and woolly-headed
Early June	*Cottonwood*, leatherwood, soft maple, willows: Bebb's, heart-leaved and hoary
Mid-June	Black willow
Late June	Canada soapberry, pin cherry, red elderberry, willows: peach-leaved, sandbar and shining
Early July	Serviceberry
Mid-July	Honeysuckle, red mulberry
Late July	Currants and gooseberries
Early August	Alternate-leaved dogwood, prickly-ash, sassafras
Mid-August	White spruce, choke cherry
Late August	Paper birch, white cedar, black cherry, silky dogwood, elderberry, sumac, white pine
Early September	Ash, buckeye, balsam fir, chinquapin and dwarf chinquapin oaks, hazelnut, hobble-bush, hop-tree, sweet fern, tamarack, tulip tree, wild plum
Mid-September	Bayberry, gray and red osier dogwood, white pine, mountain ash, wild raisin, cherry birch, river birch, gray birch, swamp birch, Virginia creeper, grapes,
Late September	Flowering dogwood, basswood, hickory, butternut, beech, New Jersey tea, winterberry, round-leaved dogwood, running strawberry bush, spicebush, crabapple, ninebark, spirea, chokeberry, white oaks
Early October	Hawthorn, hackberry, buckeye, sugar maple, ironwood, musclewood, burning bush, rough-leaved dogwood, witch hazel, sweet chestnut, junipers, magnolia, red oaks, red pine
Mid-October	Bittersweet, Manitoba maple, pawpaw, black walnut, mountain maple, clematis, tupelo, sweet gum, potentilla, autumn willow
Late October	Buttonbush, roses, choke berry, trumpet creeper, redbud, honey locust, bladdernut, sycamore
November	Catalpa, alder, Kentucky coffee tree, black locust

This table gives you a relative date for the first option to obtain fruit from the species. There is a very short window of time for some species and they are in *italics*. Check the species description and Appendix III to determine the duration of access to seed of the species. Willow seed is accessible for only a few days and sumac for several months.

Sowing Depth

One of the most critical aspects of seed germination is the amount of energy that a germinating seed has to break through the soil and mulch covering. In natural systems most seeds germinate when they fall into or are covered by a mulch of decaying vegetation. In horticulture the general practice is to use a soil or sand covering of one to two times the seed thickness. With a mulch added, the soil or sand covering should be reduced by about half. If a heavy mulch of leaves or straw is used, it must be removed around early April, prior to the seed germination. You can experiment with coarse sawdust, wood shavings and recycled wood that has been run through a hammer mill. Most of these wood mulches can be applied thinly enough for the seedlings to germinate through, with the added advantage that weeds will be suppressed.

Seed Treatment Guide Codes

V	Viability is reduced if seed dries out.
DB	Store as dried whole fruit; soak 24 hours to remove fruit later.
F	Allow the fruit to ferment for 2 to 3 days to soften the fruit if needed, prior to cleaning.
N	No pretreatment is needed for complete germination.
M	Moist soil surface must be maintained constantly until germination has started.
L	Light is needed for the seed to germinate; do not bury the seed. Cover the flat with plastic or glass.
A	Sulfuric acid pretreatment, immersion time in minutes * [never add water to acid] * Note that the seed coat must be **very dry** before you add the acid.
S	Soak stored seed in room temperature water for 24 hours prior to treatment.
H	Cover the seed with boiling water and let stand, cooling for 24 hours; legume seeds will swell.
W	Warm stratification 68°F (20°C), # = duration in days.
C	Cold stratification 35–39°F (2–4 °C) (but not freezing); # = duration in days.
G	Will germinate near the end of the cold stratification and should be planted soon. When root growth occurs during the warm treatment, the seed must still have cold treatment for the shoot.
Y	Harvest this fruit at the mature stage (yellow) and plant the seed immediately.
#	Seed may germinate the first spring, but often not till the next year.
O	Seed is best planted in protected outdoor seedbed. The information enclosed in brackets — ( ) — represents an alternative to direct seedbed sowing.
P	Seed will only germinate well in a cool, damp location with no direct sun.

Flowering Dates, Seed Dispersal Period and Seed Treatment

Name	Seed crop interval (yrs)	Flower dates	Seed dispersal	Treatment guidelines
Abies balsamea, **balsam fir**	2	May	Sept. Dec.	C 21, P
Acer negundo, **Manitoba maple**	1	Apr.	Oct. Feb.	S, C 60, G
Acer rubrum, **red maple**	1	Apr.	Late May	V, N
Acer pensylvanicum, **striped maple**	?	May June	Oct.	C 120
Acer saccharinum, **silver maple**	1	Apr.	Early June	V, N
Acer saccharum, **sugar maple**	3–5	Apr.	Oct.	S, C 90, G
Acer spicatum, **mountain maple**	1	June	Oct.	S, C 150, G
Aesculus glabra, **Ohio buckeye***	1	May	Sept.	O (C 120)
Alnus rugosa, **speckled alder**	1–3	Apr.	Dec. Mar.	C 60, M
Amelanchier spp., **serviceberry**	1	May	June July	DB, S, C 120, G
Aronia prunifolia, **chokeberry**	1	May June	Oct.	DB, S, C 150, G
Asimina triloba, **pawpaw***	2	May	Oct.	F, S, C120
Betula papyrifera, **paper birch**	1–2	Apr. May	Aug Mar	L, M
Betula alleghaniensis, **yellow birch**	2	Apr. May	Sept. Feb.	C 30, M
Campsis radicans, **trumpet creeper***	1	July Aug.	Oct. Mar.	N, L
Carpinus caroliniana, **blue beech**	3–5	Apr. May	Oct.	S, W 60, C 80, P
Carya spp., **hickories**	1–3	May June	Sept. Oct.	V, O (C 120)
Castanea dentata, **chestnut***	2–3	May June	Oct.	V, O (C 150)
Catalpa speciosa, **northern catalpa**	1	Late May	Nov. Mar.	N

Name	Seed crop interval (yrs)	Flower dates	Seed dispersal	Treatment guidelines
Ceanothus americanus, New Jersey tea*	1	July Sept.	Oct.	H, C 60
Celastrus scandens, bittersweet	1	May June	Oct. Mar.	DB, S, C 110–150
Celtis occidentalis, hackberry	1	May	Oct. Mar.	DB, S, C 90
Celtis tenuifolia, dwarf hackberry*	1	May	Oct. Mar.	DB, S, C 150
Cephalanthus occidentalis, buttonbush	1	July	Oct. Mar.	N
Cercis canadensis, redbud*	1	Late May	Nov. Feb.	H
Clematis occidentalis, rock clematis*	1	July Aug.	Oct.	W 30, C 60
Clematis virginiana, virgin's bower	1	July Aug.	Oct.	W 30, C 60
Cornus alternifolia, pagoda dogwood	1	June	July Aug.	S, W 60, C 90
Cornus drummondii, rough-leaved dogwood*	1	June	Oct. Nov.	DB, S, W 60, C 60
Cornus florida, flowering dogwood	1	May	Sept. Oct.	DB, S, C 150, G
Cornus amomum (obliqua), silky dogwood	1	July	Sept.	DB, S, W 60, C 90–120
Cornus racemosa, gray dogwood	1	June July	Sept. Oct.	DB, S, W 60, C 90
Cornus rugosa, round-leaved dogwood	1	June July	Sept. Oct.	S, C 180
Cornus stolonifera, red osier dogwood	1	June July	Sept. Oct.	S, W 60, C 120
Corylus spp., hazelnut	1	Apr.	Aug. Sept.	O (C 150)
Crataegus crus galli, cockspur hawthorn	1	May June	Sept. Oct	F, A 120, W 30, C 150 #
Crataegus mollis, downy hawthorn	1	May June	Sept. Oct.	F, S, W 30, C 150, G #
Crataegus succulenta, fleshy hawthorn	1	May June	Sept. Oct.	F, A 30, C 150 #
Diervilla lonicera, honeysuckle diervilla	1	June July	Oct. Mar.	N
Dirca palustris, leatherwood	1	Apr.	June	W 90, C 180 G
Euonymus obovata, strawberry bush	1	May June	Sept. Oct.	S, W 20–50 (expand) G, C120
Euonymus atropurpurea, burning bush*	1	Late June	Oct. Dec	DB, S, W 90, C 180, G # try DB, S, C 110, G
Fagus grandifolia, beech	2–3	May	4 Sept. Oct.	V, O (C 90–180, G)
Fraxinus americana, white ash	3–5	Apr. May	Sept. Feb.	Y, O # (W 30, C 60)
Fraxinus nigra, black ash	5–7	May June	Sept. Oct.	Y, O # (W 60, C 90)
Fraxinus pensylvanica, red ash	1	Mar. Apr.	Sept. Feb.	Y, O # (W 60, C 180)
Fraxinus profunda, pumpkin ash*	2–3	Mar. Apr.	Oct. Nov.	N
Fraxinus quadrangulata, blue ash*	3–4	Mar. Apr.	Sept. Feb.	Y, O # (S, W 60, C 90,)
Gleditsia triacanthos, honey locust*	1	May June	Oct. Mar.	H, O
Gymnocladus dioica, Kentucky coffee tree	1	May June	Nov. Mar.	Scratch surface, S, O
Hamamelis virginiana, witch hazel	1	Oct.	Oct.	S, W 60, C 120, G
Hypericum spp., St. John's-wort	1	July Aug.	Oct. Mar.	N, M
Ilex verticilata, winterberry	1	Late May	Sept. Dec.	S, W 60,C 60
Juglans spp., walnuts	1	May	Sept. Oct.	V, O (C 90–120)
Juniperus communis, common juniper	1	May June	Oct Mar.	S, W 90, C 90–120, #
Juniperus virginiana, red cedar	1	May June	Oct. Feb.	S, W 100, C 85, #
Larix laricina, tamarack	1–3	Apr. May	Aug. Oct.	C 60
Lindera benzoin, spicebush	1	Apr.	Sept. Oct.	DB, S, W 50, C 160, G
Liquidambar styraciflua, sweet gum	1	May	Oct. Feb.	N
Liriodendron tulipifera, tulip tree	1	June	1 Sept. Oct.	Y, O, # (S, C 150, #)
Lonicera dioica, honeysuckle	1	May June	July Aug.	S, C 70
Lonicera canadense, fly honeysuckle	1	May June	July Aug.	N
Lonicera involucrata, northern honeysuckle	1	June	Aug.	S, C 90
Magnolia acuminata, cucumber tree*	1	June July	Early Oct.	DB, S, C 210
Malus coronaria, crab apple	1	May June	4 Sept. Oct.	S, C 130, G
Morus rubra, red mulberry*	1–2	Apr. May	July Aug.	N, M
Myrica pensylvanica, bayberry	1	May June	Sept. Nov.	DB, S, C 90
Nyssa sylvatica, black gum*	1	May June	Mid Oct.	S, C 90
Ostrya virginiana, hop hornbeam	1	Apr. May	Sept. Oct.	S, W 60, C 150
Parthenocissus spp., Virginia creeper	1	June	Sept. Oct	S, C 90 also F 10, W 15, G
Physocarpus opulifolius, ninebark	1	June July	Sept. Apr.	C 60
Picea glauca, white spruce	2–3	May	Aug. Feb.	S, N
Pinus resinosa, red pine	3–7	May June	Oct. Nov.	C 60
Pinus strobus, white pine	1–3	Late May	Aug. Sept.	S, C 60
Platanus occidentalis, sycamore	1–2	Apr.	Nov. Mar.	N
Populus spp., aspens and cottonwood	1	Apr.	May June	V, N, M
Potentilla fruticosa, shrubby cinquefoil	1	June Sep.	Oct. Feb.	N, M
Prunus spp., cherries	1–3	June	July Sept.	O (S, C 120–150, G,)

Name	Seed crop interval (yrs)	Flower dates	Seed dispersal	Treatment guidelines
Prunus spp., **plums**	1–2	May	Sept.	O (S, C 120–150, G,)
Ptelea trifoliata, **hop-tree***	1	June	Sept.	S, W 30, C 90
Quercus alba, **white oak** (group)	2–4	Apr. May	2 Sept. 1 Oct.	V, O (V, W 15, G, C 90–120)
Quercus muehlenbergii, **chinquapin oak** (group)*	1–3	Apr. May	2 Sept. 1 Oct.	V, O (V, W 3–10, G, C 90–120)
Quercus rubra, **red oak** (group)	1–3	Apr. May	2 Sept. 3 Oct.	V, O (V, S, C 90–120)
Rhus aromatica, **fragrant sumac**	1	Early May	Aug. Apr.	H, V
Rhus copallina, **shining sumac***	1	July	Sept. Mar.	H, V
Rhus typhina, **staghorn sumac**	1	May July	Sept. June	H, V
Ribes spp., **currants and gooseberries**	1	May June	4 July 2 Aug.	A ca. 10, C 150, M, #
Rosa spp., **roses**	1	June July	Oct. Mar.	DB, S, C 120–150, G
Rubus spp., **raspberries and blackberries**	1	May Aug.	July Sept.	W 90, C 90
Salix spp., **willows**	1	Apr. June	4 May 3 June	V, N, M
Salix serissima, **autumn willow**	1	June	Mid Oct.	V, N, M
Sambucus canadensis, **elderberry**	1	July.	4 Aug. Sept.	S, W 60, C 140,G, M
Sambucus pubens, **American red elderberry**	1	Mid May	1 July Aug.	A ca. 10, W 60, C 120, M
Sassafras albidum, **sassafras**	1	Mid May	Mid Aug.	S,C 150
Shepherdia canadensis, **Canada soapberry**	1	Early May	4 June 1 July	A 20, C 90
Sorbus americana/decora, **mountain ash**	1	June	Sept. Dec.	DB,S,C 120–150,G
Spiraea alba, **meadowsweet**	1	July Aug.	Sept. Mar.	N,M
Staphylea trifolia, **bladdernut**	1	May	Sept. Mar.	W ca. 60, C 120, #
Symphoricarpos albus, **snowberry**	1	July	Sept. Dec.	A 70, W 45, C 180, M,
Taxus canadensis, **yew**	1	Mid May	3 July Aug.	Y, O, #
Thuja occidentalis, **eastern white cedar**	1–3	May	Sept. Feb.	N, M
Tilia americana, **basswood**	1–2	June July	3 Sept. Dec.	Y, O, #
Tsuga canadensis, **eastern hemlock**	2–3	May	3 Oct. Dec.	C 120, P, M
Ulmus spp., **elms**	1–3	Apr.	4 May 1 June	N
Viburnum dentatum, **southern arrowwood**	1	May June	Sept. Oct.	S, W 180, C 120
Viburnum lentago, **nannyberry**	1	May	Aug. Sept.	S, W 180, C 120, #
Viburnum trilobum, **highbush cranberry**	1	June July	Aug. Sept.	S, W ca. 60, G, C 60
Vitis spp., **grape**	1	Early June	3 Sept. 3 Oct.	S, C 180
Zanthoxylum americanum, **prickly-ash**	1	Apr. May	Aug.	S, C 150

* *These species may be rare or even absent from portions of the range, both at present and historically, and therefore are not appropriate for restoration plantings. Further, it is best to check with the local conservation agency to work with existing programs and to avoid confusion with recovery actions that may be underway on these species.*

The treatments listed were developed for small-scale operations; however, they can be adjusted for large-scale production or direct outdoor sowing in protected nursery beds.

Conversion Factors: Imperial system to metric system

To change:	Into:	Multiply by:
Inches	Millimeters	25.4
Inches	Centimeters	2.54
Feet	Meters	0.305
Yards	Meters	0.914
Miles	Kilometers	1.609
Square inches	Square centimeters	6.452
Square feet	Square meters	0.093
Square yards	Square meters	0.836
Cubic inches	Cubic centimeters	16.39
Cubic feet	Cubic meters	0.028
Cubic yards	Cubic meters	0.765

Appendix IV – Further Resources and Reading

Websites

American Association of Botanical Gardens and Arboreta
www.publicgardens.org

Bugwood Network
www.bugwood.org

Canadian Forest Service
www.cfs.nrcan.gc.ca

The Gypsy Moth Handbook
United States Department of Agriculture
www.na.fs.fed.us/fhp/gm/online_info/gm/gmhb.htm

Natural Heritage Information Centre
Ontario Ministry of Natural Resources
Information on native species and plant communities of conservation concern.
http://nhic.mnr.gov.on.ca/nhic_.cfm

Natural Resources Canada
Plant Hardiness Website
www.planthardiness.gc.ca

NatureServe Canada
A network connecting science with conservation.
www.natureserve-canada.ca

NatureServe Explorer, Online encyclopedia of life.
www.natureserve.org/explorer

New York Botanical Garden
International Plant Science Center
http://sciweb.nybg.org/science2/ScienceHome.asp

Plants National Database
Information about plants growing in the U.S.
http://plants.usda.gov/

Society for Ecological Restoration (SER)
www.ser.org

Tree Canada
www.treecanada.ca

Books, Including Books Cited

The Complete Trees of North America
by Thomas Elias, Van Nostrand Reinhold, 1980.

The Culture of Nature: North American Landscape From Disney to the Exxon Valdez by Alex Wilson, Toronto: Between the Lines, 1991.

Deciduous Forests of Eastern North America
by Lucy Braun, Macmillan, 1974.

Heart of the Land, Assembled by
The Nature Conservancy, Vintage, 1996.

Landscaping with Native Trees by Guy Sternberg and Jim Wilson, Chapters Publishing, 1995.

Loon Laughter: Ecological Fables and Nature Tales
by Paul Leet Aird.

Michigan Trees by Burton Barnes and Warren Wagner, University of Michigan Press, 1981.

Native Trees, Shrubs, and Vines for Urban and Rural America by Gary Hightshoe, Van Nostrand Reinhold, 1980.

River Notes by Barry Lopez , Avon Books, 1980.

Seeds of Woody Plants in the United States
by C. Schopmeyer, Agriculture Handbook 450, Forest Service, United States Department of Agriculture, 1989.

Shrubs of Ontario by James Soper and Margaret Heimburger, Royal Ontario Museum, 1982.

Silvics of North America by Russell Burns and Barbara Honkala, Agriculture Handbook 654, Forest Service, United States Department of Agriculture, 1990.

Trees in Canada by John Laird Farrar, Canadian Forest Service, 1995.

Trees of the Carolinian Forest: A Guide to Species, Their Ecology and Uses by Gerry Waldron, Boston Mills Press, 2003.

Trees of the Central Hardwood Forests of North America
by Donald Leopold, William McComb and Robert Muller, Timber Press, 1998.

The Woody Plants of Ohio
by Lucy Braun, Ohio State University Press, 1989.

Glossary

alien a plant introduced from elsewhere and naturalized in its new habitat

allele one of the usually two forms of a gene that occupies the same relative position on a chromosome

alvar a flat limestone area with little or no soil and a unique cover of drought-tolerant plants

aril an extra seed covering, often colored and hairy or fleshy

boreal (adjective, not a noun) of the north or northern regions

bioregion a region defined by climate, soil, moisture and its natural ecological community

bract a modified and sometimes brightly colored leaf with a flower or branch of a flower in its axil

brownfield land previously used for industrial purposes

chlorosis a reduction or loss of the normal green coloration of plants

clone a genetically identical plant produced asexually from one ancestor

cultivar a cultivated variety, as in a garden selection that is grown commercially

cutting a piece cut from a plant for propagation

dioecious having male and female organs on separate plants

dormant not actively growing; in a resting state

embryo the precursor to a plant contained within a seed

exotic introduced from or originating in or existing in a foreign or distant place

gene the basic unit of heritable characteristics in a chromosome

genotype the genetic constitution of an individual

genus (plural – **genera**) a grouping of organisms having common characteristics distinct from those of other genera, usually containing several or many species

glabrous lacking hair or down; not pubescent

glaucous covered with a dull grayish or bluish powdery bloom

graft to join a bud or branch from one plant (the scion) to the lower part of another plant (the rootstock)

hybrid the offspring of two plants or animals of different varieties or species

indigenous originating naturally in a region; native; sometimes implying locally native

inflorescence the complete flower head of a plant including stems, stalks, bracts and flowers

layer a shoot that naturally takes root while attached to the parent plant, or by inducing rooting

lenticel raised pores on woody stems that allow gas exchange

mesic (of a habitat) containing a moderate amount of moisture

monoecious with unisexual male and female organs on the same plant

native originating naturally in a region; indigenous; at times implying a wider area such as a continent

naturalize to introduce a plant to another region where it thrives in the wild; make an area natural again

ovule the part of a flower that becomes the seed after fertilization

peduncle the stalk of an inflorescence, fruit or fruit cluster

petiole the stalk joining a leaf to a stem

prickle a sharp outgrowth on young stems, and also on leaves, as in *Rosa* and *Rubus*

pubescence a soft down of short, fine hairs on the leaves and stem of plants

raceme a flower cluster with the separate flowers attached by short equal stalks along a central stem

radicle that portion of an embryo that develops into a root

rhizome an underground root-like stem bearing both roots and shoots

riparian of or on the bank of a marsh, pond, river or lake

samara a dry fruit in which part is extended to form a wing, as in elm, ash and maple

scarify scratch or make an incision in a hard seed coat

self-fertilization the fertilization of plants by their own pollen, not from others

shrub a woody plant having short, multiple stems that branch near the ground

spine a modified stipule or sharp branchlet found in a leaf axil, as in *Ribes* and *Robinia*

sprout begin to grow, put forth shoots; new growth developing from a bud, seed or other part

stratify to pretreat seeds to simulate winter conditions so that germination may occur

stipule a small leaf-like appendage at the base of a petiole

sucker a shoot springing from a root or from the lower part of the stem

tissue culture the growth of plant tissue in an artificial medium

thorn a sharp outgrowth from a stem other than at a node; a modified stem as in hawthorn

xeric (of a habitat) droughty; having little moisture available for growth

Afterword

The gathering, storage, sowing and germination of seed,
and the planting and tending
of newly germinated plants, requires us
to think like a seed.

— Henry Kock —

Henry Kock (1952–2005), horticulturist, naturalist, conservationist, author: "Born in May of 1952 with the opening of spring flowers and on the largest recorded songbird migration in North American history. From age 14–28, I did the pesticide applications on the farm. Almost every pesticide I used then is now banned as carcinogenic." Henry died of brain cancer on December 25, 2005, in Guelph, Ontario.

Shortly before he died, his partner, Anne Hansen, asked him, "Would it be all right with you if some of your friends finished your book?" Henry, struggling with brain damage that made it difficult for him to express himself, replied clearly, "Too many cooks spoil the broth."

Henry may still feel the same way, but our mutual bond with Henry and nature inspired us to complete the book he had largely finished himself. We undertook this project so as to rightfully honor Henry's work, his life and what he believed in, and his memory. Throughout, we have worked to keep Henry's "voice" as we made additions or modifications to his excellent manuscript. Keeping Henry's cautioning response in mind, we have gently stirred his broth, with perhaps a little alteration in the seasonings.

During Henry's 23 years working at the University of Guelph Arboretum in Guelph, Ontario, he germinated woody plant seeds collected throughout North America. While this book presents procedures he found successful for propagating the trees, shrubs and woody vines of the Great Lakes bioregion, it also applies to these and many other species growing outside this region.

"To think like a seed, we must first listen to the seed's story as written by the land and water where the seed formed, and the wind, water and animals that distribute it. It means respecting the natural heritage of each bioregion by growing its native species from seeds derived from the local plants — plants that have survived over time to flourish and evolve together." — *Henry Kock.*

Paul Aird,
Inglewood, Ontario

John Ambrose,
Guelph and Pelee Island,
Ontario

Gerry Waldron,
Amherstburg, Ontario

Acknowledgments

This book would not have developed without my involvement with the Arboretum at the University of Guelph, Ontario, its supportive staff and Arboretum volunteers. The book is a summary of my 20-plus years' experience collecting and germinating seeds at the Arboretum — much of the early years with the assistance of Susan Feryn-Perkin, and latterly with the assistance of Sean Fox.

My visits to the diverse natural areas of southern Ontario took place initially with the guidance of Arboretum curator Dr. John Ambrose and Arboretum botanist Steven Aboud, who both have very exacting plant identification and documentation skills. I was occasionally accompanied by well-known and insightful field botanists including Gus Yaki, Gerry Waldron and others.

Understanding the maturation and movement of fruits was not my way of seeing plants until I began to study the wildlife associated with them. Insights into the interplay between flora and fauna were often revealed by the contagious enthusiasm of Arboretum naturalists Alan Watson and Chris Earley.

On an early visit with my grandfather, Jan Kock, he declared that the art of growing plants was in my blood. Later in life, while I was visiting him in Holland, he summed up the reason that working with plants was so rewarding. He was examining a germinating seed in his backyard nursery in Hengelo and made a remark that, in translation, means "Plants are the basis of life." This simple statement has resounded over and over in my mind as the essential motivation for restoring diverse vegetation cover and sustaining the diversity of life.

I am indebted to the many hundreds of people who, over the years, participated in my *Growing Native Plants from Seed* and *Woody Plant Identification* workshops offered each fall at the Guelph Arboretum. Your collective appreciation and comments were a constant source of inspiration to me to continue refining my writing and to bring this information into published form. Workshop participants have been very successful at establishing native plants in their gardens, schoolyards and parks. Several have established thriving native plant nurseries, and I have been inspired by those managed by Cathy Dueck, Jeff Scott, Mary Gartshore and Peter Carson, Solomon Boye and Ahren Hughes. A special thanks to Mathis Natvik for sharing insights gained from the pit-and-mound construction in the expansion of Clear Creek Forest.

No book can progress without editorial help. I am indebted to Gerry Waldron for taking a first crack (in a pub, mind you) at the introductory chapters, to Ric Jordan, Arboretum Manager, for valuable comments, and to the staff at Firefly Books, who were enthusiastic from first contact — especially my editor, Laurie Coulter, who was insightful and bold enough to guide me along.

Staff at the Ontario Ministry of Natural Resources deserve credit. In discussions with Denis Joyce, I was introduced to large-scale species fluctuation models. Promoting a new awareness of gene flow, climate influence and concerns about seed sources remains the tireless and passionate work of Cathy Nielson and Barb Boysen; it has been so rewarding to work with them.

The late Alexander Wilson first placed in my mind the idea of an ecosystem book for the Great Lakes region. Mendelson Joe was straightforward enough to start me writing and Lorraine Johnson insisted that I keep writing, along with many others who shall remain nameless. Thanks also to Barb Quinlan, Frances Tutt and Chrystel Herick with their yoga, Bowen and Alexander technique that reduced back pain while sitting to write for extended periods.

Miguel Francisco Soto Cruz, founder of Arbofilia in Costa Rica, provided long-lasting inspiration. After noticing a warbler and a thrush in my urban backyard wilderness, Miguel remarked, "I know these birds," and turned to me, saying, "We live in the same forest." Unfortunately, the neotropical migrant birds visit each other more regularly than we do.

I thank my partner, Anne Hansen, for teaching me how to identify the birds and observe their activities. I am so grateful to her for her unbelievable patience, waiting as my hand reached out to make yet another seed collection during our many walks in the wilds of Ontario, and for her undivided support in many other ways.

I would be remiss if I did not take off my hat to the Forest Gene Conservation Association of Ontario (FGCA) and the Centre for Land and Water Stewardship (CLAWS) at the University of Guelph for providing financial assistance. They had faith in me before I even met a publisher.

I have long forgotten the source of some of the ideas and insights expressed in this book. To those who see their words and ideas here without recognition, I apologize, and hope that you will be pleased that your ideas are given a wider audience through me.

Addendum, December 2007 — Henry's partner, Anne Hansen, wishes to thank the following tramping and camping friends who helped Henry along the way with the exchange of plants, seeds, humor, wine and wisdom: Oxanna and Warren Adams, Lauren Baker, Robert Brandstetter, Anika Kew Brandstetter, Darius Kew Brandstetter, Karen Kew, Aimee Charbonneau, Ann Clark, Cam Collyer, Ron Kelly, Dan McDermott, Linda Pim, Andrea Rauser, Erick Rauser, Nicole Rauser, Rachel Rauser, Joan Rentoul, Sue Richards, Lenore Ross, Helen Rykens, Doug Steel, Trudy Watts, Clover Woods.

Index

Page numbers in bold type indicate a color photograph

D

E

F

Credits

t=top, tl=top left, tr=top right
cl=center left, cr=center right
b=bottom, bl=bottom left, br=bottom right

Illustrations

William Band: p. 8, p. 13, p. 15 t, p. 16, p. 20, p. 25, p. 26, p. 28 b, p. 30, p. 31, p. 39, p. 43, p. 44, p. 47, p. 55, p. 56, p. 57, p. 58, p. 66, p. 67, p. 79 bl cr, p. 81, p. 85, p. 89, p. 93, p. 96 t, p. 109, p. 110 b, p. 113 b, p. 121, p. 124 b, p. 156, p. 161, p. 166, p. 172, p. 178 tl, p. 187, p. 188, p. 196, p. 197, p. 213, p. 224, p. 225, p. 230

Kye Shuett: p. 3, p. 4, p. 5, p. 10, p. 11, p. 12, p. 15 b, p. 18, p. 24, p. 28 t, p. 29, p. 32, p. 33, p. 34, p. 36, p. 40, p. 45, p. 60, p. 63, p. 65, p. 71, p. 73, p. 74, p. 77, p. 79 t tl cl br, p. 80, p. 82, p. 83, p. 84, p. 90, p. 92, p. 94, p. 96, p. 97 c b, p. 98, p. 100, p. 102, p. 103, p. 104, p. 106, p. 107, p. 110 t, p. 111, p. 112, p. 113 t c, p. 116, p. 117, p. 118, p. 119, p. 120, p. 124 t, p. 126, p. 127, p. 128, p. 132, p. 133, p. 134, p. 136, p. 137, p. 138, p. 139, p. 141, p. 142, p. 143, p. 145, p. 146, p. 147, p. 149, p. 150, p. 153, p. 158, p. 159, p. 160, p. 163, p. 164, p. 167, p. 168, p. 170, p. 171, p. 174, p. 176, p. 177, p. 178 tr, p. 179, p. 180, p. 181, p. 182, p. 183, p. 184, p. 185, p. 190, p. 192, p. 193, p. 194, p. 199, p. 200, p. 201, p. 204, p. 205, p. 206, p. 207, p. 208, p. 209, p. 210, p. 212, p. 214, p. 216, p. 218, p. 220, p. 222, p. 226, p. 227, p. 228, p. 229, p. 231, p. 232

Photos

Front cover: Photowood Inc./CORBIS
Back cover, front flap and spine: Henry Kock
Back flap: Courtesy of Anne Hansen

Steven Aboud: p. 250 bl, p. 252 tl

Ted Bodner, Southern Weed Science Society, Bugwood.org: p. 248 bl, p. 250 tr, p. 258 tl br

John D. Byrd, Mississippi State University, Bugwood.org: p. 243 tr

William M. Ciesla, Forest Health Management International, Bugwood.org: p. 236, p. 252 tr

Bill Cook, Michigan State University, Bugwood.org: p. 245 bl

The Dow Gardens Archive, Dow Gardens, Bugwood.org: p. 240 tl, p. 243 b, p. 245 tl, p. 247 tl, p. 248 tl tr, p. 249 bl br, p. 263

Boris Hrasovec, Faculty of Forestry, Bugwood.org: p. 260 tl

Steven Katovich, USDA Forest Service, Bugwood.org: p. 246

Henry Kock: p. 236 tl, p. 240 tr, p. 244 br, p. 245 tr br, p. 247 tr, p. 249, p. 249 tl tr, p. 250 tl, p. 252 br, p. 253 tr cr br, p. 254 b, p. 255 bl br, p. 256, p. 257, p. 260 tr, p. 262 tl tr bl.

Minnesota Department of Natural Resources Archive, Minnesota Department of Natural Resources, Bugwood.org: p. 260 b

Walter Muma: p. 234, p. 235, p. 236 bl br, p. 237, p. 238 tl tr, p. 239, p. 240 b, p. 241, p. 242, p. 243 tl, p. 244 tl tr bl, p. 247 bl br, p. 248 br, p. 250 tr, p. 251 t, p. 253 bl, p. 254 t, p. 255 t, p. 258 tr bl, p. 259, p. 261 t, p. 262 br, p. 264

Joseph O'Brien, USDA Forest Service, Bugwood.org: p. 261 b

Barbara Tokarska-Guzik, University of Silesia, Bugwood.org: p. 252 bl

USDA Forest Service – North Central Research Station Archive, USDA Forest Service, Bugwood.org: p. 251 b

Paul Wray, Iowa State University, Bugwood.org: p. 238 cl, cr, bl, br, p. 253 tl